Chaotic Worlds: From Order to Disorder in Gravitational N-Body Dynamical Systems

NATO Science Series

A Series presenting the results of scientific meetings supported under the NATO Science Programme.

The Series is published by IOS Press, Amsterdam, and Springer in conjunction with the NATO Public Diplomacy Division.

Sub-Series

I. Life and Behavioural Sciences	IOS Press
II. Mathematics, Physics and Chemistry	Springer
III. Computer and Systems Science	IOS Press
IV. Earth and Environmental Sciences	Springer

The NATO Science Series continues the series of books published formerly as the NATO ASI Series.

The NATO Science Programme offers support for collaboration in civil science between scientists of countries of the Euro-Atlantic Partnership Council. The types of scientific meeting generally supported are "Advanced Study Institutes" and "Advanced Research Workshops", and the NATO Science Series collects together the results of these meetings. The meetings are co-organized by scientists from NATO countries and scientists from NATO's Partner countries — countries of the CIS and Central and Eastern Europe.

Advanced Study Institutes are high-level tutorial courses offering in-depth study of latest advances in a field.
Advanced Research Workshops are expert meetings aimed at critical assessment of a field, and identification of directions for future action.

As a consequence of the restructuring of the NATO Science Programme in 1999, the NATO Science Series was re-organized to the four sub-series noted above. Please consult the following web sites for information on previous volumes published in the Series.

http://www.nato.int/science
http://www.springer.com
http://www.iospress.nl

Series II: Mathematics, Physics and Chemistry – Vol. 227

Chaotic Worlds: From Order to Disorder in Gravitational N-Body Dynamical Systems

edited by

B. A. Steves

Glasgow Caledonian University,
U.K.

A. J. Maciejewski

University of Zielona Gora,
Poland

and

M. Hendry

University of Glasgow,
U.K.

Springer

Published in cooperation with NATO Public Diplomacy Division

Proceedings of the NATO Advanced Study Institute on
Chaotic Worlds: From Order to Disorder in Gravitational N-Body Dynamical Systems
Cortina, Italy
8-20 September 2003

A C.I.P. Catalogue record for this book is available from the Library of Congress.

ISBN-10 1-4020-4705-3 (PB)
ISBN-13 978-1-4020-4708-3 (PB)
ISBN-10 1-4020-4704-5 (HB)
ISBN-13 978-1-4020-4704-6 (HB)
ISBN-10 1-4020-4706-1 (e-book)
ISBN-13 978-1-4020-4706-0 (e-book)

Published by Springer,
P.O. Box 17, 3300 AA Dordrecht, The Netherlands.

www.springer.com

Printed on acid-free paper

Table of Contents

Preface

Research in N-body gravitational dynamics has focused on analytical and numerical studies of ordered systems, with chaotic behaviour or disorder being the fascinating anomaly. Recently, however, scientists have begun to forge the link between order and disorder: designing new methods to recognise chaos signature and developing new theories on the sources of chaotic behaviour. They are applying these techniques successfully to solve real-world problems in planetary, stellar and galactic dynamics.

This textbook is the culmination of the latest NATO Advanced Study Institute (ASI) in the Cortina series on Celestial Mechanics and Dynamical Astronomy entitled Chaotic Worlds: From Order to Disorder in Gravitational N-Body Dynamical Systems. Based on the lectures of this NATO ASI delivered by internationally renowned scientists, the book contains invaluable teaching on the latest methods of analysis for understanding and investigating ordered and chaotic behaviour in N-Body dynamics and their application to real motion in planetary systems. The book is written for researchers at PhD level in N-body gravitational dynamics, providing a systematic development from the fundamental mathematics which underpin modern studies of ordered and chaotic behaviour in N-Body dynamics to a clear view of the most recent developments and applications.

In the recent past, analytical research in Celestial Mechanics has centred on KAM theory and its applications to the dynamics of low dimensional Hamiltonian systems. Results were used to interpret observed solutions to three body problems. Order was expected and chaos or disorder the exception. Researchers turned to the curious exception, designing analytical models to study the chaotic behaviour at resonances and the effects of resonant overlaps. Numerical simulations were completed with ever longer integration times, in attempts to explore the manifestations of chaos. These methods improved our understanding but left much unexplained phenomena.

Today, chaos is the norm, order the exception. Real-world dynamics is being modeled by considering a greater number of forces acting on the bodies. These forces may drive bodies across barriers fictitiously placed by simpler models. Thus, in the context of asteroids for example, the puzzle is no longer to find a chaotic evolution able to drive an asteroid across the asteroid belt, but to find those asteroids that, like the Hildas, are able to remain in a bounded domain of the belt for times longer than the age of the Solar System (Ferraz-Mello). The twentieth century view that chaos is too difficult for our science to understand is being challenged.

We are beginning to understand the sources of Chaos. Strong chaos, though uncommon, has dramatic results, being produced by collisions (singularities in gravitational systems) or close approaches between celestial bodies. Close encounters are a major factor in the evolution of Earth crossing asteroids, the ejection of comets and asteroids from the Solar System and the evolution of many body clusters of stars. Weak chaos, the predominant form of chaos, is due to the interplay between resonances and results in the slow diffusion of bodies. Nekhoroshev's theorem is one of the most outstanding results in this area, as it provides global results valid for every initial condition in a given domain, while KAM theory provides only local results. In particular, the geometric construction of Nekhoroshev's theorem has played a major role in furthering our understanding of the complicated structure of the resonance web. Different numerical tools have been developed in order to visualize such structure with the aim of providing a global study of dynamical systems.

We are beginning to understand chaotic structure in a way not seen before. Numerical methods of measuring chaotic and regular behaviour such as Fast Liapunov Indicators, sup-maps, twist-angles, Frequency Map Analysis, fourier spectal analysis are providing lucid maps of the global dynamical behaviour of multidimensional systems. Fourier spectral analysis of orbits looks to be a powerful tool for the study of Nekhoroshev type stability. Identification of the main resonances and measures of the diffusion of trajectories can be found easily and quickly. Applied to the full N-body problem without simplification, use of these tools is beginning to explain the observed behaviour of real physical systems.

There is clearly now a diverse and exciting range of applications of these tools for understanding and measuring order and disorder in N body gravitational dynamical systems - spanning the fields of planetary, stellar and galactic astronomy. In the Solar System context alone, current applications include the study of: the dynamics and long-term evolution of massless particles such as the Kuiper belt objects; the diffusion of asteroids, motivated by the need to detect and understand the evolution of Earth- crossing asteroids; the chaotic routes of comets and meteoroids; the chaotic obliquity of planets, motivated by the desire to explain the final rotation states of Earth, Mars, and Venus; the chaotic behavior resulting from resonance between spin and orbital motion to explain behaviour such as that of Miranda and the effect of resonance on the tidal heating of satellite surfaces such as that of Europa. Bridging the gap from our own solar system to the stars is the study of stability and chaotic behaviour of extrasolar planetary systems. In stellar dynamics, the challenge is to understand how clusters of stars evolve through close encounters and resultant ejection of stars, while on a galactic scale the developing theory of chaotic motion in celestial mechanics has serious application in the relaxation of galactic structures.

The eleventh NATO ASI in the Cortina series held September 8-20, 2003, saw the return of the Institute to its origins in Cortina dAmpezzo, Italy a famous mountain resort town in the heart of the Dolomitti. The two week long ASI brought together around eighty scientists from more than thirty countries. The School once again proved to be highly successful and popular, enabling the excellent sharing of knowledge and demonstrating the wonderful spirit of the Celestial Mechanics community.

We would like to especially thank Dr Andras Széll (Glasgow Caledonian University) and Dr Shoaib Afridi (Glasgow Caledonian University) for their dedication and hard work which contributed so much to the success of the School and the final textbook. We are also indebted to the NATO Scientific Affairs Division for their valuable help and sponsorship. Further information on the NATO ASI and the Cortina series can be found at: http://www.astro.gla.ac.uk/users/martin/nato/cortina03.html. We hope you enjoy this text and look forward to meeting new friends and old at the next Cortina School in 2007.

Bonnie A Steves
Andrzej J Maciejewski
Martin Hendry

Lecturers

Alessandra Celletti	Università di Roma Tor Vergata, Italy
Sylvio Ferraz-Mello	Universidade de São Paulo, Brazil
Claude Froeschlé	Observatoire de Nice, France
Antonio Giorgilli	Università di Milano Bicocca, Italy
Richard Greenberg	University of Arizona, USA
Massimiliano Guzzo	Università degli Studi di Padova, Italy
John Hadjedemetriou	University of Thessaloniki, Greece
Ugo Locatelli	Università di Roma Tor Vergata, Italy
Andrzej J Maciejewski	University of Zielona Gora, Poland
Christian Marchal	General Scientific Direction, ONERA, Chatillon, France
Philip James Message	University of Liverpool, UK
Andrea Milani	Università di Pisa, Italy
Carles Simó	Universitat de Barcelona, Spain
Bonnie A Steves	Glasgow Caledonian University, UK
Andras Széll	Glasgow Caledonian University, UK
Giovanni Valsecchi	Istituto di Astrofisica Spaziale e Fisica Cosmica, Roma, Italy
Jörg Waldvogel	Swiss Federal Institute of Technology ETH, Zurich, Switzerland

I

TOOLS OF ORDER

CANONICAL PERTURBATION THEORY FOR NEARLY INTEGRABLE SYSTEMS

Introduction to the canonical perturbation theory for nearly integrable systems

Antonio Giorgilli
Università di Milano Bicocca, Dipartimento di Matematica e Applicazioni, Via degli Arcimboldi 8, 20126 Milano, Italy.

Ugo Locatelli
Università di Roma "Tor Vergata", Dipartimento di Matematica, Via della Ricerca Scientifica 1, 00133 Roma, Italy.

Abstract These lectures are devoted to the main results of classical perturbation theory. We start by recalling the methods of Hamiltonian dynamics, the problem of small divisors, the series of Lindstedt and the method of normal form. Then we discuss the theorem of Kolmogorov with an application to the Sun-Jupiter-Saturn problem in Celestial Mechanics. Finally we discuss the problem of long-time stability, by discussing the concept of exponential stability as introduced by Moser and Littlewood and fully exploited by Nekhoroshev. The phenomenon of superexponential stability is also recalled.

Keywords: Hamiltonian systems, perturbation theory, normal forms, KAM theory, Nekhoroshev's theorem, planetary systems

1. Ouverture

"Nous sommes donc conduit à nous proposer le problème suivant: Étudier les équations canoniques

$$\frac{dx_i}{dt} = \frac{\partial F}{\partial y_i}, \quad \frac{dy_i}{dt} = -\frac{\partial F}{\partial x_i},$$

en supposant que la function F peut se développer suivant les puissances d'un paramètre très petit μ de la manière suivante:

$$F = F_0 + \mu F_1 + \mu^2 F_2 + \dots,$$

B.A. Steves et al. (eds.), Chaotic Worlds: from Order to Disorder in Gravitational N-Body Dynamical Systems, 1–41.

en supposant de plus que F_0 ne dépend que des x et est indépendent des y; et que F_1, $F_2, \ldots$ sont des fonctions périodiques de période 2π par rapport aux y."

According to Poincaré, this is *Le problème général de la dynamique*, as stated in §13 of *Les méthodes nouvelles de la Mécanique Céleste, Tome I.*

In more usual notations and words, the problem is to investigate the dynamics of a canonical system of differential equations with Hamiltonian

$$H(p, q, \varepsilon) = H_0(p) + \varepsilon H_1(p, q) + \varepsilon^2 H_2(p, q) + \ldots, \tag{1}$$

where $p \equiv (p_1, \ldots, p_n) \in \mathcal{G} \subset \mathbf{R}^n$ are action variables in the open set $\mathcal{G}$, $q \equiv (q_1, \ldots, q_n) \in \mathbf{T}^n$ are angle variables, and ε is a small parameter.

Unless explicitly stated, every function in these lectures will be assumed to be analytic in the canonical variables p, q and in the perturbation parameter ε.

2. The Classical theory

The aim of this section is to recall some fundamental theorems of Classical Mechanics.

We first recall a few notions concerned with the canonical formalism that are needed in the rest of the lecture. We shall always assume that the Hamiltonian is autonomous, i.e., time independent.

2.1 Integrable canonical systems

We shall place ourselves in the Hamiltonian framework. We consider a $2n$-dimensional phase space $\mathcal{F}$ endowed with canonical coordinates $q_1, \ldots, q_n, p_1, \ldots, p_n$. The flow in the phase space is determined by a smooth Hamiltonian function $H(q, p, t)$ via the Hamilton's equations

$$\dot{q}_j = \frac{\partial H}{\partial p_j}, \quad \dot{p}_j = -\frac{\partial H}{\partial q_j}, \quad j = 1, \ldots, n\,.$$

The time derivative of a smooth function $\Phi(q, p)$ defined on the phase space is

$$\dot{\Phi} = \{\Phi, H\},$$

where $\{\cdot, \cdot\}$ denotes the Poisson bracket, defined as

$$\{f, g\} = \sum_{j=1}^{n} \left(\frac{\partial f}{\partial q_j} \frac{\partial g}{\partial p_j} - \frac{\partial f}{\partial p_j} \frac{\partial g}{\partial q_j} \right).$$

A time-independent function Φ is said to be a *first integral* in case $\dot{\Phi} = 0$ for the flow determined by the Hamiltonian H. Thus, a smooth first integral needs to satisfy the equation $\{\Phi, H\} = 0$. A coordinate transformation $q_j = q_j(Q,P)\,,\; p_j = p_j(Q,P)\;\; j = 1,\ldots,n$ is said to be canonical in case it preserves the Hamiltonian form of the equations. The following sufficient condition will be used below: *let the generating function* $S(P,q)$ *satisfy* $\det\left(\frac{\partial^2 S}{\partial P_j \partial q_k}\right) \neq 0$. *Then the transformation implicitly defined by* $Q_j = \frac{\partial S}{\partial P_j}(P,q)\,,\; p_j = \frac{\partial S}{\partial q_j}(P,q)$ *is canonical.* This will be enough for our purposes, so we skip the discussion concerning time dependent canonical trasformations or transformation that can not be expressed in the form above.

Let us now come to the property of integrability in Liouville's sense. To this end we need the following definition: a set $\Phi_1,\ldots,\Phi_n$ of independent functions is said to be a *complete involution system* in case it satisfies $\{\Phi_j,\Phi_k\} = 0$ for every pair j,k. A classical theorem due to Liouville is the following:

THEOREM 1 *Let the Hamiltonian $H(q,p)$ possess n independent first integrals that form a complete involution system. Then the system is integrable by quadratures.*

The process of integration goes through the construction of a generating function $S(\Phi,q) = \int \sum_j p_j(\Phi,q)\,dq_j$ of a canonical transformation $(\Phi,\alpha) = \mathcal{C}(q,p)$ such that the transformed Hamiltonian depends only on the new momenta $\Phi_1,\ldots,\Phi_n$. Thus, the solutions are written as

$$\Phi_j(t) = \Phi_{j,0}, \quad \alpha_j(t) = \alpha_{j,0} + t\frac{\partial H}{\partial \Phi_j}\bigg|_{(\Phi_{1,0},\ldots,\Phi_{n,0})}, \qquad j = 1,\ldots,n,$$

with $\alpha_{j,0}$ and $\Phi_{j,0}$ determined by the initial data.

Although the solution furnished by Liouville is actually complete, it does not bring into light the geometry of the orbits. This is better enlighted by the more recent theorem due to Arnold (1963a) and Jost (1968), essentially based on the existing theory of adiabatic invariants. The theorem essentially exploits the fact that the choice of the first integrals to be used in Liouville's construction is rather arbitrary. Actually, in most interesting cases there are good first integrals that generate a set of so called *action-angle variables.*

THEOREM 2 *Let the system be integrable in Liouville's sense, and assume that for some $c \in \mathbf{R}^n$ the level surface determined by the equations $\Phi_1(q,p) = c_1,\ldots,\Phi_n(q,p) = c_n$ contains a compact and connected component M_c. Then in a neighborhood $U(M_c)$ of M_c there are canonical* *action-angle coordinates* I, ϑ *mapping* $\mathcal{G} \times \mathbf{T}^n$ *to* $U(M_c)$*, where* $\mathcal{G} \in \mathbf{R}^n$ *is an open set, such that the Hamiltonian depends only on* $I_1, \ldots, I_n$*; the corresponding flow is*

$$\vartheta_j(t) = \vartheta_{j,0} + t\omega_j(I_{1,0}, \ldots, I_{n,0}), \quad I_j(t) = I_{j,0}, \quad j = 1, \ldots, n,$$

where $\vartheta_{j,0}$ *and* $I_{j,0}$ *are the initial data, and* $\omega_j = \frac{\partial H}{\partial I_j}$.

Thus, the orbits in the domain $\mathcal{G} \times \mathbf{T}^n$ of phase space lie on invariant tori parameterized by the action variables $I_1, \ldots, I_n$, and the motion on each torus is a Kronecker flow with frequencies $\omega_1(I), \ldots, \omega_n(I)$.

2.2 The theorem of Poincaré on non-existence of first integrals

Let us now come back to the general problem of dynamics, i.e., eq. (1), renaming the action variables as $p_1, \ldots, p_n$ and the angles as $q_1, \ldots, q_n$. For $\varepsilon = 0$ the unperturbed system with Hamiltonian $H_0(p)$ is clearly integrable in the sense above. The first naive attempt is to find first integrals for the perturbed system, i.e., the system with $\varepsilon \neq 0$. However, this is program will likely fail, in view of the following negative result due to Poincaré (1892, Tome 1, §13).

THEOREM 3 *Let us assume that the unperturbed Hamiltonian* H_0 *is non-degenerate, i.e.,*

$$\det\left(\frac{\partial^2 H_0}{\partial p_j \partial p_k}\right) \neq 0;$$

assume also the genericity condition that no coefficient $h_k(p)$ *of the Fourier expansion of* $H_1(p,q)$ *is identically zero on the manifold* $\langle k, \omega(p)\rangle = 0$*. Then there is no analytic first integral independent of* H.

It is interesting to recall the scheme of proof; the interested reader will find the details in the *Méthodes Nouvelles.* One tries to construct a first integral $\Phi(p,q,\varepsilon) = \Phi_0(p,q) + \varepsilon\Phi_1(p,q) + \varepsilon^2\Phi_2(p,q) + \ldots$ as a series expansion in the perturbation parameter ε by solving the equation $\{H, \Phi\} = 0$. This gives the recursive system

$$\begin{aligned}
\partial_\omega \Phi_0 &= 0, \\
\partial_\omega \Phi_1 &= -\{H_1, \Phi_0\}, \\
&\ldots\ldots, \\
\partial_\omega \Phi_s &= -\{H_1, \Phi_{s-1}\} - \ldots - \{H_s, \Phi_0\}, \\
&\ldots\ldots,
\end{aligned} \tag{2}$$

where $\partial_\omega = \sum_j \omega_j(p)\frac{\partial}{\partial q_j}$. The first equation is solved by any function $\Phi_0(p)$ independent of the actions, and in view of the nondegeneracy condition there are no other solutions. Thus, the problem is how to solve the equation $\partial_\omega \Phi_s = \Psi_s$ where $\Psi_s(p,q)$ is a known function. This is done as follows. Expand $\Psi_s(p,q)$ and $\Phi_s(p,q)$ in Fourier series,

$$\Psi_s(p,q) = \sum_{k\in\mathbf{Z}^n} \psi_k(p)\exp(i\langle k,q\rangle), \quad \Phi_s(p,q) = \sum_{k\in\mathbf{Z}^n} \varphi_k(p)\exp(i\langle k,q\rangle),$$

with known coefficients $\psi_k(p)$ and unknown $\varphi_k(p)$. Then the coefficients $\varphi_k(p)$ must satisfy the equation

$$i\langle k,\omega(p)\rangle\,\varphi_k(p) = \psi_k(p).$$

The formal solution is

$$\varphi_k(p) = -i\frac{\psi_k(p)}{\langle k,\omega(p)\rangle},$$

but is valid only if $\psi_k(p) = 0$ whenever $\langle k,\omega(p)\rangle = 0$. As shown by Poincaré , this happens only if $\Phi_0(p)$ is not independent of H_0, which in turn implies that Φ is not independent of H.

Thus, we clearly see that the heart of the problem of solving the general problem of dynamics is connected with the appearance of the so called *small divisors* $\langle k,\omega(p)\rangle$.

2.3 The Duffing's model and the series of Lindstedt

The problem of small divisors is related to another well known problem that shows up in classical perturbation theory, namely the problem of secular terms. Let us illustrate the problem with a very simple example.

We consider the Duffing's equation

$$\ddot{x} + \omega^2 x = \varepsilon(\cos\nu t + x^3), \tag{3}$$

describing a non-linear oscillator subjected to a small perturbation which is $\frac{2\pi}{\nu}$-periodic in time. The equation may be derived from the time-dependent Hamiltonian

$$H(x,y,t) = \frac{1}{2}\left(y^2 + \omega^2 x^2\right) - \varepsilon\left(x\cos\nu t + \frac{x^4}{4}\right).$$

where y is the momentum conjugated to the coordinate x. Let us fix the initial conditions $x(0) = a$, $\dot{x}(0) = 0$, and look for a solution in the form of a power series in the parameter ε, namely

$$x(t) = x_0(t) + \varepsilon x_1(t) + \varepsilon^2 x_2(t) + \dots.$$

By substitution in (3) we get the recursive system of equations

$$\begin{aligned}
\ddot{x}_0 + \omega^2 x_0 &= 0, \\
\ddot{x}_1 + \omega^2 x_1 &= \cos\nu t + x_0^3, \\
\ddot{x}_2 + \omega^2 x_2 &= 3x_0^2 x_1, \\
&\ldots, \\
\ddot{x}_s + \omega^2 x_s &= \psi_s(x_0, \ldots, x_{s-1}), \\
&\ldots,
\end{aligned}$$

where $\psi_s(x_0, \ldots, x_{s-1})$ is a known function. The first equation has the well known solution $x_0(t) = a\cos\omega t$. Replacing this in the second equation we get

$$\ddot{x}_1 + \omega^2 x_1 = \cos\nu t + \frac{3a^3}{4}\cos\omega t + \frac{a^3}{4}\cos 3\omega t,$$

which is easily solved as

$$x_1 = \frac{1}{\omega^2 - \nu^2}\cos\nu t + \frac{3a^3}{8\omega} t\sin\omega t - \frac{a^3}{32\omega^2}\cos 3\omega t,$$

provided $\omega \neq \nu$. However, this solution contains a non-periodic term with a coefficient which is linear in t; this is a *secular term.* A moment's thought will allow one to realize that the function ψ_s on the r.h.s. of the equation for the coefficient of order s in ε will be a trigonometric polynomial with coefficients that are polynomials of degree $s-1$ in t, and so $x_s(t)$ will contain coefficients of degree s.

The elimination of secular terms from the power series expansion of the solution is achieved by the method of Lindstedt. The underlying idea is to pick a fixed frequency μ, and to look for a quasi-periodic solution with basic frequencies μ and ν. This is actually the same thing as looking for a quasi-periodic orbit on an invariant 2-dimensional torus. The process of solution is the following. Write the Duffing's equation as

$$\ddot{x} + \mu^2 x = \varepsilon(\cos\nu t + x^3 - \delta x).$$

where $\mu^2 = \omega^2 + \varepsilon\delta$, and look again for a solution in power series of ε. This gives the recursive system of equations

$$\begin{aligned}
\ddot{x}_0 + \mu^2 x_0 &= 0, \\
\ddot{x}_1 + \mu^2 x_1 &= \cos\nu t + x_0^3 - \delta x_0, \\
\ddot{x}_2 + \mu^2 x_2 &= 3x_0^2 x_1 - \delta x_1, \\
&\ldots,
\end{aligned}$$

$$\ddot{x}_s + \mu^2 x_s = \psi_s(x_0, \ldots, x_{s-1}),$$
$$\ldots,$$

The first equation has the simple solution $x_0 = a_0 \cos \mu t$ with an arbitrary coefficient a_0 (let's forget the initial phase). Replacing it in the second equation we get

$$\ddot{x}_1 + \mu^2 x_1 = \cos \nu t + \frac{3a_0^3}{4} \cos \mu t - \delta a_0 \cos \mu t + \frac{a_0^3}{4} \cos 3\mu t$$

Here comes the crucial point: we determine the unperturbed amplitude a_0 as a solution of the equation $\frac{3a_0^2}{8} - \delta a_0 = 0$, so that the resonant term in the equation for x_1 is removed. That is, the amplitude of the oscillation is determined as a function of the frequency μ. Thus, we get the solution

$$x_1 = \frac{1}{\mu^2 - \nu^2} \cos \nu t - \frac{a^3}{32\mu} \cos 3\mu t + a_1 \cos(\mu t + \varphi_1),$$

with undetermined constants a_1, φ_1. The solution does not contain secular terms; however, the frequency μ must satisfy $\mu \pm \nu \neq 0$. The process can be repeated at order $2, 3, \ldots$ using the arbitrary solution of the homogeneous equation in order to remove the resonant terms. However, it is not difficult to check that the r.h.s. of the equation for x_s is a trigonometric polynomial containing terms of the form $\frac{\sin}{\cos}(j\mu + k\nu)t$, where j, k are arbitrary integers such that $|j| \leq 2s + 1$ and $|k| \leq s$. Each of these terms introduces in the solution a coefficient with a denominator $\mu^2 - (j\mu + k\nu)^2$, which introduces further conditions on the frequency μ. Letting $s \to \infty$, one realizes that the ratio μ/ν must be at least irrational, in order to avoid vanishing divisors; on the other hand, this does not forbid the denominators to become arbitrarily small, thus raising serious doubts on the convergence of the series.

The problem of the convergence of Lindstedt's series has been accurately investigated by Poincaré (1892, tome II, chap. XIII, § 148–149). However, he was unable to decide this question (here, n_1, n_2 play the role of our frequencies μ, ν):

"*... les séries ne pourraient-elles pas, par example, converger quand ... le rapport n_1/n_2 soit incommensurable, et que son carré soit au contraire commensurable (ou quand le rapport n_1/n_2 est assujetti à une autre condition analogue à celle que je viens d'énoncer un peu au hasard)? Les raisonnements de ce chapitre ne me permettent pas d'affirmer que ce fait ne se présentera pas. Tout ce qu'il m'est permis de dire, c'est qu'il est fort invraisemblable.*"

The challenging question of the convergence of the series of Lindstedt has been solved indirectly by Kolmogorov (1954). A direct proof of the convergence has revealed to be a major challenge, due to compensations among fast growing terms generated by the algorithm of Lindstedt. This phenomenon has been understood only three decades later, with the work of Eliasson (1988) and some successive works (see, e.g., Gallavotti 1994; Chierchia and Falcolini 1996).

2.4 The method of normal form

Let us go back to the general problem of dynamics. The goal is to perform a near the identity canonical transformation that gives the Hamiltonian a suitable form, that we shall generically call a *normal form*. We shall use the method based on composition of Lie series. Let us briefly recall how the method works. A near the identity transformation is produced by the canonical flow at time ε of a generating function $\chi(p,q)$, and takes the form

$$\begin{aligned} p &= \exp(\varepsilon L_\chi)\, p' = p' + \varepsilon \left.\frac{\partial\chi}{\partial q}\right|_{p',q'} + \frac{\varepsilon^2}{2} L_\chi \left.\frac{\partial\chi}{\partial q}\right|_{p',q'} + \ldots \\ q &= \exp(\varepsilon L_\chi)\, q' = q' + \varepsilon \left.\frac{\partial\chi}{\partial p}\right|_{p',q'} + \frac{\varepsilon^2}{2} L_\chi \left.\frac{\partial\chi}{\partial p}\right|_{p',q'} + \ldots, \end{aligned}$$

where $L_\chi \cdot = \{\cdot, \chi\}$.

The method presents two main advantages with respect to the usual method based on generating functions in mixed variables. Firstly, no function inversion is required. Secondly, the transformation of a function may be performed without substitutions of variables, in view of the formula

$$f(p,q)\Big|_{(q,p)=\exp(\varepsilon L_\chi)(q',p')} = \big(\exp(\varepsilon L_\chi)\, f\big)(q',p'). \tag{4}$$

That is, we don't need to make substitutions; we should simply apply the exponential operator $\exp(\varepsilon L_\chi)$ to the function, and change the name of the variables. The algorithm is effectively represented in a graphic form. Let $f = f_0 + \varepsilon f_1 + \ldots$ the function to be transformed, and assume that we want to calculate the power series expansion in ε of the transformed function $g = \exp(\varepsilon L_\chi)\, f = g_0 + \varepsilon g_1 + \ldots$. then it is enough to calculate the *Lie triangle*

$$
\begin{array}{cccccc}
g_0 & f_0 & & & & \\
 & \downarrow & & & & \\
g_1 & L_\chi f_0 & f_1 & & & \\
 & \downarrow & \downarrow & & & \\
g_2 & \frac{1}{2}L_\chi^2 f_0 & L_\chi f_1 & f_2 & & \\
 & \downarrow & \downarrow & \downarrow & & \\
g_3 & \frac{1}{3!}L_\chi^3 f_0 & \frac{1}{2}L_\chi^2 f_1 & L_\chi f_2 & f_3 & \\
 & \downarrow & \downarrow & \downarrow & \downarrow & \\
\vdots & \vdots & \vdots & \vdots & \vdots & \ddots
\end{array}
$$

The term g_s is the sum of all contribution that appear on the same row.

The drawback of the method is that not every near the identity transformation can be represented as a Lie series with a fixed generating function. However, this is not a big disadvantage: just use a *composition* of Lie series with a sequence of generating functions.

The reduction of the Hamiltonian to a normal form may be performed precisely via a sequence of transformations. A general scheme based on expansions in the small parameter ε is the following. Starting with a Hamiltonian of the form (1) we look for a (sequence of) near the identity canonical transformation(s) that produces the normal form up to a finite order r, i.e., a Hamiltonian of the form

$$H^{(r)}(p,q) = H_0(p) + \varepsilon Z_1(p,q) + \ldots + \varepsilon^r Z_r(p,q) + \varepsilon^{r+1}\mathcal{R}^{(r)}(p,q,\varepsilon).$$

Here, the functions $Z_1, \ldots, Z_r$ are in normal form in the sense that they satisfy some nice condition to be invented, and $\mathcal{R}^{(r)}(p,q,\varepsilon)$ is a unnormalized remainder. The upper index recalls the fact that the normal form is determined up to order r. Our dream is to allow $r \to \infty$. Let us see how to do it with a recursive procedure. Let us assume that the Hamiltonian has been given the normal form up to order $r-1$; this is true, e.g., for $r = 1$, just considering the Hamiltonian (1) with no functions Z. Thus, we have determined the Hamiltonian as

$$H^{(r-1)}(p,q) = H_0(p) + \varepsilon Z_1(p,q) + \ldots + \varepsilon^{r-1} Z_{r-1}(p,q) + \varepsilon^r H_r^{(r-1)}(p,q) + \ldots$$

We look for a generating function $\chi_r(p,q)$ such that

$$\exp(\varepsilon^r L_{\chi_r}) H^{(r-1)} = H_0 + \varepsilon Z_1 + \ldots + \varepsilon^r Z_r + \ldots$$

with Z_r in normal form (whatever it means). By isolating the the coefficient of ε^r we get the condition $H_r^{(r-1)} + L_{\chi_r} H_0 = Z_r$. Thus, the problem is to solve the equation

$$\partial_\omega \chi_r + Z_r = H_r^{(r-1)}, \quad \partial_\omega = L_{H_0} = \sum_j \omega_j(p) \frac{\partial}{\partial q_j} \tag{5}$$

for χ_r and Z_r. Assuming that we know how to do it, we determine $H^{(r)} = \exp(\varepsilon^r L_{\chi_r}) H^{(r-1)}$ by constructing the appropriate Lie triangle. An explicit formula for the transformed Hamiltonian is

$$writtH_{sr+m}^{(r)} = \frac{1}{s!} L_{\chi_r}^s Z_m + \sum_{k=0}^{s-1} \frac{1}{k!} L_{\chi_r}^k H_{(s-k)r+m}^{(r-1)},$$

for $r \geq 2$, $s \geq 1$, and $1 \leq m < r$, and

$$H_{sr}^{(r)} = \frac{1}{(s-1)!} L_{\chi_r}^{s-1} \left(\frac{1}{s} Z_r + \frac{s-1}{s} H_r^{(r-1)} \right) + \sum_{k=0}^{s-2} \frac{1}{k!} L_{\chi_r}^k H_{(s-k)r}^{(r-1)},$$

for $r \geq 1$ and $s \geq 2$. This is the wanted normal form up to order r.

Thus, the procedure that gives the Hamiltonian a normal form is well established in all details but one: nothing has been said on what the normal form actually is. The point is that the properties that characterize the normal form depend on what one is looking for. On the other hand, the restrictions are dictated by eq. (5), which must admit a solution.

2.5 The normal form of Birkhoff

The classical and most common attempt, already discussed by Poincaré, consists in trying to remove all dependencies on the angles from the Hamiltonian. This is usually called *the normal form of Birkhoff*. Such a normal form turns out to be particularly useful when the unperturbed Hamiltonian is linear, i.e., in (1) we have $H_0(p) = \langle \omega, p \rangle$ with some $\omega \in \mathbf{R}^n$. Therefore we illustrate the theory in the latter case. However, with some caveat, the method is useful also in the general case: see Section 4.3 on Nekhoroshev's theorem.

To the frequency vector ω we associate the *resonance module* $\mathcal{M}_\omega$ defined as

$$\mathcal{M}_\omega = \{ k \in \mathbf{Z}^n \ : \ \langle k, \omega \rangle = 0 \}; \tag{6}$$

This is a subgroup of $\mathbf{Z}^n$. The dimension $m = \dim \mathcal{M}_\omega$ is called the *multiplicity* of the resonance of ω; this is the number of independent integer vectors $k \in \mathcal{M}_\omega$ satisfying $\langle k, \omega \rangle = 0$. The frequency vector ω is said to be *non-resonant* in case $\dim \mathcal{M}_\omega = 0$.

We say that the function $Z(p,q)$ is in normal form with respect to $\mathcal{M}_\omega$ in case its Fourier expansions contains only Fourier modes belonging to $\mathcal{M}_\omega$, i.e., it has the form

$$Z_s = \sum_{k\in\mathcal{M}_\omega} z_k(p)\exp(i\langle k,q\rangle).$$

If $\dim\mathcal{M}_\omega = 0$ then the normal form depends only on the action variables p. Therefore, the system in normal form turns out to be integrable, because it possesses the n independent first integrals $p_1,\ldots,p_n$. If instead $1 \le \dim\mathcal{M}_\omega < n$ then the normal form possesses $n - \dim\mathcal{M}_\omega$ independent first integrals of the form

$$\Phi_\nu = \langle \nu, p\rangle \quad \text{with } \mathcal{M}_\omega \perp \nu \in \mathbf{R}^n\,.$$

All these results hold true in formal sense, i.e., disregarding the problem of convergence of the series.

Let us show how the equation (5) for the generating function and the normal form may be solved. Expand the known function $H_r^{(r-1)}$ in Fourier series as

$$H_r^{(r-1)} = \sum_{k\in\mathbf{Z}^n} h_k(p)\exp(i\langle k,q\rangle);$$

then determine the corresponding coefficients $z_k(p)$ of Z_r and $c_k(p)$ of χ_r as

$$\begin{aligned} z_k(p) &= h_k(p), & c_k(p) &= 0 & &\text{for } k\in\mathcal{M}_\omega \\ z_k(p) &= 0, & c_k(p) &= -i\frac{h_k(p)}{\langle k,\omega\rangle} & &\text{for } k\notin\mathcal{M}_\omega. \end{aligned}$$

The definition of $\mathcal{M}_\omega$ assures that no denominator in c_k vanishes, thus making the construction consistent. However, the denominators may assume arbitrarily small values, thus raising doubts on the convergence of the series[1]. It may happen that the frequencies ω are not resonant, but quite close to a resonant value. Thus, it is desirable not to include in the generating function terms that have a very small denominator. In this case the normal form is conveniently defined with reference to a resonance module $\mathcal{M}$ such that[2] $\mathcal{M}_\omega \subset \mathcal{M}$. The latter condition excludes zero denominators from the generating function, thus assuring the consistency of the whole procedure.

---

[1]The reader will notice that for $\dim\mathcal{M}_\omega = n-1$ the denominators are bounded away from zero, being all multiples of a fixed quantity. However, this is not enough to guarantee the convergence of the series.

[2]A resonance module may be defined as a subgroup $\mathcal{M}\subset\mathbf{Z}^n$ satisfying the condition $\mathbf{Z}^n \cap \mathrm{span}(\mathcal{M}) = \mathcal{M}$, where $\mathrm{span}(\mathcal{M})$ is the subspace of $\mathbf{R}^n$ spanned by $\mathcal{M}$, and $\mathbf{Z}^n$ is considered as immersed in $\mathbf{R}^n$.

As we know from theorem 3, due to Poincaré, in the general case of a non-degenerate unperturbed Hamiltonian $H_0(p)$ the construction of the normal form can not be completely performed in a consistent manner. However, we are still allowed to perfom a suitable construction in domains where some resonances may be excluded, provided we perform a Fourier cutoff on the perturbation. We shall discuss this method in connection with Nekhoroshev's theorem.

3. Invariant tori

This section is devoted to the statement of the theorem of Kolmogorov on the persistence of Invariant tori. We first state the theorem and recall in some detail the formal method invented by Kolmogorov. Then we show how the original scheme can be rearranged in the form of a constructive algorithm, suitable for an explicit calculation via algebraic manipulations. Finally we shall apply this method to the problem of three bodies.

3.1 The theorem of Kolmogorov

A first application of the normal form method is the proof of the theorem of Kolmogorov on the persistence of invariant tori. We outline here the basic scheme proposed in Kolmogorov's original paper.

We shall say that a Hamiltonian system is in Kolmogorov's normal form in case the Hamiltonian function is written as

$$H(p,q) = \sum_{j=1}^{n} \omega_j p_j + \mathcal{R}(p,q), \tag{7}$$

where $\omega \in \mathbf{R}^n$ and $\mathcal{R}(p,q)$ is at least quadratic in the actions p. The usefulness of this normal form is transparent if one writes the Hamilton's equations, namely

$$\dot{q}_j = \omega_j + \frac{\partial \mathcal{R}}{\partial p_j}, \quad \dot{p}_j = -\frac{\partial \mathcal{R}}{\partial q_j}, \quad j = 1, \ldots, n,$$

and chooses the initial conditions as $p(0) = 0$, with arbitrary initial phases $q(0)$. Indeed, in this case it is immediately seen that the solution is

$$q_j(t) = \omega_j t + q_j(0), \quad p_j(t) = 0, \quad j = 1, \ldots, n,$$

since the derivatives of $\mathcal{R}$ in the r.h.s. of the equations do vanish for $p = 0$. Thus, the Hamiltonian (7) possesses an invariant torus $p = 0$, carrying quasi-periodic motions.

The question is whether we can give the Hamiltonian (1) the normal form (7). Here is a formal statement.

THEOREM 4 *Consider a canonical system with Hamiltonian*

$$H(p,q) = h(p) + \varepsilon f(p,q). \tag{8}$$

Let us assume that the unperturbed part of the Hamiltonian is non-degenerate, i.e., $\det\left(\frac{\partial^2 h}{\partial p_j \partial p_k}\right) \neq 0$, *and that* $p^* \in \mathbf{R}^n$ *is such that the corresponding frequencies* $\omega = \frac{\partial h}{\partial p}(p^*)$ *satisfy a diophantine condition, i.e.,*

$$|\langle k, \omega\rangle| \geq \gamma |k|^{-\tau} \quad \forall\, 0 \neq k \in \mathbf{Z}^n$$

with some constants $\gamma > 0$ *and* $\tau \geq n-1$. *Then for* ε *small enough the Hamiltonian (8) admits an invariant torus carrying quasi-periodic motions with frequencies* ω. *The invariant torus lies in a* ε*-neighborhood of the unperturbed torus* $\{(p,q) : p = p^*,\ q \in \mathbf{T}^n\}$.

The theorem was first stated by Kolmogorov in a very short note (Kolmogorov 1954), were the essential hints for working out the proof are also given. The first published proofs are due to Moser (1962), who proved the theorem for the case of an area preserving mapping of an annulus, and Arnold (1963b and 1963c), who introduced a scheme of proof different from that proposed by Kolmogorov. We recall here only the formal aspect of the proof.

Let us forget for a while the expansion in a parameter ε and write the Hamiltonian in the simple form

$$H(p,q) = h(p) + f(p,q),$$

where f is a small perturbation, i.e., $|f(p,q)| \leq \varepsilon$ at all points of phase space. The flow of the unperturbed Hamiltonian $h(p)$ is quasi-periodic on invariant tori parameterized by the action variables p. The non-degeneracy condition assures that the frequencies $\omega(p) = \frac{\partial h}{\partial p}$ may be used as local coordinates in place of p. Thus we may select an unperturbed torus p^* with frequency ω satisfying the diophantine condition; this reminds the method of Lindstedt, where the frequency is fixed in advance. Expanding the Hamiltonian in Taylor series around the torus p^* we get

$$\begin{aligned} H = \eta + \langle \omega, p - p^*\rangle + \tfrac{1}{2}\langle C(q)(p-p^*), (p-p^*)\rangle + \\ A(q) + \langle B(q), (p-p^*)\rangle + O((p-p^*)^3), \end{aligned}$$

where η is a constant that may be ignored, and

$$C(q) = \left(\frac{\partial^2 h}{\partial p \partial p} + \frac{\partial^2 f}{\partial p \partial p}\right)\bigg|_{p=p^*}, \quad A(q) = f(p^*, q), \quad B(q) = \frac{\partial f}{\partial p}\bigg|_{p=p^*}.$$

By a translation of the action variables we may always set $p^* = 0$, thus getting the Hamiltonian

$$H = \langle \omega, p \rangle + \tfrac{1}{2}\langle C(q)p, p \rangle + A(q) + \langle B(q), p \rangle + O(p^3), \qquad (9)$$

Thus, in order to give the Hamiltonian the normal form (7) we should remove the unwanted terms $A(q)$, $\langle B(q), p \rangle$. Remark that these terms are small provided f is small. Following Kolmogorov, we look for a canonical transformation with generating function

$$\chi(p, q) = X(q) + \langle Y(q), p \rangle + \langle \xi, q \rangle.$$

where $\xi \in \mathbf{R}^n$ and X, $Y(q)$ are functions to be determined. The first term $X(q)$ of the generating function introduces a *deformation* of the torus depending on the angles, since a straighforward application of the Lie series algorithm gives

$$p_j = p'_j + \frac{\partial X}{\partial q_j}, \quad q_j = q'_j, \quad j = 1, \ldots, n.$$

The last term $\langle \xi, q \rangle$ causes a translation of the torus by ξ. The term $\langle Y(q), p \rangle$ introduces a small change of both the actions and the angles.

By calculating the first term of the Lie transform of the Hamiltonian it is easily seen that the unwanted terms $A(q)$, $\langle B(q), p \rangle$ are removed if the generating function satisfies the equations

$$\partial_\omega X = A(q), \quad \partial_\omega Y = B(q) - C(q)\left(\xi + \frac{\partial X}{\partial q}\right).$$

The first equation is solved by following the procedure sketched in Section 2.2; no zero denominators can occur in view of the diophantine property of the frequency ω. In the second equation the real vector ξ is determined so as to compensate the average $\overline{B(q)}$ over the angles. This requires solving the linear equation $\overline{C(q)}\,\xi = \overline{B(q)}$, which admits a solution since the real matrix $\overline{C(q)}$ (i.e., the average of $C(q)$ over the angles) is assumed to be non-degenerate. Then the equation for $Y(q)$ is solved again with the method of Section 2.2. The transformed Hamiltonian turns out to have again the form (9), but with new functions $A'(q')$ and $B'(q')$ that are hopefully smaller than $A(q)$ and $B(q)$. Thus, the process may be iterated.

The problem is to prove that the iteration actually converges, thus ensuring the existence af an invariant torus. This requires several technical tools that can not fit into the present short review. A complete proof along the lines sketched above may be found, e.g., in Benettin et al. (1984). It should be stressed that the theorem of Kolmogorov proves that the perturbed invariant tori characterized by strongly non-resonant frequencies are not destroyed by small perturbations, being just deformed and possibly translated by a little amount. A better description of the set of invariant tori, as results form Arnold's version of the proof, is that they form a closed, nowhere dense set in the phase space, the relative measure of which tends to 1 as $\varepsilon \to 0$.

The existence of invariant tori may be used as an argument of stability for systems with 2 degrees of freedom, because, in this case, the invariant tori of Kolmogorov are invariant 2-dimensional surfaces that split the surface of invariant energy in disconnected regions. This implies that an orbit with initial point in the gap between two invariant tori is perpetually confined there. For systems with more than 2 degrees of freedom the complement of the invariant tori is connected, so that there is no known mechanism of confinement. Therefore, an orbit with initial point in the complement of invariant tori could move anywhere in that subset of the energy surface. Such a phenomena has been named *Arnold diffusion*, in view of the simple example proposed by Arnold (1964). However, a proof that diffusion generically occurs is still to be invented.

3.2 Expansion of the Hamiltonian

We come now to reformulating the algorithm of Kolmogorov in a form suitable for an explicit calculation via algebraic manipulation. This is the constructive method that we have actually used in order to prove the existence of KAM tori for the secular part of the problem of three bodies, with a computer-assisted method.

The first step is to identify a suitable form for the expansion of the Hamiltonian. Let us be a bit pedantic. Following Kolmogorov, we should expand the Hamiltonian (1) in Taylor series around an unperturbed torus p^* with a given frequency ω. Thus, forgetting unessential constants, we have

$$\begin{aligned} H_0(p) &= \langle \omega, p - p^* \rangle + \frac{1}{2} \langle C(p - p^*), (p - p^*) \rangle + \dots \\ H_s(p, q) &= A_s(q) + \langle B_s(q), (p - p^*) \rangle + \dots, \qquad s = 1, 2, \dots \end{aligned} \tag{10}$$

where $C = \frac{\partial^2 H_0}{\partial p \partial p}(p^*)$ is a $n \times n$ real matrix that is assumed to be non-degenerate, and the function $A_s(q)$, $B_s(q)$, ... are expressed as derivatives of $H_s(p, q)$ evaluated at $p = p^*$. Thus, we have a double expansion in powers of the small parameter ε and of $p - p^*$. With a translation on the action space we may actually consider $p^* = 0$, and so we shall do.

However, the expansion above is still impractical for our purposes, because the functions $A_s(q)$, $B_s(q)$, ... still need to be expanded in an infinite Fourier series of the angles; e.g., we should write $A(q) = \sum_{k\in\mathbf{Z}^n} a_k \exp(i\langle k,q\rangle)$. It is more convenient to work with trigonometric polynomials, so that every part of the expansion contains only a finite number of terms. To this end, we introduce a Fourier cutoff by splitting every function of the angles in an infinite number of slices that contain only a finite number of Fourier modes. This may be done in many arbitrary ways, so let us illustrate just one method. We choose an arbitray integer K, ad write, e.g.,

$$A_s(q) = \underbrace{\sum_{0\le|k|\le K} a_k\, e^{i\langle k,q\rangle}}_{A_{s,1}(q)} + \underbrace{\sum_{K<|k|\le 2K} a_k\, e^{i\langle k,q\rangle}}_{A_{s,2}(q)} + \ldots + \underbrace{\sum_{(m-1)K<|k|\le mK} a_k\, e^{i\langle k,q\rangle}}_{A_{s,m}(q)} + \ldots$$

That is, we collect in $A_{s,m}(q)$ the part of the Fourier expansion of $A(q)$ containing modes k between $(m-1)K+1$ and mK; the average (i.e., the mode $k=0$) is added to $A_{s,1}(q)$. The same splitting is applied to the functions $B_s(q)$ and to any other function of the angles resulting from the expansion (10). This is fully justified in view of the known property of analytic functions, namely that the coefficient a_k of the Fourier expansion decays to zero as $e^{-mK\sigma}$, with some positive constant σ. This assures that the size of the functions $A_{s,m}(q)$ decreases as $e^{-mK\sigma}$, thus introducing a third parameter in the expansion. The actual choice of K is rather arbitrary. In a practical calculation it may be determined by making a suitable compromise between the wanted precision (which requires a number of coefficients large enough) and the computer power (which may impose strong limitations on the number of terms that can actually be handled). A typical method, although not always the best one, is to choose K so that $e^{-K\sigma} \sim \varepsilon$. In a theoretical approach the best way is to leave K as a free parameter, and to determine it at the end.

Now we collect together all terms of the perturbation that have: (i) the same polynomial degree in p, and (ii) the same global degree in ε and $e^{-K\sigma}$. This still includes a finite number of terms. We shall denote such a term by $f_l^{(0,m)}(p,q)$, meaning that it is a homogeneous polynomial of degree m in the action variables p and a trigonometric polynomial of degree mK in the angles. The extra upper index 0 simply reminds us that this is the Hamiltonian at zeroth step of the process of normalization;

the index will be increased step-by-step. E.g., we set $f_0^{(0,1)} = \varepsilon A_{1,1}(q)$, $f_0^{(0,2)} = \varepsilon^2 A_{2,1}(q) + \varepsilon A_{1,2}(q)$, $f_1^{(0,1)} = \varepsilon \langle B_{1,1}(q), p\rangle$, and so on.

We emphasize that the properties of $f_l^{(m)}$ of being a homogeneous polynomial of degree l in the actions p and a trigonometric polynomial of degree mK in the angles q is preserved by the operation of Poisson bracket. This makes this scheme particularly suitable for performing canonical transformations via the algorithm of Lie series.

3.3 The algorithm

Having rearranged the initial Hamiltonian $H^{(0)}$ according to the scheme of the previous section, let us represent it as a diagram

$$
\begin{array}{cccccc}
 & \vdots & \vdots & \vdots & \vdots & \cdots \\
H^{(0)} \ : & h_2 & f_2^{(0,1)} & f_2^{(0,2)} & f_2^{(0,3)} & \cdots \\
 & \langle \omega, p\rangle & f_1^{(0,1)} & f_1^{(0,2)} & f_1^{(0,3)} & \cdots \\
 & 0 & f_0^{(0,1)} & f_0^{(0,2)} & f_0^{(0,3)} & \cdots
\end{array}
$$

where terms of the same perturbation order are aligned by columns, and terms of the same degree in p are aligned on rows. The vertical dots represent terms of degree higher than 2 in p. We may always assume that the functions in the lowest row have zero average, because this would just contribute an innocuous constant to the Hamiltonian. Recall that $h_2(p) = \frac{1}{2}\langle Cp, p\rangle$, where C is a real symmetric $n \times n$ matrix which is non-degenerate in view of the nondegeneracy condition on the Hamiltonian.

We perform a first canonical transformation with generating function $\chi_1^{(1)}(q) = X^{(1)}(q) + \langle \xi^{(1)}, q\rangle$, where the function $X(q)$ and the real vector ξ are determined from the equations

$$\{X^{(1)}, \langle \omega, p\rangle\} + f_0^{(0,1)} = 0, \quad \overline{\{\chi_1^{(1)}, h_2\} + f_1^{(0,1)}} = 0$$

where the overline denotes average with respect to the angles. The equations are better written in the form

$$\partial_\omega X^{(1)} = f_0^{(0,1)}, \quad \langle C\xi^{(1)}, p\rangle = \overline{f_1^{(0,1)}}. \tag{11}$$

The latter equation is justified by simply calculating the Poisson bracket

$$\{\chi_1^{(1)}, h_2\} = \{\langle \xi^{(1)}, q\rangle, h_2\} + \{X^{(1)}, h_2\} = \langle C\xi^{(1)}, p\rangle + \{X^{(1)}, h_2\},$$

and remarking that the average of the second term over the angles is clearly zero. The first equation (11) is solved using the Fourier expansion of $X^{(1)}(q)$, and following the procedure of Section 2.2. Thus, $X^{(1)}(q)$ is a trigonometric polynomial of degree K. The second equation is linear in p, and reduces to a linear equation for ξ that admits a unique solution because C is non-degenerate.

We now perform the transformation, thus constructing an intermediate Hamiltonian $\hat{H}^{(1)} = \exp(L_{\chi_1^{(1)}})H^{(0)}$. This means that we should apply the operator $\exp(L_{\chi_1^{(1)}})$ to every term in the diagram above for $H^{(0)}$. Here we must pay a little attention to the action of the Lie derivative $L_{\chi_1^{(1)}}$ on a function $f_l^{(0,m)}(p,q)$. Since $\chi_1^{(1)}$ is independent of p, the Poisson bracket decrements by one degree on p; on the other hand, since $\chi_1^{(1)}$ is a trigonometric polynomial of degree K it increments by K the trigonometric degree. This is illustrated in the following diagram:

$$
\begin{array}{cccccccccc}
 & \vdots & & \vdots & & \vdots & & \vdots & & \dots \\
 & & \searrow & & \searrow & & \searrow & & \searrow & \\
 & h_2 & & \hat{f}_2^{(0,1)} & & \hat{f}_2^{(0,2)} & & \hat{f}_2^{(0,3)} & & \dots \\
\hat{H}^{(1)}\ : & & \searrow & & \searrow & & \searrow & & \searrow & \\
 & \langle\omega,p\rangle & & \hat{f}_1^{(0,1)} & & \hat{f}_1^{(0,2)} & & \hat{f}_1^{(0,3)} & & \dots \\
 & & \searrow & & \searrow & & \searrow & & \searrow & \\
 & 0 & & 0 & & \hat{f}_0^{(0,2)} & & \hat{f}_0^{(0,3)} & & \dots
\end{array}
$$

A few comments are mandatory, here. E.g., take the term h_2 in the initial Hamiltonian, and transform it. This gives $\exp(L_{\chi_1^{(1)}})h_2 = h_2 + L_{\chi_1^{(1)}}h_2 + \frac{1}{2}L^2_{\chi_1^{(1)}}h_2$ and nothing else. The first term is kept in its place; the second term is added to $f_1^{(0,1)}$ and the result is renamed $\hat{f}_1^{(0,1)}$; the third term is added to $f_0^{(0,2)}$. Remark that in view of equation (11) $\hat{f}_1^{(0,1)}$ has zero average, because ξ has been determined so as to kill the average of $f_1^{(0,1)}$. Similarly, we calculate $\exp(L_{\chi_1^{(1)}})f_1^{(0,1)} = f_1^{(0,1)} + L_{\chi_1^{(1)}}f_1^{(0,1)}$ and add the second term to $f_0^{(0,2)}$. This exhausts all contributions to $f_0^{(0,2)}$, and we rename the final result $\hat{f}_0^{(0,2)} = f_0^{(0,2)} + \frac{1}{2}L^2_{\chi_1^{(1)}}h_2 + L_{\chi_1^{(1)}}f_1^{(0,1)}$. When we transform $\langle\omega,p\rangle$ we have $\exp(L_{\chi_1^{(1)}})\langle\omega,p\rangle = \langle\omega,p\rangle + L_{\chi_1^{(1)}}\langle\omega,p\rangle$, and the second term kills $f_0^{(0,1)}$ in view of equation (11); thus we put 0 in its place. All the rest of the diagram is constructed with the same procedure.

Let's now go to the second step. We determine a second generating function $\chi_2^{(1)}(p,q) = \langle Y^{(1)}(q), p\rangle$ by solving the equation

$$\partial_\omega \chi_2^{(1)} = \hat{f}_1^{(0,1)}. \tag{12}$$

This results in a set of n equations for the vector function $Y^{(1)}(q)$, namely $\partial_\omega Y^{(1)} = \frac{\partial \hat{f}_1^{(0,1)}}{\partial p}$, the second term having zero average. Thus, all the equations may be solved with the procedure of Section 2.2. Finally, we construct the new Hamiltonian $H^{(1)} = \exp(L_{\chi_1^{(1)}})\hat{H}^{(1)}$, thus getting the new diagram

$$\begin{array}{rccccccccc}
 & \vdots & & \vdots & & \vdots & & \vdots & & \ldots \\
H^{(1)}\ : & h_2 & \rightarrow & h_2^{(1)} & \rightarrow & f_2^{(1,2)} & \rightarrow & f_2^{(1,3)} & \rightarrow & \ldots \\
 & \langle\omega,p\rangle & \rightarrow & 0 & \rightarrow & f_1^{(1,2)} & \rightarrow & f_1^{(1,3)} & \rightarrow & \ldots \\
 & 0 & \rightarrow & 0 & \rightarrow & f_0^{(1,2)} & \rightarrow & f_0^{(1,3)} & \rightarrow & \ldots
\end{array}$$

Let us add a few remarks in this case too. The generating function $\chi_2^{(1)}$ is linear in p, and is a trigonometric polynomial of degree K. Hence, the Lie derivative $L_{\chi_2^{(1)}}$ keeps the degree in p and increments by K the trigonometric degree. This means that in the diagram above the terms generated by the Lie derivative propagate along horizontal rows. E.g., the transformation of h_2 gives $\exp(L_{\chi_2^{(1)}})h_2 = h_2 + L_{\chi_2^{(1)}}h_2 + \frac{1}{2}L^2_{\chi_2^{(1)}}h_2 + \ldots$ (this is actually an infinite sum). The first term keeps its place. The second term adds to $\hat{f}_2^{(0,1)}$, and we rename the result $h_2^{(1)}$ because from now on it will not change anymore. The third term adds to $\hat{f}_2^{(0,2)}$, together with the contribution of $L_{\chi_2^{(1)}}\hat{f}_2^{(0,1)}$; we rename the result $f_2^{(1,2)}$, and so on. Similarly, we calculate $\exp(L_{\chi_2^{(1)}})\langle\omega,p\rangle = \langle\omega,p\rangle + L_{\chi_2^{(1)}}\langle\omega,p\rangle + \ldots$; the second term kills $\hat{f}_1^{(0,1)}$ thanks to (12), and the rest of the expansion is added to the appropriate term in the same row.

This completes the first normalization step, ending up with a diagram similar to the one for $H^{(0)}$, but with the relevant difference that in the second column the terms of degree zero and one in p have been canceled out.

The procedure may be iterated *ad infinitum*. After $s-1$ normalization steps we shall obtain a Hamiltonian that can be represented by the diagram

$$
\begin{array}{lccccccc}
 & \vdots & \vdots & \dots & \vdots & \vdots & \dots \\
H^{(s-1)}: & h_2 & h_2^{(1)} & \dots & h_2^{(s-1)} & f_2^{(s-1,s)} & \dots \\
 & \langle\omega,p\rangle & 0 & \dots & 0 & f_1^{(s-1,s)} & \dots \\
 & 0 & 0 & \dots & 0 & f_0^{(s-1,s)} & \dots
\end{array}
$$

The columns $1,\ldots,s-1$ do not contain any term independent of or linear in p, and the terms which are quadratic or more are denoted with h instead of f in order to remind us that they will not change any more. The perturbation is actually relegated to the columns $s,\ldots$. Here too, we may cancel out the average of $f_0^{(s-1,s)}$, because it is just a constant. We say that $H^{(s-1)}$ is in Kolmogorov's normal form up to order $s-1$ In order to do the next step we determine two generating functions $\chi_1^{(s)}(q)=X^{(s)}(q)+\langle\xi^{(s)},q\rangle$ and $\chi_2^{(s)}(p,q)=\langle Y^{(s)}(q),p\rangle$, where the function $X^{(s)}$, the real vector $\xi^{(s)}$ and the vector function $Y^{(s)}(q)$ are determined by solving the equations

$$
\begin{aligned}
\partial_\omega X^{(s)} &= f_0^{(s-1,s)}, \quad \langle C\xi^{(s)},p\rangle = \overline{f_1^{(s-1,s)}} \\
\partial_\omega \chi_2^{(s)} &= L_{\chi_1^{(s)}} h_2 + f_1^{(s-1,s)} - \overline{f_1^{(s-1,s)}}.
\end{aligned}
\tag{13}
$$

The solution of this equations is worked out as in the first step, and we calculate the new Hamiltonian as $H^{(s)}=\exp(L_{\chi_2^{(s)}})\circ\exp(L_{\chi_1^{(s)}})H^{(s-1)}$. There is only one point to be noticed: the generating functions $\chi_1^{(s)}$ and $\chi_2^{(s)}$ are trigonometric polynomials of degree sK; thus the Lie derivatives $L_{\chi_1^{(s)}}$ and $L_{\chi_2^{(s)}}$ increase the trigonometric degree by sK. Moreover, they should be considered as being of order s in ε and $e^{-K\sigma}$. Therefore the action of the Lie derivative affects the column s places at right with respect to the term it is acting on. By the way, this remark is crucial in order to construct an appropriate scheme of estimates that leads to the proof of the convergence of the whole procedure. For a detailed explanation of this point see, e.g., Giorgilli and Locatelli (1999). After having performed this transformation, we get a new Hamiltonian which still has the form (13), but s has been increased by 1. Letting s go to infinity we find a Hamiltonian

$$
H^{(\infty)} = \ldots\circ\exp(L_{\chi_2^{(s)}})\circ\exp(L_{\chi_1^{(s)}})\circ\ldots\circ\exp(L_{\chi_2^{(1)}})\circ\exp(L_{\chi_1^{(1)}})H^{(0)}
$$

which has been given the normal form (7) of Kolmogorov. For the proof of convergence of the algorithm see Giorgilli and Locatelli (1997a and 1997b).

3.4 Construction of the invariant torus

What is not immediately evident from the discussion above is how to write the equation for the Invariant torus. This is hidden in the discussion of Section 3.1. We should recall that the Lie series actually defines a coordinate transformation, although thanks to the formula (4) we can perform the whole normalization procedure without even mentioning it. Actually, denoting by $p^{(s)}, q^{(s)}$ the coordinates that give the Hamiltonian the normal form up to order s, i.e., the form analogous to that represented in the diagram of $H^{(s-1)}$ above, we can calculate the transformation in explicit form as

$$\begin{aligned} p &= \exp(L_{\chi_2^{(s)}}) \circ \exp(L_{\chi_1^{(s)}}) \circ \ldots \circ \exp(L_{\chi_2^{(1)}}) \circ \exp(L_{\chi_1^{(1)}})\, p^{(s)}, \\ q &= \exp(L_{\chi_2^{(s)}}) \circ \exp(L_{\chi_1^{(s)}}) \circ \ldots \circ \exp(L_{\chi_2^{(1)}}) \circ \exp(L_{\chi_1^{(1)}})\, q^{(s)}, \end{aligned} \tag{14}$$

where all functions in the r.h.s. must be considered as expressed in terms of the new coordinates $p^{(s)}, q^{(s)}$. In order to calculate the inverse transformation we should invert the relations above, thus writing

$$\begin{aligned} p^{(s)} &= \exp(-L_{\chi_1^{(1)}}) \circ \exp(-L_{\chi_2^{(1)}}) \circ \ldots \circ \exp(-L_{\chi_1^{(s)}}) \circ \exp(-L_{\chi_1^{(s)}}) p, \\ q^{(s)} &= \exp(-L_{\chi_1^{(1)}}) \circ \exp(-L_{\chi_2^{(1)}}) \circ \ldots \circ \exp(-L_{\chi_2^{(s)}}) \circ \exp(-L_{\chi_2^{(s)}}) q, \end{aligned} \tag{15}$$

Here too, letting s go to infinity we may construct the complete transformation. If the transformation to normal form is analytic, being expressed as an absolutely convergent sequence of analytic transformations, so are the transformation of coordinates.

From a practical point of view, we shall never be able to perform that whole normalization process for a generic Hamiltonian. However, we can perform a finite number, r say, of steps, and consider the Hamiltonian $H^{(r)}$ truncated at the column r of the diagram above as the approximate normal form that we are interested in. Let us call $\overline{H}^{(r)}(p^{(r)}, q^{(r)})$ the truncated Hamiltonian. Then the canonical equations for $\overline{H}^{(r)}(p^{(r)}, q^{(r)})$ admit the simple solution

$$p^{(r)} = 0, \quad q^{(r)} = \omega t + q_0^{(r)}, \tag{16}$$

where $q_0^{(r)}$ are the values of the phases at time zero. If we denote $(p, q) = \mathcal{C}(p^{(r)}, q^{(r)})$ the canonical transformation (14) truncated at order r, then we can recover the orbit in the original coordinates by simply substituting the solution (16) in the transformation.

The argument above provides us a method for checking the reliability of our calculation: we first determine the value of $q_0^{(r)}$ corresponding to the initial conditions (by using (15) with $s = r$); then we calculate the solution (16) expressed in the original coordinates (through the canonical transformation $\mathcal{C}$); finally, we compare this "semi-analytical" integration with a numerical one (still in the original coordinates).

3.5 Invariant tori in the neighborhood of an elliptic equilibrium

Let us consider a canonical system with Hamiltonian

$$H(x,y) = \frac{\nu_1}{2}(x_1^2 + y_1^2) + \ldots + \frac{\nu_n}{2}(x_n^2 + y_n^2) + \sum_{l\geq 3} \mathcal{P}_l(x,y), \tag{17}$$

where $(x,y) \in \mathbf{R}^{2n}$ are the canonical coordinates, $\nu_1, \ldots, \nu_n$ the frequencies, and $\mathcal{P}_l(x,y)$ is a homogeneous polynomial of degree l. This is a typical form for a Hamiltonian system in the neighborhood of an elliptic equilibrium point, that here is at the origin. The series representing the Hamiltonian is assumed to be a convergent in a neighborhood of the origin. The problem is to find invariant tori in that neighborhood.

The immediate difficulty is that the unperturbed Hamiltonian is degenerate. In fact, introducing action-angle variables via the canonical transformation

$$x_j = \sqrt{2I_j}\cos\varphi_j, \quad y = \sqrt{2I_j}\sin\varphi_j,$$

the Hamiltonian takes the form

$$H(I,\varphi) = \langle \nu, I\rangle + \mathcal{P}_3(I,\varphi) + \mathcal{P}_4(I,\varphi) + \ldots \tag{18}$$

where the unperturbed part is linear in the action, and so it is degenerate. In order to remove the degeneration we perform a preliminary reduction of the Hamiltonian to a Birkhoff's normal form up to a finite order. If the unperturbed frequencies ν satisfy the condition

$$\langle k, \nu\rangle \neq 0 \text{ for } |k| \leq 4\,,\ k \in \mathbf{Z}^n\,,$$

then we can give the Hamiltonian the form

$$H(I,\varphi) = \langle k, \nu\rangle + h_2(I) + \mathcal{P}_5(I,\varphi) + \mathcal{P}_6(I,\varphi) + \ldots$$

with h_2 quadratic in the actions I, and with a remainder $\mathcal{P}_5(I,\varphi), \ldots$ which is at least of degree 5 in the original coordinates x, y. On need, we can also perform a Birkhoff normalization up to a higher order s, provided the frequencies satisfy the non-resonance condition above for $|k| \leq s$. Thus, we can consider $\langle k, \nu \rangle + h_2(I)$ as the unperturbed system, and we should check that this is non-degenerate.

Assuming that we have succeeded in constructing a non-degenerate system, we pick a good frequency ω (e.g., satisfying a diophantine condition) for which we can solve the linear equation in the unknown I

$$\nu + \frac{\partial h_2}{\partial I}(I) = \omega.$$

Let I^* be a solution of this equation. Then we can perform a power series expansion around I^*, and give the Hamiltonian the form of the diagram for $H^{(0)}$ which is suitable for applying the Kolmogorov's normalization algorithm . It may be useful to remark that the functions $\mathcal{P}_5(I, \varphi), \ldots$ are already trigonometric polynomials in the angles φ, which makes the separation of terms definitely easier. The reader may be puzzled by the fact that there is apparently no small parameter. However, there is a hidden parameter: it is the ratio $|I^*|/\varrho$, where ϱ is the convergence radius of the Hamiltonian in Birkhoff's normal form.

3.6 An application to the Sun-Jupiter-Saturn system

Applying the theorem of Kolmogorov to a real system is not an easy matter. Due to the very strong requests on the smallness of the parameter ε (that we have not reported for simplicity in the statement of theorem 4), even using the best available analytical estimates it is typical to end up with ridiculous results. Thus, we resort to the computer-assisted methods of proof. To this end, we consider the so called secular dynamics for the motion of Jupiter and Saturn: this will be our model. We proceed in four steps, with the aid of an algebraic manipulator.

(i) We determine the expansion of the secular Hamiltonian in Poincaré variables, thus describing the system with a model having two degrees of freedom.

(ii) We determine via frequency analysis the angular frequencies of the orbit with the initial data of Jupiter and Saturn in our secular model.

(iii) We construct analytically an invariant torus with angular frequencies that approximate very well the numerically calculated ones.

(iv) For an interval of values of energy that contains the current value in the secular model of the Sun-Jupiter-Saturn system we determine two families of invariant KAM tori that bound the orbit on every energy surface.

The conclusion is that at least in the approximation of the secular system the orbit of Jupiter and Saturn is confined forever between two KAM invariant tori, thus assuring the perpetual stability of this system. Below we resume in brief the method and the result. For more details the reader is referred to Locatelli and Giorgilli (2000).

Having performed the reduction of the centre of mass and of the angular momentum, the Hamiltonian of the problem of three bodies can be written in heliocentric coordinates and using the Poincaré variables as

$$H_{\mathrm{tb}} = -\frac{1}{2}\left(\frac{\mu_1^2\beta_1^3}{\Lambda_1^2} + \frac{\mu_2^2\beta_2^3}{\Lambda_2^2}\right) - \mathcal{G}\frac{m_1 m_2}{\Delta} + T. \tag{19}$$

Here, $\mathcal{G}$ is the gravitational constant, $\beta_j = \frac{m_j m_0}{m_j + m_0}$ for $j = 1, 2$, m_0 being the mass of the Sun and m_1 , m_2 the masses of the planets, Δ is the distance between the planets and T a term coming from the expression of the kinetic energy in heliocentric coordinates. The action variables for the two planets are Λ_1 , Λ_2, with conjugate angles λ_1 , λ_2. We recall that the Poincaré variables are

$$\begin{cases} \Lambda_j = \beta_j\sqrt{\mu_j a_j}, & \xi_j = \sqrt{2\Lambda_j}\sqrt{1-\sqrt{1-e_j^2}}\cos\omega_j, \\ \lambda_j = l_j + \omega_j, & \eta_j = -\sqrt{2\Lambda_j}\sqrt{1-\sqrt{1-e_j^2}}\sin\omega_j, \end{cases} \qquad j = 1, 2,$$

with the usual notations a_j , e_j , l_j and ω_j for the semi-major axes, the eccentricities, the mean anomalies and the perihelion arguments, respectively.

Using the classical methods of Celestial Mechanics, we can expand the distance Δ in the Hamiltonian (19) as a function of the Poincaré variables, and we can calculate the so called *secular system at order two in the masses* (used, for instance, in Laskar 1988 in a model with 8 planets, to study the long term evolution of the solar system). In the secular system the dependency on the angles λ_1 , λ_2 (which evolve much faster than the other Poincaré variables) is dropped out by simply averaging the Hamiltonian over the angles themselves. Thus, the actions Λ_1 , Λ_2 are first integrals for the secular system, which are replaced with their numerical values corresponding to the data for the real system Sun-Jupiter-Saturn at a fixed initial time. Therefore, we can actually expand the secular Hamiltonian as a power series in the form

$$\mathcal{H}_{\mathrm{sec}}(\xi_1, \xi_2, \eta_1, \eta_2) = \sum_{s\geq 1} \sum_{\substack{i_1+i_2+\\ j_1+j_2=2s}} c_{i_1,i_2,j_1,j_2}\, \xi_1^{i_1}\xi_2^{i_2}\eta_1^{j_1}\eta_2^{j_2}. \tag{20}$$

Let us remark that only terms of even polynomial degree occur in the previous expansion. We did calculate the coefficients of the expansion in (20) up to terms of degree 6. This defines the model. The table of the coefficients may be found in Locatelli and Giorgilli (2000).

The Hamiltonian (20) has an elliptic equilibrium at the origin. Therefore, we may apply the scheme outlined in Section 3.5 in order to find a torus in the neighborhood of the equilibrium. Such an invariant torus is characterized by angular frequencies (ω_1, ω_2). In order to approximate as much as possible the real secular dynamics of Jupiter and Saturn, we choose $(\omega_1\,,\,\omega_2)$ with the following criterion: by using the frequency analysis method (see, e.g., Laskar 1995), we calculate the two main frequencies related to the motion induced by the canonical flow of (20), starting from the initial condition $(\xi_1(0), \xi_2(0), \eta_1(0), \eta_2(0))$ that corresponds to the data for the real system Sun-Jupiter-Saturn at the initial time. The frequency analysis method provides us, of course, two numerical values of $(\omega_1\,,\,\omega_2)$ that are correct up to a finite number of digits, but we should consider Diophantine frequencies in order to apply the KAM theorem. Therefore, we modify the values of ω_1 and ω_2 by a very small quantity so that the continued fraction of the ratio ω_1/ω_2 is a noble number (i.e., a number the continued fraction of which ends with a infinite sequence of ones, as happens for the golden ratio).

The approximate construction of the torus is a rather lengthy calculation, but may be performed with the computers available nowadays. The reliability of the construction may be appreciated by plotting the distance between the orbit calculated via numerical integration of the original system and the orbit calculated via the normal form of Kolmogorov. The result is reported in Figure 1 for different orders of approximation of the torus. One sees that the successive applications of the algorithm of Kolmogorov give a better and better agreement, until the small difference between the frequencies of the integrated orbit and the fixed frequencies of the motion on the torus shows up, causing a very slow drift.

However, this is not yet rigorous. In order to prove the existence of a torus we should actually prove that the *infinite* sequence of transformation to Kolmogorov normal form is convergent. Since pushing the actual calculation of the normal form is impractical, we may resort to a simple evaluation of the size of the generating function. This may be done up to quite high orders using recursion formulæ; however, the estimates become worse. The result may be appreciated by looking at Figure 2. In this case the normal form has been explicitly calculated up to order 33. Then the norms have been estimated via recursive formulæ up to order 2000. As one sees, the clear indication is that the process converges, although the recursive estimates appear to be definitely pessimistic.

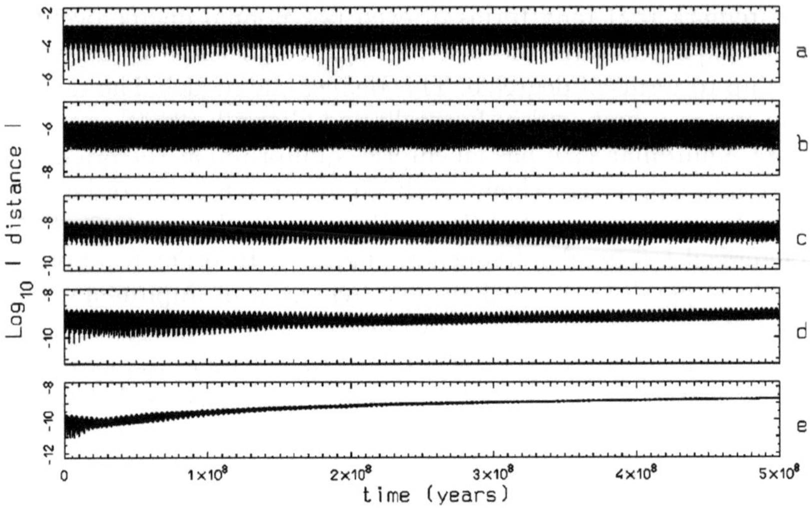

Figure 1. The distance $d(t)$ between the numerically integrated orbit and the approximated motion $(\tilde{\xi}_r(t), \tilde{\eta}_r(t))$ calculated via the approximation of the invariant torus. The curves a–e refer to the step r of Kolmogorov with $r = 1, 5, 9, 11, 13$, respectively. The convergence may be appreciated by looking at the vertical scale. The drift effect in Figure 1e is due to the error in the determination of the frequencies.

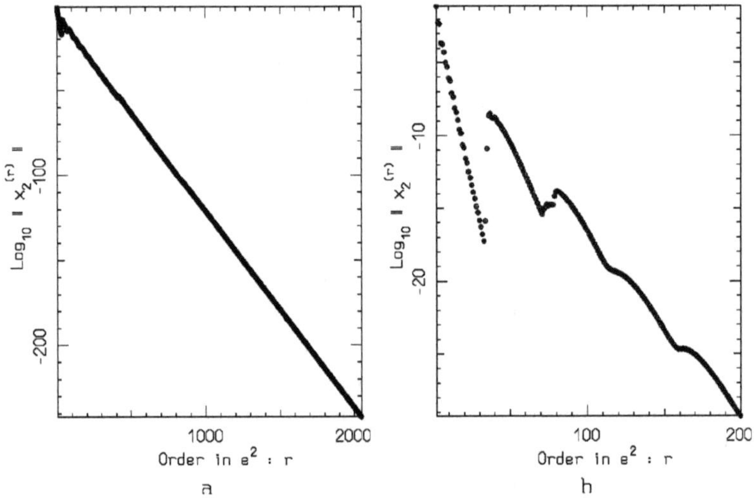

Figure 2. Decrease of the generating functions defined by the Kolmogorov normalization algorithm. Figure 2a has been enlarged in Figure 2b, where we can appreciate the change of the slope occurring when the calculation of the norms is no longer made starting with the coefficients of the expansions, but instead only by iterating the estimates (in our case for $r = 33$).

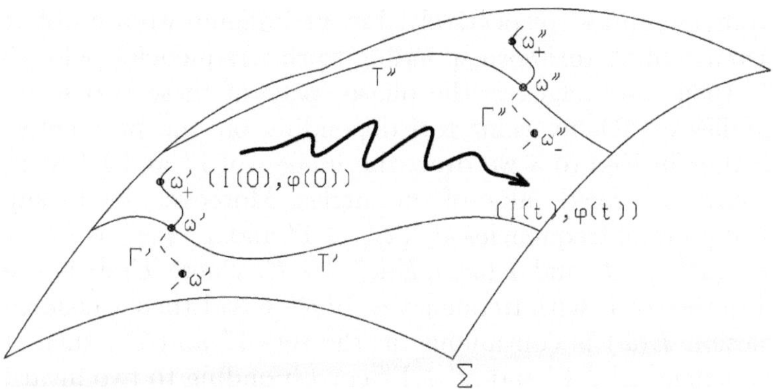

Figure 3. Illustrating the topological confinement of the orbit in the 4D phase space. The continuous curves Γ' and Γ'' represent two sets of 2D invariant tori that intersect transversally an energy surface. An orbit with initial datum in the gap between two tori will be eternally trapped in the same region (see text).

The convergence of the normalization algorithm may be proved if one is able to estimate the perturbation left after a given number of normalization step. This may be done with analytical estimates. Thus, we can apply a formal statement of KAM theorem to a Hamiltonian with a remainder dramatically reduced, thanks to both explicit calculation of the expansion and recursive estimates. A fully rigorous result may be achieved by performing all the calculations using interval arithmetics, so that we have full control on the propagation of roundoff errors.

The procedure above allows us to prove only that there is an invariant torus *close* to the initial conditions of the Sun-Jupiter-Saturn system, not that the orbit of the system actually lies on a torus. Since we can not exclude the possibility of Arnold's diffusion, this is not enough to prove the perpetual stability of the orbit of the secular system. Therefore, we make a more accurate analysis in order to prove that the orbit is actually confined in a gap between two invariant tori. The procedure is illustrated in Figure 3.

Let us write the Hamiltonian in action-angle variables I, φ in the usual form $h(I)+\varepsilon f(I,\varphi)$, where ε is the perturbation parameter. Consider the initial conditions $(I(0),\varphi(0))$ and let us denote by $\bar{E}$ the corresponding value of the energy. After having assumed suitable non-degeneracy conditions on the unperturbed Hamiltonian $h(I)$, we can uniquely identify the invariant tori surviving the perturbation by their angular frequencies. Making reference to the frequency plane (ω_1,ω_2), we consider two straight lines $\omega_1/\omega_2 = \text{const}'$, const'' of frequencies satisfying a Diophantine condition and we choose two segments Γ' and Γ'' lying on each of the previous lines, respectively. Let us imagine we are able to prove the existence of all tori corresponding to the frequencies belonging to Γ' and Γ''; then the images in the phase space of these two segments are two families of 2D invariant tori depending on one parameter. Let us remark that in Figure 3 we drew the images of Γ' and Γ'' which correspond only to a fixed value of the angles. Moreover, let us suppose to know two pairs of frequencies ω'_{-}, $\omega'_{+} \in \Gamma'$ and ω''_{-}, $\omega''_{+} \in \Gamma''$, such that $E(\omega'_{-}), E(\omega''_{-}) < \bar{E}$ and $E(\omega'_{+}), E(\omega''_{+}) > \bar{E}$, where $E(\omega)$ is the energy related to the torus with frequency ω. Since KAM theory ensures us that the function $E(\omega)$ is continuous on the sets Γ' and Γ'', then there are two frequencies $\omega' \in \Gamma'$ and $\omega'' \in \Gamma''$ corresponding to two invariant tori, say T' and T'' respectively, such that they belong to the energy surface Σ related to the level $\bar{E}$. Thus, it is enough to check that the initial data belong to the gap between T' and T'' on the surface Σ in order to assure that the orbit will be trapped there forever.

The interested reader may find a detailed exposition of this result in Locatelli and Giorgilli (2000).

4. Long time stability

Stability in mathematical sense is usually associated with some property that is satisfied for all times. E.g., an equilibrium point is said to be stable if every neighborhood V of the point is invariant for all orbits with initial point in a second neighborhood $U \subset V$. However, stability in strict sense is hard to prove. On the other hand, we can look for weaker statements that are still meaningful for the real world.

For instance, in the case of an equilibrium we may consider a ϱ-neighborhood of the equilibrium Δ_ϱ, with arbitrary ϱ, and look for the following property:

(a) Consider all orbits $(x(t), y(t))$ with initial point $(x(0), y(0)) \in \Delta_{\varrho_0}$ for some positive ϱ_0. Choose $\varrho > \varrho_0$ (e.g., let $\varrho = 2\varrho_0$) and prove that $(x(t), y(t)) \in \Delta_\varrho$ for $|t| \le T(\varrho_0)$ with some "large" $T(\varrho_0)$, e.g., increasing to infinity as $\varrho_0 \to 0$. An even stronger question that applies to the more general case of a Hamiltonian $H(p, q) = h(p) + \varepsilon f(p, q)$ of the form (1) is the following:

(b) Prove that for every orbit $(p(t), q(t))$ the actions p satisfy $|p(t) - p(0)| < \varepsilon^b$ $(0 < b \le 1)$ for $|t| \le T(\varepsilon)$ with some "large" $T(\varepsilon)$, e.g., increasing to infinity as $\varepsilon \to 0$.

The request above may be meaningful if we take into consideration some characteristics of the dynamical system that is (more or less accurately) described by our equations. In this case "large" should be interpreted as *large with respect to some characteristic time of the physical system, or comparable with the lifetime of it.* For instance, for modern accelerators, a characteristic time is the period of revolution of a particle of the beam and the typical lifetime of the beam during an experiment may be a few days, which may correspond to some 10^9 revolutions; for the solar system the lifetime is the estimated age of the universe, which corresponds to some 10^9 revolutions of Jupiter; for a galaxy, we should consider that the stars may perform a few hundred revolutions during a time as long as the age of the universe, which means that a galaxy does not really need to be much stable in order to exist.

From a mathematical viewpoint the word "large" is more difficult to explain, since there is no typical lifetime associated to a differential equation. Hence, in order to give the word "stability" a meaning in the sense above it is essential to consider the dependence of T on ε (or on ϱ_0). In this respect the continuity with respect to initial data does not help too much. For instance, if we consider the trivial example of the differential equation $\dot{x} = x$ one will immediately see that if $x(0) = x_0 > 0$ is the initial point, then we have $x(t) > 2x_0$ for $t > T = \ln 2$ no matter how small is x_0; hence T may hardly be considered to be "large", since it remains constant as x_0 decreases to 0.

The question about finite but large stability times may be answered in many different ways, depending on how we choose $T(\varepsilon)$. Here is a short list.

(i) *Adiabatic invariance:* $T(\varepsilon) \sim 1/\varepsilon$.

(ii) *Complete stability* according to Birkhoff: $T(\varepsilon) \sim 1/\varepsilon^r$, for some $r > 1$.

(iii) *exponential stability* according to Littlewood, Moser and Littlewood: $T(\varepsilon) \sim \exp(1/\varepsilon^a)$, for some $0 < a \leq 1$.

(iv) *Superexponential stability:* $T(\varepsilon) \sim \exp\left[\exp(1/\varepsilon^a)\right]$.

(v) *Perpetual stability:* $T(\varepsilon) = +\infty$.

Perpetual stability is a quite strong property that can be proved only in some special cases. The cases of an integrable system or of a system with two degrees of freedom that satisfies the hypotheses of KAM theorem have already been discussed. We can add the case of an elliptic equilibrium, which turns out to be stable in case the equilibrium corresponds to a minimum (or a maximum) of the Hamiltonian; this result goes back to Dirichlet.

The rest of this section is devoted to the discussion of the theory of complete stability, exponential stability and superexponential stability.

4.1 Complete stability

The concept of complete stability is due to Birkhoff (1927), who introduced it in connection with the study of the normal form of a Hamiltonian system in the neighborhood of an equilibrium. However, the same method works properly also if one considers the neighborhood of an invariant torus, as we are going to discuss. We shall use a direct construction of the first integrals, which turns out to be quite simple.

Let us recall that in the case of an elliptic equilibrium the typical form of the Hamiltonian is

$$H(x,y) = \sum_{l=1}^{n} \frac{\omega_l}{2}(x_l^2 + y_l^2) + H_1(x,y) + H_2(x,y) + \ldots \tag{21}$$

where $\omega \in \mathbf{R}^n$ is the vector of the real frequencies, and $H_s(x,y)$ is a homogeneous polynomial of degree $s+2$. We shall assume here that the frequencies are non-resonant, and even diophantine. We have already observed that the canonical transformation

$$x_j = \sqrt{2p_j}\cos q_j\,, \quad y = \sqrt{2p_j}\sin q_j\,, \tag{22}$$

gives the Hamiltonian the form

$$H(p,q) = \langle \omega, p\rangle + H_1(p,q) + H_2(p,q) + \ldots \tag{23}$$

i.e., the unperturbed part is linear. It is an easy matter also to verify that $H_s(p,q)$ turns out to be a homogeneous polynomial of degree $s+2$ in $p_1^{1/2},\ldots,p_n^{1/2}$, and a trigonometric polynomial of degree $s+2$ in the angles; this is the result of the transformation (22) when acting on a homogeneous polynomial in x,y. The fact that the transformation introduces a singularity at the origin is a bit annoying, but it turns out to be essentially harmless as long as we are interested only in the formal aspect of the calculations, as we shall do in this section. The relevant fact is that in a ϱ-neighborhood of the equilibrium the unperturbed Hamiltonian is $O(\varepsilon^2)$ with $\varepsilon = \varrho^{1/2}$, and the perturbation decreases as $|H_s(p,q)| < C\varepsilon^{s+2}$. Thus, H_s has relative size $O(\varepsilon^s)$ with respect to the unperturbed part of the Hamiltonian.

As a matter of fact, in a neighborhood of an unperturbed KAM torus the Hamiltonian may be given a form which is very similar to (23). Indeed, applying the Kolmogorov's normalization algorithm as illustrated in Section 3 we conclude that in the neighborhood of an invariant torus with Diophantine frequencies ω the Hamiltonian may be given the Kolmogorov's normal form (7). As a byproduct of the proof we obtain also that $\mathcal{R}(p,q)$ (let us recall that it is at least quadratic in the actions p)

is analytic. With an argument similar to that of Section 3.2 we can split the remainder $\mathcal{R}(p, q)$ into a convergent series $H_1(p, q) + H_2(p, q) + \ldots$ where $H_s(p, q)$ is a (non-homogeneous) polynomial of degree $s+1$ in the actions p and a trigonometric polynomial of degree sK in q, for some $K \geq 1$. Moreover, thanks to the analyticity of $\mathcal{R}(p, q)$ we may assure that the size of $H_s(p, q)$ decreases geometrically, i.e., in a ϱ-neighborhood of the invariant torus we have $|H_s(p, q)| < C\varepsilon^{s+1}$ with $\varepsilon \sim \varrho$, while the unperturbed Hamiltonian is $O(\varepsilon)$. Again, H_s has relative size $O(\varepsilon^s)$ with respect to the unperturbed part of the Hamiltonian.

Taking into account only the formal aspect, a direct construction of first integrals may be performed as in Section 2.2, namely with the same procedure used by Poincaré in order to prove the non-existence of first integrals. The apparent paradox is removed by remarking that the unperturbed Hamiltonian does not fulfil the hypotheses of the theorem of Poincaré : this remark has been first made by Whittaker (1916); later this method was used by Cherry (1924a and 1924b) and Contopoulos (1960 and 1963). Thus, we recursively solve the system of equations (2), namely

$$\begin{aligned}
\partial_\omega \Phi_0 &= 0, \\
\partial_\omega \Phi_1 &= -\{H_1, \Phi_0\}, \\
&\ldots\ldots, \\
\partial_\omega \Phi_s &= -\{H_1, \Phi_{s-1}\} - \ldots - \{H_s, \Phi_0\}, \\
&\ldots\ldots,
\end{aligned}$$

by setting Φ_0 to coincide with any of the actions. Precisely, for every $s \geq 1$ we solve the equation $\partial_\omega \Phi_s = \Psi_s$, where Ψ_s is the r.h.s. of the equation for Φ_s, and may be written as

$$\Psi_s(p, q) = \sum_{k \in \mathbf{Z}^n} \psi_k(p) \exp(i\langle k, q\rangle),$$

with known coefficients $\psi_k(p)$. As we have seen in Section 2.2 we obtain the solution by simply setting $\Phi_s(p, q)$ to have the same form with coefficients

$$\varphi_k(p) = -i\psi_k(p)/\langle k, \omega\rangle. \tag{24}$$

We stress that this solution is valid only if $\Psi_s(p, q)$ has zero average over the angles, i.e., if its trigonometric expansion does not contain a term which is independent of the angles. Using the normal form of Birkhoff it may be proven that this is always the case if the frequencies are non-resonant (see, e.g., Celletti and Giorgilli 1991). Remark also that at every order s the known term Ψ_s on the r.h.s. is a trigonometric polynomial of finite degree. With a bit of patience, the reader will be able to check that the expansion of the first integral has the same characteristics as the expansion of the Hamiltonian. That is, in the case of an elliptic equilibrium, Φ_s is a homogeneous polynomial of degree $s+2$ in $p_1^{1/2}, \ldots, p_n^{1/2}$ and a trigonometric polynomial of degree $s+2$ in the angles; in the case of the neighborhood of an invariant torus, Φ_s is a non-homogeneous polynomial of degree $s+1$ in p and a trigonometric polynomial of degree sK in q. Moreover, in both cases the relative size of the term Φ_s with respect to the first term is $O(\varepsilon^s)$. More precisely, in a ϱ-neighborhood of the equilibrium (of the invariant torus) we have $|\Phi_s(p,q)| \leq B_s \varepsilon^{s+\lambda}$ with with some (possibly big) constant B_s depending on s, and with $\lambda = 2$ for the case of the elliptic equilibrium and $\lambda = 1$ for the case of the invariant torus.

According to the scheme above we may construct n independent formal integrals $\Phi^{(l)} = p_l + \Phi_1^{(l)}(p,q) + \ldots$, for $l = 1, \ldots, n$. What is doubtful here is the convergence of the series so generated. However, we may just decide to truncate the construction after a finite number r of steps, thus constructing approximate first integrals

$$\Phi^{(l,r)} = p_l + \Phi_1^{(l)}(p,q) + \ldots + \Phi_r^{(l)}(p,q).$$

By construction, in a ϱ-neighborhood of $p = 0$ we have $|\Phi_s^{(l)}(p,q)| \leq B_s \varepsilon^{s+\lambda}$. Therefore, for ε small enough we also have

$$|p_l - \Phi^{(l,r)}| < 2B_r \varepsilon^{1+\lambda}. \tag{25}$$

Here comes the argument of Birkhoff, that we can formulate as follows. By the triangle inequality, we have

$$\begin{aligned} |p_l(t) - p_l(0)| \;&\leq\; |p_l(t) - \Phi^{(l,r)}(t)| + |\Phi^{(l,r)}(t) - \Phi^{(l,r)}(0)| \\ &\quad + |\Phi^{(l,r)}(0) - p_l(0)|. \end{aligned} \tag{26}$$

The first and the last term in the r.h.s. are $O(\varepsilon^{1+\lambda})$, in view of (25). In order to estimate the second term we calculate the time derivative of $\Phi^{(l)}$ as

$$\dot{\Phi}^{(l,r)} = \{\Phi^{(l,r)}, H\} = \{\Phi_1^{(l)}, H_r\} + \ldots + \{\Phi_r^{(l)}, H_1\} + \ldots,$$

where the dots denote terms of higher order. Therefore, we have $|\dot{\Phi}^{(l,r)}(p,q)| \sim B_r \varepsilon^{r+\lambda+1}$, which means that if we let $\varepsilon \to 0$ then the inequality

$$|\Phi^{(l,r)}(t) - \Phi^{(l,r)}(0)| \sim B_r \varepsilon^{r+1+\lambda} |t|, \tag{27}$$

holds true as long as the orbit remains in the ϱ-neighborhood. If we allow the effect of the noise to become comparable with that of the deformation we conclude, with Birkhoff,

$$\left|p_l(t) - p_l(0)\right| = O(\varepsilon^{1+\lambda}) \quad \text{for } |t| < T(\varepsilon) \sim 1/\varepsilon^r . \tag{28}$$

If we set $r = 1$ this method is equivalent to the method of averaging over the angles, namely the usual procedure leading to adiabatic invariance. Hence, the theory of complete stability due to Birkhoff is actually a strengthening of the adiabatic invariance.

Let us add a short comment on this result. Assume for a moment that $\Phi^{(l,r)}$ are exact first integrals, $(l = 1, \ldots, n)$. Then the orbit lies on an invariant surface—actually a torus—which is ε—close to the unperturbed torus $p_l = p_l(0)$. Hence the actions $p_l(t)$ are not constants, being $\dot{p}_l = O(\varepsilon)$. We say that this change is due to a *deformation* of the invariant surface, which shows up in a time $O(1/\varepsilon)$, but remains bounded. Moreover, since the invariant surface is still a torus, the time evolution of the actions is actually quasi-periodic.

Let us now take into account the fact that $\Phi^{(l,r)}$ are only approximate first integrals. Indeed, as stated by (27), the time derivative of $\Phi^{(l,r)}$ is $O(\varepsilon^{r+1+\lambda})$, so that its effect is veiled by that of the deformation, and we can observe it only after a much longer time $O(1/\varepsilon^r)$. We say that this is a very slow *noise* that is superimposed to the deformation.

4.2 Exponential stability

In its simplest formulation, exponential stability is the result of making quantitative the heuristic estimates above. Essentially, the problem is to evaluate the r-dependence of the constant B_r in (25). In this form the result has been first stated by Moser (1955) and Littlewood (1959a and 1959b).

Let us try to give a rough estimate on the constant B_r, using heuristic considerations on the recursive system (2). For definiteness, let us refer to the case of the invariant torus; the case of an elliptic equilibrium is very similar. Every step requires two operations, namely:

(i) determining the known function $\Psi_s = -\{H_1, \Phi_{s-1}\} - \ldots - \{H_s, \Phi_0\}$ by calculating Poisson brackets, and

(ii) solving the equation $\partial_\omega \Phi_s = \Psi_s$.

The first operation (i) contains derivatives of polynomials of degree at most $s + 1$ and trigonometric polynomials of degree at most sK. Hence, the size of Φ_{s-1} is multiplied by a factor $\leq s + 1$, besides other possible factors that are independent of the degree. The second operation (ii) introduces the small denominators. Since Ψ_s is a trigonometric polynomial of degree sK, the worst possible divisor is

$$\alpha_s = \min_{0<|k|\le sK} \big|\langle k,\omega\rangle\big| .$$

Recalling that H_s is $O(\varepsilon^{s+1})$ we expect that in a ϱ-neighborhood of the invariant torus the size of Φ_s may be evaluated as

$$\big|\Phi_s(p,q)\big| \sim \big[(s+1)\varepsilon/\alpha_s\big]\big|\Phi_{s-1}(p,q)\big| .$$

This is a recursive estimate that we may apply starting with $\Phi_0(p) = |p_l| < \varrho \sim \varepsilon$. Thus, for the first integrals truncated at order r we get

$$\big|\Phi_r(p,q)\big| \sim \big[(r+1)!\,\varepsilon^{r+1}\,(\alpha_1\cdot\ldots\cdot\alpha_r)^{-1}\big]. \tag{29}$$

Now, in view of the diophantine estimate

$$\big|\langle k,\omega\rangle\big| > \gamma|k|^{-\tau} \quad \text{for some } \gamma>0 \text{ and } \tau\ge n-1,$$

we may replace the constant α_s by $\gamma(sK)^{-\tau}$. This gives the final estimate

$$B_r \sim (r+1)!^{\tau+1}\,C^{r+1},$$

with some positive constant C independent of r.

Here comes the exponential estimate. According to this estimate we rewrite the bound (27) on the noise as

$$\big|\Phi^{(l,r)}(t) - \Phi^{(l,r)}(0)\big| \sim (r+1)!^{\tau+1}\,\varepsilon^{r+1}|t|, \tag{30}$$

Remark that in this formula ε is given, being related to the size of the ϱ-neighborhood of the torus, but r is arbitrary. Therefore we may look for a best choice of r as a function of ε so that the quantity $(r+1)!^{\tau+1}\,\varepsilon^{r+1}$ is minimized. It is an easy matter to check that the minimum is reached for

$$r \sim \varepsilon^{-1/(\tau+1)}.$$

Using this value for r and the Stirling's formula, we get

$$B_r\varepsilon^r \sim r!^{\tau+1}\,\varepsilon^r \sim (e^{-r}r^r)^{\tau+1}\varepsilon^r \sim e^{-r(\tau+1)}\varepsilon^{-r}\varepsilon^r = \exp(1/(a\varepsilon^a)),$$

where $a = 1/(\tau+1)$. Replacing the latter estimate in (28) we conclude

$$\big|p_l(t) - p_l(0)\big| = O(\varepsilon^{1+\lambda}) \quad \text{for } |t| < T(\varepsilon) \sim \exp(1/(a\varepsilon^a)),$$

i.e., the exponential stability of a KAM torus or of an elliptic equilibrium.

4.3 The theorem of Nekhoroshev

The theorem of Littlewood complements the Kolmogorov's theorem, since it gives information on orbits lying in an open subset of the phase space, but only over a large time. Here is a formal statement.

THEOREM 5 *Let the Hamiltonian $H(p,q) = h(p)+\varepsilon f(p,q,\varepsilon)$ be analytic in some domain $\mathcal{G} \times \mathbf{T}^n$, with $\mathcal{G} \subset \mathbf{R}^n$ an open set, and assume that the unperturbed Hamiltonian is convex, i.e.,*

$$|\langle C(p)v, v\rangle| \geq m\|v\|^2 \quad \text{for all } v \in \mathbf{R}^n\,, \quad C_{jk}(p) = \frac{\partial^2 h}{\partial p_j \partial p_k}\,.$$

Then for ε small enough the following statement holds true: if $p(0) \in \mathcal{G}$ then the corresponding motion $(p(t), q(t))$ satisfies the inequality

$$\operatorname{dist}(p(t) - p(0)) \leq \varepsilon^b \quad \forall\, t \text{ such that } |t| \leq T(\varepsilon) \sim \exp(1/\varepsilon^a),$$

with some positive coefficients $a < 1$ and $b < 1$.

In the original formulation of Littlewood the condition of convexity is replaced by a weaker one, called *steepness*.

A complete account on the proof goes far beyond the limit of the present notes. The interested reader may find an informal exposition, e.g., in Giorgilli (1995a and 1995b). Detailed proofs may be found in several articles (see, e.g., Littlewood1977; Littlewood 1979; Benettin et al. 1985; Benettin and Gallavotti 1986; Lochak 1992; Giorgilli 1992; Giorgilli 2003). However, let us stress a few relevant points.

The standard scheme of proof of the theorem is composed by a so called *analytic part*, based on the construction of a local Birkhoff normal form in suitably chosen domains, and of a *geometric part* which takes care of covering the phase space with good domains. A remarkable exception to this scheme is the paper by Lochak (1992), where the geometric part is replaced by a clever use of convexity and of the simultaneous approximations of real numbers with rationals. We shall sketch the traditional scheme.

The analytic part is the following. In the domain $\mathcal{G}$ of the actions we isolate a *non-resonance domain* $\mathcal{V}$, characterized by a resonance module $\mathcal{M}$, a real constant α and an integer N, where the following inequality holds true:

$$|\langle k, \omega(p)\rangle| > \alpha \quad \text{for all } p \in \mathcal{V},\ k \in \mathbf{Z}^n \setminus \mathcal{M} \text{ and } |k| \leq N. \tag{31}$$

In such a domain we may perform the construction of Birkhoff's normal form illustrated in Section 2.5. To this end we split again the Hamiltonian as a series

$$H(p,q) = H_0(p) + H_1(p,q) + H_2(p,q) + \ldots$$

where $H_s(p,q)$ is a trigonometric polynomial of degree sK in q with coefficients depending on p, namely

$$H_s(p,q) = \sum_{|k|\le sK} h_k(p)\exp(i\langle k,q\rangle),$$

taking care only that we must have $|H_s(p,q)| = O(\varepsilon^s)$. Again, this may be performed in many ways. In a non-resonance domain $\mathcal{V}$ with $N = rK$ we may perform r steps of the construction of the resonant normal form of Birkhoff, because, due to the condition (31), no denominator $\langle k,\omega(p)\rangle$ vanishes in $\mathcal{V}$. Hence we may construct a normal form

$$H^{(r)}(p,q) = H_0(p) + Z^{(r)}(p,q) + R^{(r)}(p,q)$$

where $|R^{(r)}(p,q)|| = O(\varepsilon^{r+1})$ in the domain $\mathcal{V}$, and

$$Z^{(r)}(p,q) = \sum_{k\in\mathcal{M}\,,\,|k|\le rK} z_k(p)\exp(i\langle k,q\rangle).$$

Let us now set for a moment $R^{(r)}(p,q) = 0$. Then, according to the general theory discussed in Section 2.5, the Hamiltonian in normal form possesses $n-\dim\mathcal{M}$ independent first integrals of the form $\Phi(p) = \langle\nu,p\rangle$, with $\nu\perp\mathcal{M}$. With elementary geometrical considerations we conclude that any orbit with initial point $p_0\in\mathcal{V}$ lies on a plane $\Pi_{\mathcal{M}}(p_0)$ through p_0 and parallel to $\mathcal{M}$; we shall call this plane the *plane of fast drift*. This is true in the coordinates of the normal form. If we look at the original coordinates then we must take into account the *deformation* due to the canonical transformations —as we already remarked while discussing the case of an elliptic equilibrium. Moreover, we must consider also the *noise* due to the remainder, but in this case too we have $|\dot{\Phi}| = O(\varepsilon^r)$, so that the noise causes only a slow drift that becomes comparable with the deformation only after a time $T(\varepsilon)\sim 1/\varepsilon^r$.

The conclusion is illustrated in Figure 4: until the orbit remains in the non-resonance domain $\mathcal{V}$ it lies in a small neighborhood of the plane of fast drift $\Pi_{\mathcal{M}}(p_0)$. However, this is not enough to assure the long-time preservation of the actions, because the analytic theory does not assure that the orbit will be confined for a long time in $\mathcal{V}$. The question is: what happens if the orbit leaves the non-resonance domain?

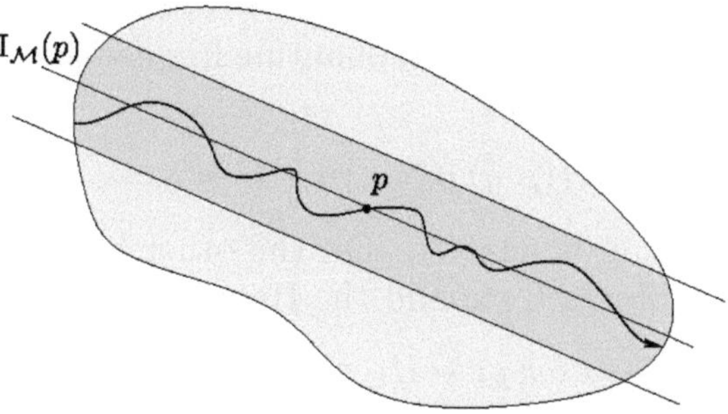

Figure 4. The local dynamics in a non-resonance domain. The orbit lies in a neighborhood of the plane of fast drift.

The answer to the question above is hidden in the geometric part of the theorem: this is indeed the clever and deep contribution of Littlewood. Essentially, the geometric construction of Littlewood is based on two facts, namely: (i) the covering of the action space by a family of non-resonant domains characterized by a finite number of resonance moduli $\mathcal{M}$, and (ii) the use of the steepness (or convexity) condition in order to avoid the phenomenon of the *overlapping of resonances*. In rough terms, the problem is to assure that if the orbit leaves in a short time a non-resonance domain characterized by a resonance module $\mathcal{M}$, then it must enter a less resonant domain, i.e., a local domain characterized by a resonance module $\mathcal{M}'$ with $\dim \mathcal{M}' < \dim \mathcal{M}$. A main role is played here by the condition of convexity (or steepness, as in the original formulation of Littlewood): it assures that the plane of fast drift is transversal to the *resonant manifold* characterized by $\langle k, \omega(p) \rangle = 0$ (which follows from convexity) or at least it presents a tangency of finite order (which follows from steepness), so that the motion along the plane of fast drift can not cause a fast diffusion along the resonant manifold. With a clever argument, this leads to the conclusion that the motion of the orbit due to the deformation remains bounded, and only the slow drift due to the noise can take the orbit far from its initial point p_0 in the action space. The exponential estimate of the stability time follows from a process of optimization similar to that of Section 4.2.

4.4 Superexponential stability

Superexponential stability is in a sense the outcome of the combination of perturbation methods. The simplest case to be considered is, again, that of an elliptic equilibrium or of the neighborhood of an invariant KAM torus. For definiteness, let us consider the latter case.

Recall that according to the theorem of Kolmogorov, in a neighborhood of an invariant torus with Diophantine frequencies we may give the Hamiltonian the normal form

$$H(p,q) = \langle \omega, p \rangle + O(p^2),$$

where the Hamiltonian is analytic. On the other hand we have already remarked that we may well expand the Hamiltonian as a series

$$H(p,q) = \langle \omega, p \rangle + H_1(p,q) + H_2(p,q) + \dots$$

where $H_s(p,q)$ is, e.g., a polynomial of degree $s+1$ in p and a trigonometric polynomial of degree sK in q, with some integer K, and with the crucial condition that in a ϱ-neighborhood of the torus we have $|H_s(p,q)| \sim \varepsilon^{s+1}$ with $\varepsilon \sim \varrho$. In Section 4.1 we have used this form of the Hamiltonian in order to illustrate the theory of complete stability of Birkhoff, and in Section 4.2 we have added a scheme of quantitative estimates that leads to exponential stability.

Let us now proceed in a different manner, which however is essentially equivalent to the procedure illustrated there. Instead of looking directly for first integrals, we go through the process of constructing a Birkhoff's normal form. This is what we have done in Section 2.5, at least formally: we have actually seen that we can construct a normal form up to an arbitrary large order. Thus, we may give the Hamiltonian the form

$$H(p,q) = H_0(p) + Z(p) + R(p,q), \tag{32}$$

where in a ϱ-neighborhood of the invariant torus we have $|Z(p)| = O(\varepsilon^2)$ and $|R(p,q)| < B_r\varepsilon^{r+1})$ for some r, and with some constant B_r. If, with a lot of patience, we add a scheme of quantitative estimates, we may see again that $B_r = C^r r!^{\tau+1}$, as we did for the formal integrals. Furthermore, by optimizing the choice of r as a function of ε we conclude again that $|R(p,q)| = O\big(\exp(-1/\varepsilon^{1/(\tau+1)})\big)$. If we stop at this point we simply recover the exponential stability of the torus.

Here come the interesting part of the game. Looking at (32) we realize that this is a Hamiltonian system with an unperturbed integrable part $h(p) = H_0(p) + Z(p)$ and a perturbation $R(p,q)$, with $|R(p,q)| = O(\eta)$, where $\eta = \exp(-1/\varepsilon^a)$, and $a = \frac{1}{\tau+1}$. Now, if $h(p)$ satisfies the hypotheses of Littlewood's theorem we can apply it, thus concluding for the exponential stability with respect to η, i.e., the length of the time of stability is estimated as $T(\eta) = \exp(1/\eta)$. Using the expression above for η as a function of ε we conclude for a stability time

$$T(\varepsilon) \sim \exp\big[\exp(1/\varepsilon^a)\big].$$

References

Arnold, V. I. (1963a). A theorem of Liouville concerning integrable problems of dynamics, *Sibirsk. Math. Zh.*, 4:471–474.

Arnold, V. I. (1963b). Proof of a theorem of A. N. Kolmogorov on the invariance of quasi-periodic motions under small perturbations of the Hamiltonian, *Usp. Mat. Nauk* 18:13–. In Russian. English translation in *Russ. Math. Surv.*, 18:9–36.

Arnold, V. I. (1963c). Small denominators and problems of stability of motion in classical and Celestial Mechanics, *Usp. Math. Nauk* 18(6):91–.In Russian. English translation in *Russ. Math. Surv.* 18(6):85–191.

Arnold, V. I. (1964). Instability of dynamical systems with several degrees of freedom, *Sov. Math. Dokl.* 5:581–585.

Arnold, V. I. (1976). *Méthodes mathématiques de la méchanique classique.* Editions MIR.

Benettin, G., Galgani, L., Giorgilli, A. and Strelcyn, J. M. (1984). A proof of Kolmogorov's theorem on invariant tori using canonical transformations defined by the Lie method. *Il Nuovo Cimento*, 79:201–216.

Benettin, G., Galgani, L., Giorgilli, A. (1985). A proof of Nekhoroshev's theorem for the stability times in nearly integrable Hamiltonian systems. *Cel. Mech.*, 37:1–25.

Benettin, G., Gallavotti, G. (1986). Stability of motions near resonances in quasi–integrable Hamiltonian systems. *Journ. Stat. Phys.*, 44:293–338.

Birkhoff, G. D. (1927). *Dynamical systems.* New York.

Born, M. (1927) *The Mechanics of the atom.* Frederick Ungar Publ. Co., New York.

Celletti, A. and Giorgilli, A. (1991). On the stability of the Lagrangian points in the spatial restricted problem of three bodies. *Cel. Mech.*, 50:31–58.

Cherry, T. M. (1924a). On integrals developable about a singular point of a Hamiltonian system of differential equations. *Proc. Camb. Phil. Soc.*, 22:325–349.

Cherry, T. M. (1924b). On integrals developable about a singular point of a Hamiltonian system of differential equations, II. *Proc. Camb. Phil. Soc.*, 22:510–533.

Chierchia, L. and Falcolini, C. (1996). Compensations in small divisor problems. *Comm. Math. Phys.*, 175:135–160.

Contopoulos, G. (1960). A third integral of motion in a Galaxy. *Z. Astrophys.*, 49:273–291.

Contopoulos, G. (1963). Resonant case and small divisors in a third integral of motion, I. *Astron. J.*, 68:763–779.

Contopoulos, G. (1965). Resonant case and small divisors in a third integral of motion, II. *Astron. J.*, 70:817–835.

Diana, E., Galgani, L., Giorgilli, A. Scotti, A. (1975). On the direct construction of formal integrals of a Hamiltonian system near an equilibrium point. *Boll. Un. Mat. It.*, 84–89.

Eliasson, L. H. (1988). Absolutely convergent series expansion for quasi–periodic motions. report 2–88, Dept. of Math., Univ. of Stockolm; later published in *MPEJ*, 2:1–33 (1996).

Gallavotti, G. (1994). Twistless KAM tori. *Comm. Math. Phys.*, 164:145–156.

Giorgilli, A. and Galgani, L. (1978). Formal integrals for an autonomous Hamiltonian system near an equilibrium point. *Cel. Mech.*, 17:267–280.

Giorgilli, A. Rigorous results on the power expansions for the integrals of a Hamiltonian system near an elliptic equilibrium point. *Ann. Ist. H. Poincaré*, 48(4):423–439.

Giorgilli, A. and Zehnder, E. (1992). Exponential stability for time dependent potentials. *ZAMP*, 43(5):827–855.

Giorgilli, A. (1995a). Quantitative methods in classical perturbation theory. In Roy, A. E. and Steves, B. A., editors. *Proceedings of the Nato ASI school: "From Newton to chaos: modern techniques for understanding and coping with chaos in N–body dynamical systems"*. Plenum Press, New York.

Giorgilli, A. (1995b). Methods of complex analysis in classical perturbation theory. In Benest, D. and Froeschlé, C., editors. *Chaos and diffusion in Hamiltonian systems.* Editions Frontières.

Giorgilli, A. and Locatelli, U. (1997a). Kolmogorov theorem and classical perturbation theory. *ZAMP*, 48:220–261.

Giorgilli, A. and Locatelli, U. (1997b). On classical series expansions for quasi–periodic motions. *MPEJ*, 3(5).

Giorgilli, A. and Morbidelli, A. (1997). Invariant KAM tori and global stability for Hamiltonian systems. *ZAMP*, 48:102–134.

Giorgilli, A. and Locatelli, U. (1999). A classical self–contained proof of Kolmogorov's theorem on invariant tori. In Simó, C., editor. *Hamiltonian systems with three or more degrees of freedom*, NATO ASI series C, 533. Kluwer Academic Publishers, Dordrecht–Boston–London.

Giorgilli, A. (2003). Notes on exponential stability of Hamiltonian systems. *Pubblicazioni della Classe di Scienze, Scuola Normale Superiore, Pisa, Centro di Ricerca Matematica "Ennio De Giorgi"*.

Jost, R. (1968). Winkel–und Wirkungsvariable für allgemeine mechanische Systeme. *Helvetica Physica Acta.* 41:965–968.

Kolmogorov, A. N. (1954). Preservation of conditionally periodic movements with small change in the Hamilton function. *Dokl. Akad. Nauk SSSR*, 98:527–530. In Russian. English translation in Casati, G. and Ford, J., editors. *Lecture Notes in Physics*, 93:51–56 (1979).

Laskar, J. (1988). Secular evolution over 10 million years. *Astronomy and Astrophysics*, 198:341–362.

Laskar, J. (1995). Introduction to frequency map analysis. In Simó, C., editor. *Hamiltonian systems with three or more degrees of freedom*, NATO ASI series C, 533. Kluwer Academic Publishers, Dordrecht–Boston–London.

Littlewood, J. E. (1959). On the equilateral configuration in the restricted problem of three bodies. *Proc. London Math. Soc. (3)*, 9:343–372.

Littlewood, J. E. (1959). The Lagrange configuration in Celestial Mechanics. *Proc. London Math. Soc. (3)*, 9:525–543.

Locatelli, U. and Giorgilli, A. (2000). Invariant tori in the secular motions of the three-body planetary systems. *Cel. Mech.*, 78:47–74.

Lochak, P. (1992). Canonical perturbation theory via simultaneous approximations, *Usp. Math. Nauk.*, 47:59–140. English translation in *Russ. Math. Surv.*

Morbidelli, A. and Giorgilli, A. (1995). Superexponential stability of KAM tori. *J. Stat. Phys.*, 78:1607–1617.

Moser, J. (1955). Stabilitätsverhalten kanonisher differentialgleichungssysteme. *Nachr. Akad. Wiss. Göttingen, Math. Phys. Kl IIa*, 6:87–120.

Moser, J. (1962). On invariant curves of area–preserving mappings of an annulus. *Nachr. Akad. Wiss. Gött,. II Math. Phys. Kl 1962*, 1–20.

Nekhoroshev, N. N. (1977). Exponential estimates of the stability time of near–integrable Hamiltonian systems. *Russ. Math. Surveys*, 32:1–65.

Nekhoroshev, N. N. (1979). Exponential estimates of the stability time of near–integrable Hamiltonian systems, II. *Trudy Sem. Petrovs.*, 5:5–50.

Poincaré, H. (1892). *Les méthodes nouvelles de la mécanique céleste.* Gauthier–Villars, Paris.

Siegel, C. L. (1941). On the integrals of canonical systems. *Ann. Math.*, 42:806–822.

Whittaker, E. T. (1916). On the adelphic integral of the differential equations of dynamics, *Proc. Roy Soc. Edinburgh, Sect. A*, 37:95–109.

PERIODIC ORBITS IN GRAVITATIONAL SYSTEMS

John D. Hadjedemetriou
Department of Physics, University of Thessaloniki, Greece

Abstract The periodic orbits play an important role in the study of the stability of a dynamical system. The methods of study of the stability of a periodic orbit are presented both in the general case and for Hamiltonian systems. The Poincaré map on a surface of section is presented as a powerful tool in the study of a dynamical system, especially for two or three degrees of freedom. Special attention is given to nearly integrable dynamical systems, because our solar system and the extra solar planetary systems are considered as perturbed Keplerian systems. The continuation of the families of periodic orbits from the unperturbed, integrable, system to the perturbed, nearly integrable system, is studied.

Keywords: Periodic orbits, resonances, stability

1. The gravitational N-Body problem

The Newtonian gravitational force is the dominant force in the N-Body systems in the universe, as for example in a planetary system, a planet with its satellites, or a multiple stellar system. The long term evolution of the system depends on the topology of its phase space and on the existence of ordered or chaotic regions. The topology of the phase space is determined by the position and the stability character of the periodic orbits of the system (the fixed points of the Poincaré map on a surface of section). Islands of stable motion exist around the stable periodic orbits. chaotic motion appears at unstable periodic orbits. This makes clear the importance of the periodic orbits in the study of the dynamics of such systems.

B.A. Steves et al. (eds.), Chaotic Worlds: from Order to Disorder in Gravitational N-Body Dynamical Systems, 43–79.

In many cases there is only one massive body whose gravitational attraction provides the dominant force; this is the case with a planetary system, where the Sun is the main attracting body, or a planet surrounded by satellites. In this case the motion of the small bodies (planets or satellites) follow Keplerian orbits, perturbed by the gravitational interaction between the small bodies. This is a nearly integrable dynamical system. In these systems resonances exist between the small bodies in their motion around the massive body. These correspond to periodic motion, and this makes clear the importance of the resonances in the dynamical properties of a nearly integrable system.

The simplest model of a gravitational system is a system of two bodies moving in Keplerian orbits around their common center of mass. This is an integrable system. We consider now a hierarchy of models, starting from the above mentioned integrable system and adding more bodies to the system. We have different models, which are used to study particular systems.

- **The restricted three-body problem**: Two bodies of finite masses, called *primaries*, revolve around their common center of mass in *circular* orbits and a third body with *negligible* mass moves under their gravitational attraction, but does not affect the orbits of the two primaries. In most astronomical applications the second primary has a small mass compared to the first primary, and consequently the motion of the third, massless, body is a perturbed Keplerian orbit. This is a model for the study of an asteroid (Jupiter being the second primary), a trans-Neptunian object (Neptune being the second primary) or an Earth-like planet in an extrasolar planetary system.

- **The general three-body problem**: Three bodies with finite masses moving under their gravitational attraction. This is a model for a triple stellar system. In many astronomical applications one of the three bodies has a large mass and the other two bodies have small, but not negligible masses. This is a model for an extrasolar planetary system, or a system of two satellites moving around a major planet. In the latter two cases the two small bodies move in perturbed Keplerian orbits.

We will begin our study with systems of two degrees of freedom, and then extend the results to three or more degrees of freedom. The study will be for a general dynamical system and applications will be made to gravitational systems of astronomical interest.

2. Periodic orbits in systems with two degrees of freedom: Autonomous case

2.1 Periodic orbits

Let us consider a dynamical system with two degrees of freedom, defined by the set of two second order differential equations

$$\begin{aligned}\ddot{x}_1 &= F_1(x_1, x_2, \dot{x}_1, \dot{x}_2),\\ \ddot{x}_2 &= F_2(x_1, x_2, \dot{x}_1, \dot{x}_2).\end{aligned} \tag{1}$$

The initial conditions that determine a solution are

$$x_{10},\ x_{20},\ \dot{x}_{10},\ \dot{x}_{20}, \tag{2}$$

and the corresponding solution has the form

$$\begin{aligned}&x_1(x_{10}, x_{20}, \dot{x}_{10}, \dot{x}_{20}; t),\\ &x_2(x_{10}, x_{20}, \dot{x}_{10}, \dot{x}_{20}; t).\end{aligned} \tag{3}$$

The solution is periodic, with period T, if, for every t

$$x_i(x_{10}, x_{20}, \dot{x}_{10}, \dot{x}_{20}; t+T) = x_i(x_{10}, x_{20}, \dot{x}_{10}, \dot{x}_{20}; t). \tag{4}$$

2.2 Existence of symmetric periodic orbits

We assume that the differential equations are invariant under the transformation

$$x_1 \to x_1,\ x_2 \to -x_2,\ t \to -t. \tag{5}$$

This property appears in several models that are of astronomical interest. It means that if

$$x_1(t),\ x_2(t)$$

is a solution, then

$$x_1(t),\ -x_2(-t) \tag{6}$$

is also a solution. Note that this second solution and the first solution are symmetric with respect to the x_1-axis. Consequently, if an orbit starts from the x_1 axis perpendicularly, with $\dot{x}_{10} = 0$, and crosses again the x_1 axis perpendicularly, with $\dot{x}_{10} = 0$, the orbit is closed and consequently is a *symmetric* periodic orbit with respect to the x_1-axis. This behaviour is shown in Figure 1.

The initial conditions of a symmetric periodic orbit are

$$x_{10},\ x_{20} = 0,\ \dot{x}_{10} = 0,\ \dot{x}_{20},$$

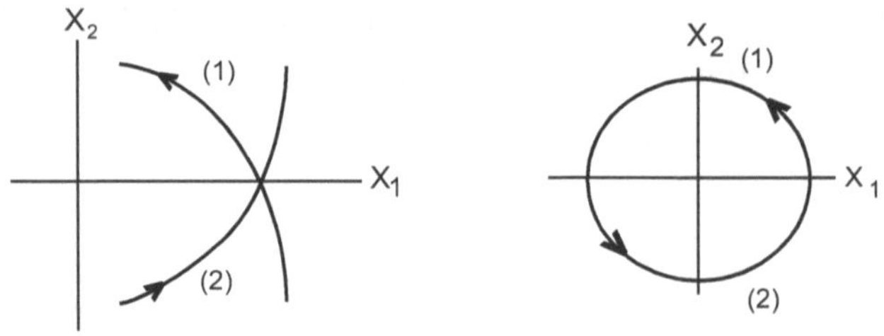

Figure 1. The two symmetric solutions (first panel) and their coincidence to a single orbit (second panel) if the periodicity conditions apply

which means that a symmetric periodic orbit is determined only by two nonzero initial conditions

$$x_{10},\ \dot{x}_{20}. \tag{7}$$

From the above we see that the *periodicity conditions* are

$$\begin{aligned} x_2(x_{10}, 0, 0, \dot{x}_{20}; T/2) &= 0, \\ \dot{x}_1(x_{10}, 0, 0, \dot{x}_{20}; T/2) &= 0, \end{aligned} \tag{8}$$

which imply that the orbit starts perpendicularly from the x-axis and crosses the x-axis again perpendicularly after a time interval equal to half the period T. We remark that the second *perpendicular* crossing may take place after several (non perpendicular) crossings from the x_1-axis.

The periodic orbits are not isolated, in general. They belong to families, along which the period varies. A family of symmetric periodic orbits is represented by a continuous curve in the space of initial conditions $x_{10}, \dot{x}_{20}$. This curve is called a *characteristic curve.*

3. Variational equations

A periodic orbit is an orbit which repeats itself for infinite time with period T. We shall study now the behaviour of the system in the vicinity of a periodic solution by considering perturbed initial conditions – i.e, initial conditions in the vicinity of the initial conditions of the periodic orbit.

We express the system of differential equations (1) as a system of four differential equations of first order,

$$\begin{aligned} \dot{x}_1 &= x_3, \\ \dot{x}_2 &= x_4, \\ \dot{x}_3 &= F_1(x_1, x_2, x_3, x_4), \\ \dot{x}_4 &= F_2(x_1, x_2, x_3, x_4), \end{aligned}$$

or, in general

$$\dot{x}_i = f_i(x_1, x_2, x_3, x_4). \quad (i = 1, ...4) \tag{9}$$

Let

$$x_i = x_i(x_{10}, x_{20}, x_{30}, x_{40}; t), \quad (i = 1, ...4) \tag{10}$$

be a solution of the system (9), non periodic in general, corresponding to the initial conditions $x_1(0), x_2(0), x_3(0), x_4(0)$. We consider now new initial conditions, in the vicinity of these initial conditions, of the form

$$x_1(0) + \xi_1(0),\ x_2(0) + \xi_2(0),\ x_3(0) + \xi_3(0),\ x_4(0) + \xi_4(0),$$

where $\xi_i(0)$ are small. The new solution can be expressed in the form

$$x'_i(t) = x_i(t) + \xi_i(t), \quad (i = 1, ...4) \tag{11}$$

where $\xi(t)$ is the deviation vector between the solution (10) and the perturbed solution (11), for the *same time* t,

$$\xi(t) = x'_i(t) - x_i(t).$$

The behaviour of the system in the vicinity of the solution (10) depends on the deviation vector $\xi(t)$.

We assume that the initial perturbation $\xi(0)$ is small, and consequently, for continuity reasons, the deviation $\xi(t)$ should be also small, at least for a finite time interval. For this reason we linearize the system of differential equations (9), to first order terms in the $\xi_i(t)$, by substituting the perturbed solution (11) into system (9) and keeping only first order terms in ξ_i. We obtain the system of *variational equations*,

$$\dot{\xi}_i = \sum_{k=1}^{4} p_{ik}\xi_k, \quad p_{ik} = \left(\frac{\partial f_i}{\partial x_k}\right)_{x_i(t)}, \quad (i = 1, ...4) \tag{12}$$

which describes the evolution of the system (9) in the neighborhood of the orbit (10), to first order terms in the deviations. The partial derivatives are computed for the solution $x_i(t)$.

The variational equations (12) are a system of four linear differential equations with time dependent coefficients. If the solution $x(t)$ is T-periodic, then the partial derivatives are also T-periodic. In this latter case the system of variational equations is a *linear system with periodic coefficients*. The theory related to the study of such systems is the *Floquet theory* and some elements of it will be presented in the following sections.

3.1 The fundamental matrix of solutions

The general solution of the linear system (12) is expressed as a linear combination of four linearly independent solutions. In particular, let us consider a 4×4 matrix $\Delta(t)$ whose columns are four linearly independent solutions corresponding to the initial conditions $\Delta(0) = I_4$, where I_4 is the 4×4 unit matrix. This matrix is called *fundamental matrix of solutions* and the general solution of the variational equations is expressed in the form

$$\xi(t) = \Delta(t)\xi(0). \tag{13}$$

A basic property of the matrix $\Delta(t)$ is the Liouville-Jacobi formula (Yakubovich and Starzhinskii, 1975; Jordan and Smith, 1988)

$$\det \Delta(t) = \det \Delta(0) \exp \int_o^t \mathrm{trace}(P) dt, \tag{14}$$

where P is the matrix of coefficients for the system of variational equations (12).

We shall prove now that the columns of the matrix $\Delta(t)$ are the partial derivatives of the solution $x_i(x_{10}, x_{20}, x_{30}, x_{40}, t)$ with respect to the initial conditions. In particular, we shall prove that the j-th column, $j = 1, ...4$, is given by

$$\begin{pmatrix} \partial x_1/\partial x_{j0} \\ \partial x_2/\partial x_{j0} \\ \partial x_3/\partial x_{j0} \\ \partial x_4/\partial x_{j0} \end{pmatrix}.$$

Proof

The solution $x_i(x_0; t)$ satisfies the system (9),

$$\frac{\partial x_i(x_0, t)}{\partial t} = f_i(x_1(x_0, t), x_2(x_0, t), x_3(x_0, t), x_4(x_0, t)). \ (i = 1, ...4)$$

If we apply to the above equations the operator $\partial/\partial x_{j0}$, $j = 1, ...4$, we obtain

$$\frac{\partial}{\partial t}\left(\frac{\partial x_i}{\partial x_{jo}}\right) = \sum_{k=1}^{4} \left(\frac{\partial f_i}{\partial x_k}\right) \frac{\partial x_k}{\partial x_{j0}}, \ (i = 1, ...4) \tag{15}$$

for each x_{j0}. We note that the system (16) is the system of variational equations (12) satisfied by the vector (15). This means that the four vectors (15), for $j = 1, ...4$, are four linearly independent solutions of the variational equations. In addition, we note that $\partial x_i/\partial x_{j0} = \delta_{ij}$ for $t = 0$. Consequently, these four vectors are the four columns of the fundamental matrix of solutions $\Delta(t)$.

3.2 Variational equations for a periodic solution

We assume now that the solution $x(t)$ is T-periodic. Then the system of variational equations corresponding to this periodic solution is a linear system with *periodic* coefficients.

We shall prove that the derivative $\dot{x}_i(t)$ of the periodic solution $x_i(t)$ is a solution of the variational equations.

Proof

The solution $x_i(t)$ satisfies the system (9):

$$\dot{x}_i = f_i(x_1, x_2, x_3, x_4). \quad (i = 1, ...4)$$

If we apply the operator d/dt we obtain

$$\frac{d}{dt}(\dot{x}_i(t)) = \sum_{j=1}^{4} \left(\frac{\partial f_i}{\partial x_j}\right)_{x_i(t)} \dot{x}_j(t).$$

This is the system of variational equations for the solution $\xi_i = \dot{x}_i(t)$. So we come to the conclusion that *the variational equations that correspond to a T-periodic orbit have always a T-periodic solution, which is the derivative $\dot{x}_i(t)$ of the periodic solution.*

4. Linear stability of a periodic orbit

Let $x_i(t)$ be a periodic orbit and $x'(t)$ a perturbed orbit, which, to a linear approximation, can be expressed in the form

$$x_i'(t) = x_i(t) + \xi_i(t),$$

where $\xi_i(t)$ is the solution of the variational equations. This latter solution is expressed in the form

$$\xi(t) = \Delta(t)\xi(0), \tag{16}$$

and for $t = T$,

$$\xi(T) = \Delta(T)\xi(0). \tag{17}$$

From this expression we obtain, by induction,

$$\xi(nT) = [\Delta(T)]^n\xi(0). \tag{18}$$

Equations (18) and (19) give the deviation, to a linear approximation, of the perturbed orbit $x'(t)$ from the periodic orbit $x(t)$ after a time interval equal to n times the period T, due to an initial deviation $\xi(0) = x'(0) - x(0)$. In fact Equation (19) is a mapping of the initial deviation

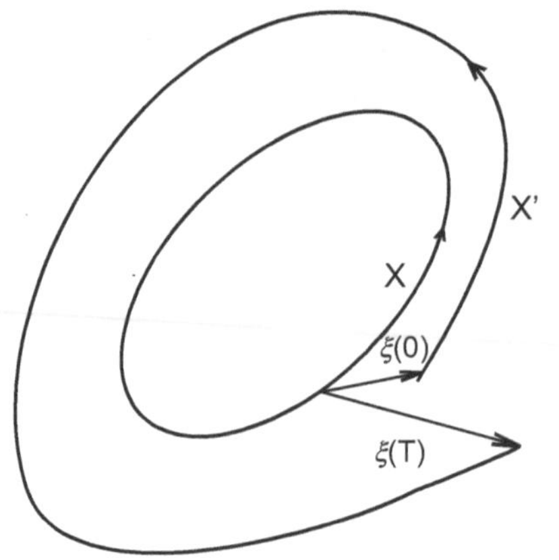

Figure 2. Mapping at integral multiples of the period

$\xi(0)$ at integral multiples of the period T (see Figure 2). This is a linear mapping defined by the matrix $\Delta(T)$. It is clear that the stability of the periodic orbit $x(t)$ depends on the properties of the mapping (19) – i.e. on the eigenvalues of the matrix $\Delta(T)$. The matrix $\Delta(T)$ is called the *monodromy matrix*.

4.1 Unit eigenvalues of the monodromy matrix $\Delta(T)$

Existence of a periodic solution $\xi(t)$

Let $\xi(t)$ be a T-periodic solution of the variational equations. We have $\xi(t+T) = \xi(t)$, for any t and consequently, for $t = 0$, $\xi(T) = \xi(0)$. Due to this latter relation, Equation (17) takes the form, for $t = T$, $\xi(0) = \Delta(T)\xi(0)$, and finally

$$(\Delta(T) - I)\,\xi(0) = 0. \tag{19}$$

Thus we come to the conclusion that *if the system of variational equations has a periodic solution, the monodromy matrix has a unit eigenvalue.*

Remark: Note that the system of variational equations that corresponds to a T-periodic solution $x(t)$, has a T-periodic solution $\xi(t) = \dot{x}(t)$. The corresponding eigenvector is $\xi(0) = \dot{x}(0)$, as is seen from Equation (20), and represents a tangent displacement along the periodic orbit (see Figure 3).

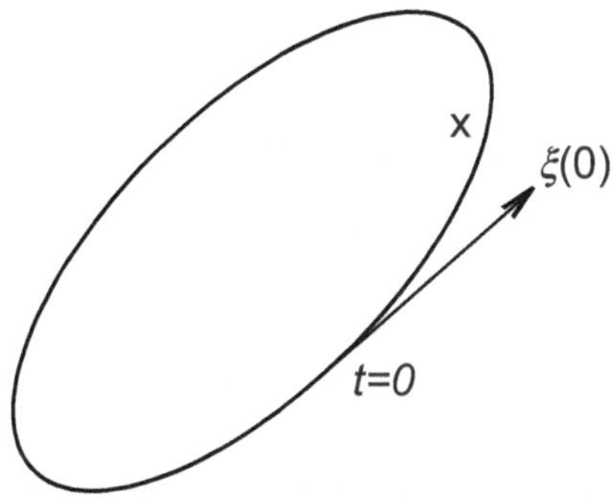

Figure 3. Tangent displacement $\xi(0) = \dot{x}(0)$

Existence of an integral of motion

Let $G(x_1, x_2, x_3, x_4) =$ constant be an integral of motion. If $x_i(x_0, t)$ is a T-periodic solution, we have the relation

$$G(x_1(x_0,t), x_2(x_0,t), x_3(x_0,t), x_4(x_0,t)) =$$
$$G(x_{10}, x_{20}, x_{30}, x_{40}).$$

We apply to the above relation the operator $\partial/\partial x_{j0}$, and we obtain

$$\sum_j \left(\frac{\partial G}{\partial x_k}\right)_t \left(\frac{\partial x_k}{\partial x_{j0}}\right)_t = \left(\frac{\partial G}{\partial x_{j0}}\right).$$

We set now $t = T$ and taking into account that

$$\left(\frac{\partial G}{\partial x_k}\right)_{t=T} = \left(\frac{\partial G}{\partial x_k}\right)_{t=0},$$

due to the fact that $x(t)$ is periodic, we obtain

$$(\Delta^\tau(T) - I)\,\nabla G = 0, \tag{20}$$

where Δ^τ is the transpose of Δ. From this relation we obtain that if $\nabla G \neq 0$ then $\Delta(T)^\tau$ has a unit eigenvalue. Thus finally, we come to the conclusion that *if the dynamical system has an integral of motion, which is not stationary along the periodic orbit, the monodromy matrix $\Delta(T)$ has a unit eigenvalue.*

Existence of a unit eigenvalue of $\Delta(T)$

We shall prove that a unit eigenvalue of $\Delta(T)$ implies a T-periodic solution of the variational equations:

Let $\xi(0)$ be the eigenvector corresponding to the unit eigenvalue. We have

$$(\Delta(T) - I)\xi(0) = 0,$$

from which we obtain

$$\xi(0) = \Delta(T)\xi(0).$$

But, according to Equation (18), $\xi(T) = \Delta(T)\xi(0)$, and finally

$$\xi(T) = \xi(0),$$

which implies that $\xi(t)$ is periodic. The initial conditions of this periodic orbit is the eigenvector of the monodromy matrix corresponding to this unit eigenvalue.

In general, if there exist two unit eigenvalues with two linearly independent eigenvectors, respectively, then the system of variational equations has two independent periodic solutions.

Remark: We shall prove later that in a Hamiltonian system there are always two unit eigenvalues, but only one eigenvector. So, there is only one periodic solution of the variational equations, corresponding to the double unit eigenvalue.

4.2 Special solutions

Let us assume that the system of variational equations has a special solution with the property

$$\xi(t+T) = \lambda\xi(t). \tag{21}$$

If we set $t = 0$ we obtain

$$\xi(T) = \lambda\xi(0), \tag{22}$$

which means that the initial deviation $\xi(0)$ is mapped, after a period T, to a multiple of it.

Equation (23) is a difference equation, and its solution is of the form

$$\xi(t) = f(t)\lambda^{t/T}, \quad f(t)\text{: } T\text{-periodic.} \tag{23}$$

From Equation (23) and Equation (18), $\xi(T) = \Delta(T)\xi(0)$, we obtain

$$\Delta(T)\xi(0) = \lambda\xi(0)$$

and finally

$$(\Delta(T) - \lambda I)\,\xi(0) = 0.$$

This means that if a solution with property (22) exists, λ is an eigenvalue of the monodromy matrix and $\xi(0)$ the corresponding eigenvector.

Let us assume now that the monodromy matrix has an eigenvalue λ and let $\xi(0)$ be the corresponding eigenvector. We shall prove that this eigenvector is the initial conditions of a solution with the property (22). We have $\Delta(T)\xi(0) = \lambda\xi(0)$ and also, because of Equation (18), $\Delta(T)\xi(0) = \xi(T)$. From these two relations we obtain

$$\xi(T) = \lambda\xi(0). \tag{24}$$

Equation (17) is valid for any t and setting $t = t + T$ we have,

$$\xi(t+T) = \Delta(t+T)\xi(0). \tag{25}$$

We note now that $\Delta(t+T)$ is a solution matrix (because the matrix p_{ij} of the variational equations is T-periodic) and consequently it satisfies relation (17), in matrix form (Meyer and Hall, 1992; Yakubovich and Starzhinskii, 1975),

$$\Delta(t+T) = \Delta(t)\Delta(T).$$

Equation (26) now takes the form

$$\xi(t+T) = \Delta(t)\Delta(T)\xi(0). \tag{26}$$

and taking into account the relations (18) and (25) we obtain

$$\xi(t+T) = \lambda\Delta(t)\xi(0),$$

and finally, using (17) we obtain

$$\xi(t+T) = \lambda\xi(t).$$

Thus finally we come to the conclusion that *to each eigenvalue λ of the monodromy matrix $\Delta(T)$ there exists a solution of the form $f(t)\lambda^{t/T}$, where $f(t)$ is T-periodic, whose initial conditions is the eigenvector $f(0)$ corresponding to the eigenvalue λ.*

Characteristic exponents

We define a parameter α as

$$\alpha = \frac{1}{T}\ln\lambda, \tag{27}$$

where the principal value of the logarithm is taken. Then the solution (23) is expressed as

$$\xi(t) = f(t)e^{\alpha t}. \tag{28}$$

The parameter α is called the *characteristic exponent* and is related to the eigenvalue λ by the relation (28). Note that for $\lambda = 1$, it is $\alpha = 0$.

The monodromy matrix $\Delta(T)$ is a 4×4 matrix and it has four eigenvalues. If there exist four independent eigenvectors to the four eigenvalues of $\Delta(T)$ (this may not be the case if some eigenvalues are multiple), then there exist four independent solutions with the property (22) and consequently of the form (29). In this case the general solution is a linear combination of solutions of the form (29).

From the above is clear that the periodic orbit $x(t)$ is linearly stable if and only if

$$\mathrm{Re}(\alpha) \leq 0. \tag{29}$$

For $\alpha = 0$ ($\lambda = 1$) the solution (29) is T-periodic.

5. Hamiltonian systems

Gravitational systems are Hamiltonian. For this reason we shall study in this section the special properties that a Hamiltonian system possesses, in addition to the general properties obtained in the previous sections. We will start with systems with two degrees of freedom.

A Hamiltonian system is defined by the Hamiltonian function

$$H(x_1, x_2, x_3, x_4), \tag{30}$$

where x_1, x_2 are the coordinates and x_3, x_4 the momenta.

The Hamiltonian equations are

$$\begin{aligned} \dot{x}_1 &= \partial H/\partial x_3, \\ \dot{x}_2 &= \partial H/\partial x_4, \\ \dot{x}_3 &= -\partial H/\partial x_1, \\ \dot{x}_4 &= -\partial H/\partial x_2, \end{aligned}$$

or

$$\dot{x} = -J\nabla H, \tag{31}$$

where ∇H is a column vector and J the 4×4 symplectic matrix

$$J = \begin{pmatrix} 0 & -I_2 \\ +I_2 & 0 \end{pmatrix}. \tag{32}$$

Note that $J^{-1} = -J$.

5.1 Variational equations of Hamiltonian systems

The variational equations of a Hamiltonian system (32) have the special form given by

$$\dot{\xi} = -JA\xi, \tag{33}$$

where the elements a_{ij} of the 4×4 matrix A are

$$a_{ij} = \frac{\partial^2 H}{\partial x_i \partial x_j}. \quad (i, j = 1, ...4) \tag{34}$$

The system (34) is called a *linear Hamiltonian system.* It is easy to see that it can be expressed in the Hamiltonian form (32) with Hamiltonian

$$H = \frac{1}{2}\xi^{\tau} A \xi = \frac{1}{2} \sum_{i,j=1}^{4} a_{ij}\xi_i\xi_j.$$

From the relations (35) we can verify that the trace of the matrix of the coefficients of a linear Hamiltonian system (34) is equal to zero. Consequently, due to the general property (14), the determinant of the fundamental matrix of solutions $\Delta(t)$ is equal to unity (see also Meyer and Hall, 1992),

$$\det \Delta(t) = 1.$$

For $t = T$ we obtain

$$\det \Delta(T) = 1, \tag{35}$$

from which we see that *the determinant of the monodromy matrix is equal to unity.*

As we proved in Section 3.1, the fundamental matrix of solutions $\Delta(t)$ is expressed in the form

$$\Delta(t) = \frac{\partial(x_1, x_2, x_{,3}, x_4)}{\partial(x_{10}, x_{20}, x_{30}, x_{40})}. \tag{36}$$

This means that the determinant of the Jacobian of the flow in phase space is equal to one. Consequently, *the volume in phase space is conserved* (the Liouville theorem).

5.2 Symplectic property

An important property of the monodromy matrix of a Hamiltonian system is the symplectic property that we shall prove now. Let ξ and ξ' be two solutions of the variational equations. The following property holds:

$$\xi_1\xi_3' + \xi_2\xi_4' - \xi_3\xi_1' - \xi_4\xi_2' = \text{constant},$$

or, in matrix form,

$$\xi'^{\tau} \, J \, \xi = \text{constant}, \tag{37}$$

where ξ, ξ' are 4-vectors and J is the symplectic matrix (33).

Proof

The proof can be made by direct substitution.

If we apply property (38) to the four solutions that are the four columns of the matrix $\Delta(t)$, we obtain the equation

$$\Delta^{\tau}(t)\, J\, \Delta(t) = J.$$

We set now $t = T$ and obtain

$$\Delta^{\tau}(T)\, J\, \Delta(T) = J. \tag{38}$$

This is an important property of the monodromy matrix of a Hamiltonian system, which is called the *symplectic* property. Thus we come to the conclusion that *the monodromy matrix of a Hamiltonian system is symplectic.*

5.3 Eigenvalues of a symplectic matrix

We shall see now that the eigenvalues of a symplectic matrix have some special properties. We express property (39) as

$$\Delta^{\tau}(T) = J\Delta^{-1}(T)J^{-1},$$

from which we see that the matrix $\Delta^{\tau}(T)$ is related to the matrix $\Delta^{-1}(T)$ by a similarity transformation. Consequently, they have the same set of eigenvalues. Thus finally, we come to the conclusion that the eigenvalues of $\Delta(T)$ are in reciprocal pairs. In addition, due to the fact that the matrix $\Delta(T)$ is real, they are also in complex conjugate pairs.

From the above we see that the four eigenvalues $\lambda_1, \lambda_2, \lambda_3, \lambda_4$ of the monodromy matrix have the property

$$\lambda_1\lambda_2 = 1, \quad \lambda_3\lambda_4 = 1. \tag{39}$$

We note now that the variational equations correspond to a periodic orbit $x(t)$. So, $\xi(t) = \dot{x}(t)$ is a periodic solution of the variational equations and according to Section 4.1, one eigenvalue is equal to one, $\lambda_1 = 1$. Using now relation (40) we come to the conclusion that *the monodromy matrix of a Hamiltonian system corresponding to a periodic orbit has a double unit eigenvalue.*

$$\lambda_1 = 1, \quad \lambda_2 = 1. \tag{40}$$

5.4 Stability

Stability is determined by the two nonzero eigenvalues λ_3, λ_4 of the monodromy matrix. As we proved, these eigenvalues are reciprocal and

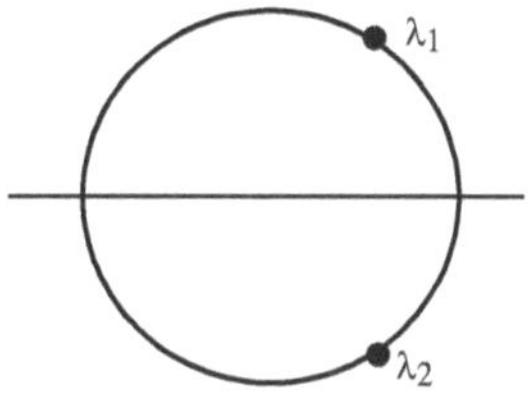

Figure 4. Eigenvalues complex conjugate on the unit circle

also complex conjugate, so they are either on the unit circle, in the complex plane, or on the real axis, one inside the unit circle and the other outside. A special case is $\lambda_3 = \lambda_4 = +1$ or $\lambda_3 = \lambda_4 = -1$. All of these cases are shown in Figures 4-6. It is evident that the orbit is stable only in the case where the two nonzero eigenvalues λ_3, λ_4 are complex conjugate, on the unit circle (Figure 4). If the eigenvalues λ_3, λ_4 are real, the orbit is unstable, because one of them will be larger than $+1$ or smaller than -1.

Stability criteria can be obtained from the elements of the monodromy matrix as follows. The eigenvalues are the roots of the characteristic equation of $\Delta(T)$ and consequently

$$\lambda_1 + \lambda_2 + \lambda_3 + \lambda_4 = \mathrm{trace}\Delta(T),$$

$$\lambda_1\lambda_2\lambda_3\lambda_4 = \det\Delta(T) = 1.$$

Taking into account that $\lambda_1 = \lambda_2 = 1$ we find that the two nonzero eigenvalues λ_3, λ_4 are the roots of the quadratic equation

$$\lambda^2 - K\lambda + 1 = 0, \tag{41}$$

where

$$K = \mathrm{trace}\Delta(T) - 2. \tag{42}$$

Stability depends on the value of K, which is called the *stability index*. Note that the stability index depends only the trace of the monodromy matrix. We have the following cases:

- $|K| < 2$, $-4 < \mathrm{trace}\Delta(T) < 4$: eigenvalues $\lambda_{3,4}$ complex conjugate on the unit circle, characteristic exponents $\alpha = \pm i\varphi$ (see Figure 4).

- $K = 2$, $\mathrm{trace}\Delta(T) = 4$: eigenvalues $\lambda_{3,4} = 1$, characteristic exponents $\alpha = 0$ (see Figure 5a).

- $K = -2$, $\mathrm{trace}\Delta(T) = 0$: eigenvalues $\lambda_{3,4} = -1$, characteristic exponents $\alpha = 0 \pm i\pi/T$ (see Figure 5b).

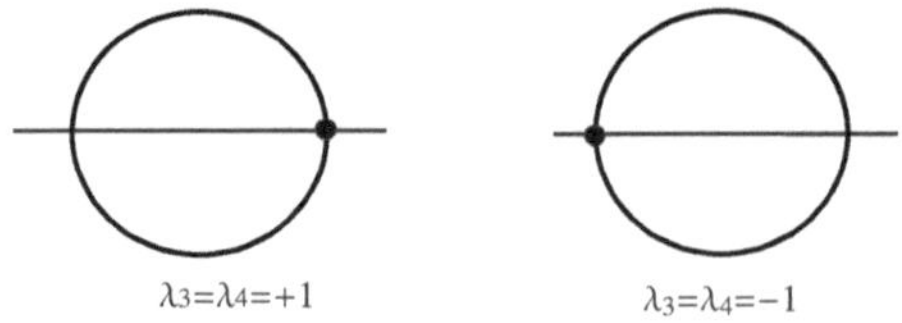

Figure 5. Eigenvalues at the critical points +1 and −1

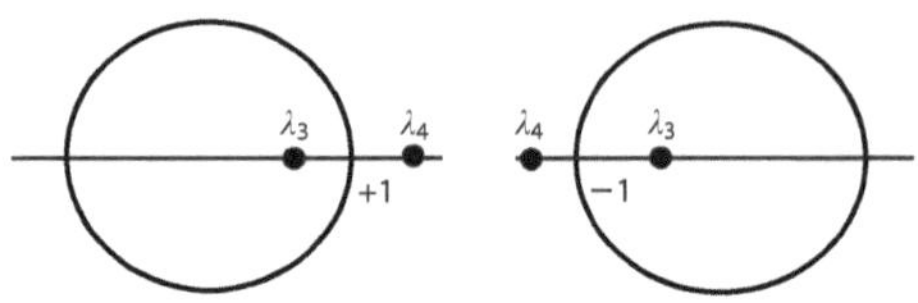

Figure 6. Real eigenvalues inside and outside the unit circle

- $K > 2$, trace$\Delta(T) > 4$: eigenvalues $\lambda_{3,4}$ real and positive, $\lambda_3 = 1/\lambda_4$, characteristic exponents $\pm\alpha$, α real (see Figure 6a).

- $K < -2$, trace$\Delta(T) < 0$: eigenvalues $\lambda_{3,4}$ real and negative, $\lambda_3 = 1/\lambda_4$, characteristic exponents $\pm(\alpha \pm i\pi/T)$, α real (see Figure 6b).

We have stability if all the eigenvalues are on the unit circle. So, we have stability only in the case that the stability index is $-2 \leq K \leq 2$.

Asymptotic stability never appears, because it is not possible for the eigenvalues λ_3, λ_4 to be both inside the unit circle. This is also a consequence of the fact that the volume in phase space is conserved.

Let us assume that a periodic orbit is stable, which implies that the eigenvalues λ_3, λ_4 are on the unit circle (Figure 4) and we assume that they are not equal to +1 or −1. If a parameter varies, then the eigenvalues λ_3, λ_4 are restricted to move *on the unit circle*, because they must be both inverse, $\lambda_3 = 1/\lambda_4$ and complex conjugate. Consequently, the stability is conserved. However, if λ_3, λ_4 meet at the points +1 or −1, then it is possible for them to go outside the unit circle, as shown in Figure 6, and thus generate instability. For this reason the orbits with $\lambda_3 = \lambda_4 = \pm 1$ are called *critical* as far as stability is concerned.

5.5 Solution in the vicinity of a T-periodic orbit $x(t)$

An orbit $x'(t)$ in the vicinity of the periodic orbit $x(t)$ is expressed in the form

$$x'(t) = x(t) + \xi(t), \tag{43}$$

to a linear approximation in the deviations, where $\xi(t)$ is the solution of the variation equations. If $\xi(t)$ is bounded then $x'(t)$ is also bounded.

The properties of the orbit $x'(t)$ in the vicinity of the periodic orbit $x(t)$ are determined from the properties of the solution $\xi(t)$ of the variational equations. In particular, if a T-periodic solution $\xi(t)$ exists, then a new T-periodic orbit $x'(t)$ exists.

We remark that the periodic solution of the variational equations $\xi(t) = \dot{x}(t)$ does not generate a new periodic orbit $x'(t)$, but the same orbit $x(t)$, with a phase shift, because $\xi(0)$ is tangent to the orbit $x(t)$ (Figure 3).

General solution of the variational equations

To each eigenvalue of $\Delta(T)$ there corresponds a solution $\xi(t)$, so we have four linearly independent solutions. However, to the double unit eigenvalue $\lambda_1 = \lambda_2 = 1$, there exists only one unit eigenvector $\xi(0) = \dot{x}(0)$. The two linearly independent solutions corresponding to the double unit eigenvalue are

$$\begin{aligned} \xi^1 &= f_1(t), \\ \xi^2 &= f_2(t) + t\, f_1(t), \end{aligned}$$

where $f_1(t)$, $f_2(t)$ are T-periodic.

For the other two eigenvalues $\lambda_3 = \lambda_4$ we have the solutions

- eigenvalues real and positive

  $\xi^{3,4} = f_{3,4}(t)\, e^{\pm \alpha t}$

- eigenvalues real and negative

  $\xi^{3,4} = f_{3,4}(t)\, e^{\pm \alpha t}\, e^{\pm i \pi t/T}$

- eigenvalues complex conjugate on the unit circle

  $\xi^{3,4} = f_{3,4}(t)\, e^{\pm i \beta t}$

where the functions $f_3(t)$, $f_4(t)$ are T-periodic.

The initial conditions for the above four special solutions are $f_i(0)$, $i = 1, \ldots 4$ and are given by

$$\begin{aligned} (\Delta(T) - I)\, f_1(0) &= 0, \\ (\Delta(T) - I) f_2(0) &= T f_1(0), \\ (\Delta(T) - \lambda_{3,4} I)\, f_{3,4}(0) &= 0, \end{aligned}$$

The general solution is a linear combination of the above four solutions:

Characteristic exponents real and positive:

$$\begin{aligned}\xi(t) &= c_1 f_1(t) + c_2(f_2(t) + t\, f_1(t)) \\ &\quad + c_3 f_3(t) e^{\alpha t} + c_4 f_4(t) e^{-\alpha t}.\end{aligned}$$

The orbit $x(t)$ is unstable.

Characteristic exponents real and negative:

$$\begin{aligned}\xi(t) &= c_1 f_1(t) + c_2(f_2(t) + t\, f_1(t)) \\ &\quad + c_3 f_3(t) e^{\alpha t} e^{i\pi t/T} + c_4 f_4(t) e^{-\alpha t} e^{-i\pi t/T}.\end{aligned}$$

The orbit $x(t)$ is unstable.

Characteristic exponents complex, on the unit circle:

$$\begin{aligned}\xi(t) &= c_1 f_1(t) + c_2(f_2(t) + t\, f_1(t)) \\ &\quad + c_3 f_3(t) e^{i\beta t} + c_4 f_4(t) e^{-i\beta t}.\end{aligned}$$

The orbit $x(t)$ is stable.

The above expressions give the totality of orbits in the vicinity of the periodic orbit $x(t)$ for all possible cases.

Isoenergetic displacements

Let us consider a periodic orbit $x(t)$, with initial conditions $x(0)$ and a nearby orbit $x'(t) = x(t)+\xi(t)$, with initial conditions $x'(0) = x(0)+\xi(0)$. The displacement vector $\xi(0)$ can be expressed as a linear combination of the four vectors $f(0)$ defined by Equations (46),

$$\xi(0) = c_1 f_1(0) + c_2 f_2(0) + c_3 f_3(0) + c_4 f_4(0), \tag{44}$$

where c_i are arbitrary constants. We shall prove that the displacement $\xi(0)$ is isoenergetic (the energy, i.e. the Hamiltonian, is not changed) if

$$c_2 = 0.$$

Proof

The Hamiltonian is constant along the periodic orbit $x(t)$,

$$H(x_1, x_2, x_3, x_4) = \text{constant}.$$

The change of H due to the displacement $\xi(0)$ is given by

$$\delta H = (\nabla H_0, \xi(0)),$$

where ∇H_0 is computed at the initial conditions $x_i(0)$.

Consider first $c_1 \neq 0$, $c_2 = c_3 = c_4 = 0$. Then $\xi(0) = c_1 f_1(0) = c_1 \dot{x}(0)$ or, due to the Hamiltonian equations, $\dot{x}_i(0) = J\nabla H_0$ and finally

$$\delta H = (\nabla H_0, c_1 J \nabla H_0) = 0. \tag{45}$$

Next, consider $c_3 \neq 0, c_1 = c_2 = c_4 = 0$, i.e. $\xi(0) = c_3 f_3(0)$. The change of the Hamiltonian is

$$\delta H = c_3(\nabla H_0, f_3(0)).$$

We take now into account that

$$\dot{x}(0) = f_1(0) = J\nabla H_0,$$

and

$$\nabla H_0 = J^{-1} f_1(0) = -J f_1(0),$$

and obtain finally

$$\delta H = -c_3(J f_1(0), f_3(0)). \tag{46}$$

We shall prove that

$$(J f_1(0), f_3(0)) = 0. \tag{47}$$

We have

$$f_1(0) = \Delta(T) f_1(0), \quad f_3(0) = \frac{1}{\lambda_3} \Delta(T) f_3(0)$$

and

$$(J f_1(0), f_3(0)) = (J f_1(0)^{\tau} . f_3(0)) =$$

$$\frac{1}{\lambda_3} f_1(0)^{\tau} \left(\Delta^{\tau}(T) J^{\tau} \Delta(T)\right) f_3(0),$$

or, using $J_{\tau} = -J$,

$$-f_1^{\tau}(0) J f_3(0) = -\frac{1}{\lambda_3} f_1^{\tau}(0) \left(\Delta^{\tau}(T) J \Delta(T)\right) f_3(0)$$

$$= \frac{1}{\lambda_3} f_1^{\tau}(0) J f_3(0).$$

From this latter relation we obtain

$$\left(1 + \frac{1}{\lambda_3}\right) f_1^{\tau}(0) J f_3(0) = 0,$$

and for $\lambda_3 \neq -1$ we have

$$(J f_1(0), f_3(0)) = 0.$$

Thus, from Equation (51) we obtain $\delta H = 0$. A similar proof holds for $\xi(0) = c_4 f_4(0)$.

Thus we come to the conclusion that

- Any perturbation which is a linear combination of $f_1(0), f_3(0),\ f_4(0)$ is isoenergetic.
- For an isoenergetic displacement, no secular term appears in the general solution, as is readily obtained from Equations(47),(48) and (49).
- In order to obtain a change of the energy, we must have a displacement along the vector $f_2(0)$.

5.6 Orbital stability

The stability that we mentioned before refers to the evolution of the deviation vector $\xi(t) = x'(t) - x(t)$ between the perturbed solution $x'(t)$ and the periodic orbit $x(t)$ *at the same time* t. If $\xi(t)$ is bounded, then the periodic orbit is stable. In this case two particles, one on the periodic orbit $x(t)$ and the other on the perturbed orbit $x'(t)$, that start close to each other at $t = 0$, would always stay close. A necessary condition is that all the eigenvalues of the monodromy matrix be on the unit circle in the complex plane. However, in a Hamiltonian system this condition is not enough for stability, because there is only one eigenvector corresponding to the double unit eigenvalue and consequently a secular term always appears in the general solution, as can be seen from Equation (49). We remark that this secular term appears if the vector of initial deviation $\xi(0) = x'(0) - x(0)$ has a component along the direction $f_2(0)$.

In order to understand the meaning of the secular term, we consider the initial conditions, for ε small,

$$x'(0) = x(0) + \varepsilon f_2(0). \tag{48}$$

The corresponding solution is

$$x'(t) = x(t) + \varepsilon f_2(0) + \varepsilon t f_1(0),$$

and using $f_1(t) = \dot{x}(t)$ we obtain

$$x'(t) = x(t) + \varepsilon t \dot{x}(t) + \varepsilon f_2(t), \tag{49}$$

and to a linear approximation in ε,

$$x'(t) = x(t + \varepsilon t) + \varepsilon f_2(t + \varepsilon t). \tag{50}$$

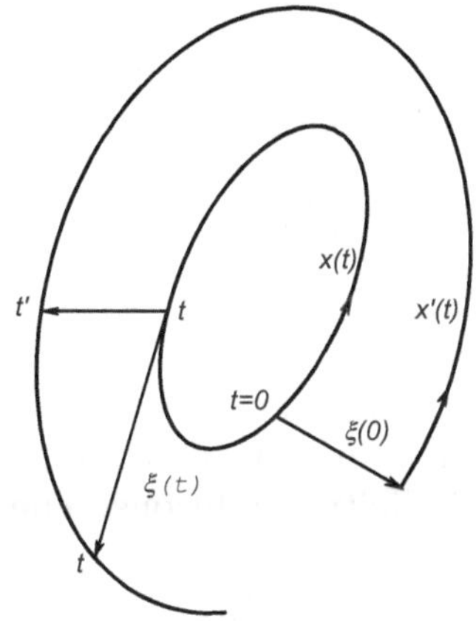

Figure 7. Orbital stability

We define the time

$$t' = t + \varepsilon t,$$

and come finally to the conclusion that

$$x'(t) - x(t') = \text{bounded}.$$

Thus we come to the conclusion that *the secular term introduces a phase shift only along the orbit*. This means that the two orbits, $x(t)$ and $x'(t)$, considered as geometrical curves, are close to each other (Figure 7). In this case we say that we have *orbital stability*, provided that the eigenvalues λ_3, λ_4 are on the unit circle and consequently the corresponding solution is bounded.

5.7 Families of periodic orbits

Consider the system of differential equations(not Hamiltonian in general)

$$\dot{x}_i = F_i(x_1, x_2, x_3, x_4), \quad (i = 1, ...4) \tag{51}$$

and let $x_{i0} = x_i(0)$ be the initial conditions corresponding to a T-periodic orbit. The solution is expressed in the form $x_i(x_{10}, x_{20}, x_{30}, x_{40}; t)$ and the periodicity conditions are

$$x_i(x_{10}, x_{20}, x_{30}, x_{40}; T) - x_{i0} = 0. \quad (i = 1, ...4) \tag{52}$$

A new periodic orbit with period $T + \delta T$, in the vicinity of the above T-periodic solution exists if the Jacobian determinant of the left-hand-side of Equation (58), computed at the initial conditions x_{i0}, is different from zero (implicit function theorem). In this case the T-periodic orbit $x(t)$ is not isolated, but belongs to a family of periodic orbits along which he initial conditions and the period vary. The Jacobian determinant is

$$|\Delta(T) - I_4|, \tag{53}$$

which however is always equal to zero, because the monodromy matrix $\Delta(T)$ has a unit eigenvalue.

In order to overcome the problem, we keep one initial condition fixed, e.g. x_{20} and vary the other three initial conditions x_{10}, x_{30}, x_{40}. We consider the first three periodicity conditions. The corresponding Jacobian matrix is

$$|\Delta_3(T) - I_3|, \tag{54}$$

where $\Delta_3(T)$ is the matrix obtained from the monodromy matrix $\Delta(T)$ by deleting the second column (because x_{20} is not considered as variable) and the fourth row (because the fourth periodicity condition is not considered). If this latter Jacobian matrix is not equal to zero (and this is the case in general), a new periodic solution exists in the vicinity of the periodic orbit $x(t)$.

The non vanishing of the matrix (59) ensures that the first three periodicity conditions are satisfied for the new initial conditions x'_{10}, x_{20}, x'_{30}, x'_{40}, with the new period T':

$$x_1(T') = x'_{10}, \quad x_2(T') = x_{20}, \quad x_3(T') = x'_{30}. \tag{55}$$

From the above we see that the first three periodicity conditions (58) are satisfied. In order for a new periodic orbit to exist, we must prove that the fourth periodicity condition is also satisfied.

We assume that an integral of motion exists

$$G(x_1, x_2, x_3, x_4) = \text{constant}.$$

Let x''_{40} be the value of x_4 after one period T'. We have

$$G(x'_{10}, x_{20}, x'_{30}, x'_{40}) - G(x'_{10}, x_{20}, x'_{30}, x''_{40}) = 0,$$

and by Taylor's theorem,

$$\left(\frac{\partial G}{\partial x_4}\right)_{x'''_{40}} (x'_{40} - x''_{40}) = 0, \tag{56}$$

where the partial derivative is computed for $(x'_{10}, x_{20}, x'_{30}, x'''_{40})$, x'''_{40} lying between x'_{40} and x''_{40}. Consequently

$$x''_{40} = x'_{40},$$

which implies that the fourth periodicity condition is also satisfied, provided that the partial derivative in (62) is not equal to zero.

We shall find now the new periodic orbit, to a linear approximation. Consider the orbit $x'(t)$, corresponding to the initial conditions

$$x'(0) = x(0) + c_2 f_2(0), \tag{57}$$

for c_2 small. The solution is given by Equation (56), and can be expressed, to a linear approximation in the small parameter c_2, as

$$x'(t) = x((1 + c_2)t) + c_2 f_2((1 + c_2)t).$$

The functions $x(t)$, $f_2(t)$ are T-periodic and consequently the new solution $x'(t)$ is $(T/(1 + c_2))$-periodic, and to first order in c_2, with period

$$T' = (1 - c_2)T.$$

From the above we conclude that a family of periodic orbits is obtained by varying the initial conditions of $x(0)$ along the vector $f_2(0)$, as is seen from Equation (63). We remark that it is along this vector that the energy (Hamiltonian) varies. The initial conditions of all the periodic orbits of the family lie on a curve that is called the *characteristic curve.* The vector $f_2(0)$ is tangent to the characteristic curve. The period and the energy vary along the family.

If the periodic orbits are symmetric with respect to the x_1-axis, then $x_{20} = \dot{x}_{10} = 0$ and the characteristic curve is a curve in the space of two initial conditions only, $x_{10}, \dot{x}_{20}$.

Remark: The procedure described in this section, for the continuation of a periodic orbit, was for a change of the initial conditions, keeping the parameters of the system fixed. In this case the period of the new periodic orbit was, in general, different from the initial period. It may happen however that the system of differential equations (57) depends on a parameter and we wish to study the continuation of the periodic orbit with respect to this parameter. The procedure is the same, but now we fix the period T and study the continuation to a new periodic orbit with the same period.

5.8 Bifurcations of families of periodic orbits in the plane

In the general solution for the displacement $\xi(t)$, there is a term of the form

$$c_3 f_3(t)e^{\sigma t} + c_4 f_4(t)e^{-\sigma t}, \tag{58}$$

where σ is either real or complex, on the unit circle (characteristic exponent). The functions $f_3(t)$, $f_4(t)$ are T-periodic.

The initial conditions that generate this component of the solution are

$$x'(0) = x(0) + c_3 f_3(0) + c_4 f_4(0). \tag{59}$$

Bifurcation with the same period

If for an orbit of the family we have $\sigma = 0$, or $\sigma = i2\pi\nu$, and we select the initial conditions (65) for $\xi(0)$, the perturbed solution

$$x'(t) = x(t) + \xi(t)$$

is also T-periodic.

A new family of periodic orbits bifurcates from the T-periodic orbit $x(t)$ of the original family, starting with the same period. Note that $\sigma = 0$ implies $\lambda_3 = \lambda_4 = 1$, which means that the bifurcation takes place at the point of the original family where the stability changes (critical points with respect to the stability).

Bifurcation with multiple period

Assume now that the characteristic exponent of a T-periodic orbit $x(t)$ of the family of periodic orbits is equal to

$$\sigma = i2\pi\frac{p}{q}.$$

If this periodic orbit $x(t)$ is described q times, then the characteristic exponent is

$$\sigma = i2\pi p \;\rightarrow\; e^{\pm\sigma} = 1.$$

This means that we also have a bifurcation of a new family of periodic orbits, from the T-periodic orbit $x(t)$ of the original family, but now the new periodic orbits start with a period equal to qT.

In the particular case $p = 1$, $q = 2$, we have $\sigma = i\pi$ and the bifurcating orbits start with a period $2T$. This type of bifurcation takes place when $\lambda_1 = \lambda_2 = -1$, which is also a critical point with respect to stability.

6. Vertical stability of planar periodic orbits

In the previous sections we studied the stability of a planar periodic orbit with respect to perturbations of the initial conditions *in the plane*. We shall study now the stability of a planar orbit with respect to perturbations of the initial conditions perpendicular to the plane of motion.

This type of stability is called *vertical stability* and completes the study of the stability of a planar periodic orbit.

Consider a dynamical system with three degrees of freedom, of the form

$$\begin{aligned} \dot{x}_1 &= f_1(x_1, x_2, x_3, \dot{x}_1, \dot{x}_2, \dot{x}_3), \\ \dot{x}_2 &= f_2(x_1, x_2, x_3, \dot{x}_1, \dot{x}_2, \dot{x}_3), \\ \dot{x}_3 &= x_3 f_3(x_1, x_2, x_3, \dot{x}_1, \dot{x}_2, \dot{x}_3). \end{aligned} \quad (60)$$

This is the form of the differential equations of many gravitational systems, for example the 3-dimensional three body problem.

It is easy to verify that equations (66) admit a planar solution, which we will assume to be periodic:

$$x_1(t),\ \ x_2(t),\ \ x_3(t) = 0, \quad (61)$$

corresponding to the initial conditions

$$x_{10},\ \ x_{20},\ \ x_{30} = 0,\ \ \dot{x}_{10},\ \ \dot{x}_{20},\ \ \dot{x}_{30} = 0.$$

We consider now a small perturbation along the x_3 axis also,

$$x_{10} + \varepsilon_1,\ \ x_{20} + \varepsilon_2,\ \ x_{30} = 0 + \varepsilon_3,\ \ \dot{x}_{10} + \varepsilon_4,\ \ \dot{x}_{20} + \varepsilon_5,\ \ \dot{x}_{30} = 0 + \varepsilon_6,$$

where ε_i are small, and we want to study the behaviour of the perturbed solution. We define new variables

$$x_4 = \dot{x}_1,\ x_5 = \dot{x}_2,\ x_6 = \dot{x}_3,$$

and a simple calculation shows that the system of variational equations for the system (66), with periodic solution (67), breaks into two uncoupled systems: a system in the planar displacements $\xi_1, \xi_2, \xi_4, \xi_5$, corresponding to the variational equations of the planar motion, and a system in the vertical displacements (along the x_3 axis) ξ_3, ξ_6. This latter system is

$$\begin{aligned} \dot{\xi}_3 &= \xi_6, \\ \dot{\xi}_6 &= f_{30}(t)\xi_3, \end{aligned}$$

where the function $f_{30}(t)$ is the T-periodic function,

$$f_3(x_1(t), x_2(t), x_3 = 0, x_4(t), x_5(t), x_6 = 0),$$

computed for the planar T-periodic solution (67). The system (68) is the system of variational equations for the displacements along the x_3 axis. The vertical stability depends on the eigenvalues λ_5, λ_6 of the monodromy matrix $\Delta_2(T)$ of this system.

If the system is Hamiltonian, $\Delta_2(T)$ is symplectic. Then, either λ_5, λ_6 are real, and of the form

$$\lambda_{5,6} = e^{\pm \alpha t},$$

and we have *vertical instability*, or else they are complex conjugate on the unit circle, of the form

$$\lambda_{5,6} = e^{\pm i\varphi t},$$

and we have *vertical stability*. The α or $i\varphi$ are the characteristic exponents for the vertical perturbations (along the x_3 axis). The solution of the system of vertical variational equations is of the form

$$\xi = c_1 f_1(t) e^{-\sigma t} + c_2 f_2(t) e^{\sigma t}, \tag{62}$$

where ξ is a vector with elements ξ_3, ξ_6, for σ equal to α (vertically unstable) or to $i\varphi$ (vertically stable).

If for a periodic orbit of the planar family we have $\sigma = 0$ (vertically critical orbit), then $\xi_3(t)$ is T-periodic. Then we have a bifurcation of a family of 3-dimensional periodic orbits from this point of the planar family, with the same period.

7. Extension to three or more degrees of freedom

All the above results concerning the eigenvalues and the stability of a periodic orbit, obtained in systems with two degrees of freedom, can be easily extended to three or more degrees of freedom.

In a Hamiltonian system the monodromy matrix is a $2n \times 2n$ symplectic matrix, and the eigenvalues are in reciprocal pairs (because of the symplectic property), and in complex conjugate pairs (because the elements of the matrix are real).

There are the following possibilities for the eigenvalues:

- Complex conjugate on the unit circle, $e^{\pm i\varphi}$ (see Figure 8): Stability
- Real, on the real axis, in reciprocal pairs (positive or negative), λ, $1/\lambda$ (see Figure 9): Instability
- Complex, inside and outside the unit circle, in reciprocal and in complex conjugate pairs (see Figure 10), $Re^{i\varphi}$, $Re^{-i\varphi}$, $R^{-1}e^{i\varphi}$, $R^{-1}e^{-i\varphi}$: Complex instability

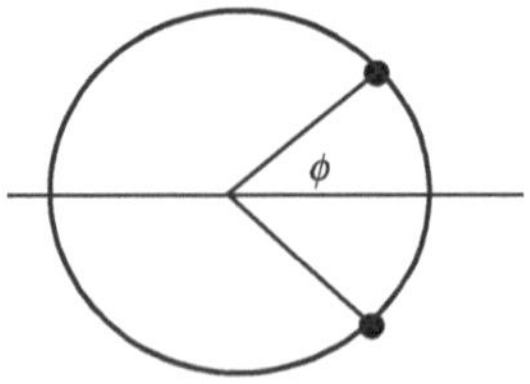

Figure 8. Position of the eigenvalues on the unit circle: stability

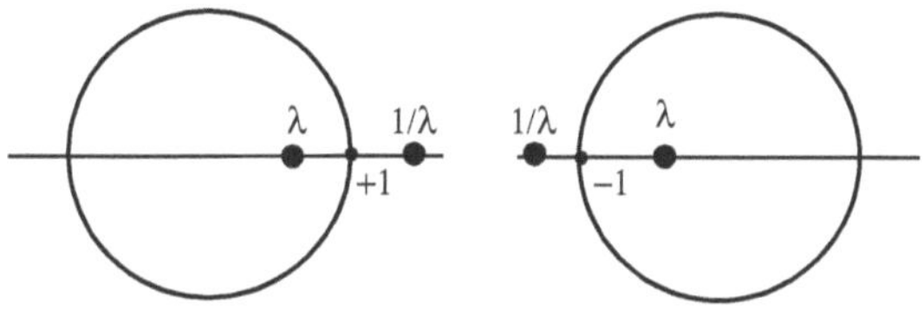

Figure 9. Position of the eigenvalues on the real axis: Instability

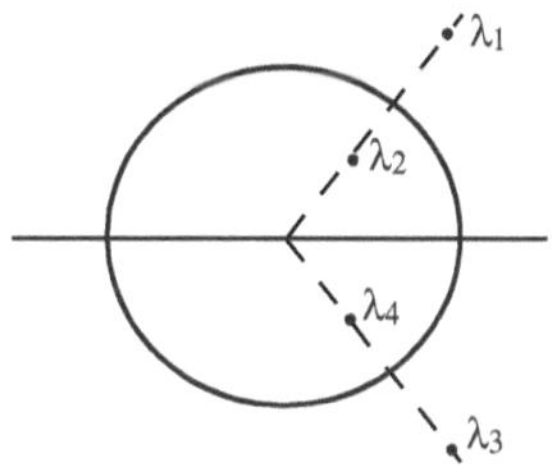

Figure 10. Position of the eigenvalues inside and outside the unit circle in reciprocal pairs and in complex conjugate pairs: Complex instability

Note that in three, or more, degrees of freedom we have a new type of instability, *complex instability*, which cannot appear in systems with two degrees of freedom.

8. The Poincaré map

The Poincaré map is a method to transform the continuous flow in n-dimensional phase space to an equivalent discrete flow (map) in a phase space of $(n-1)$-dimensions (or $(n-2)$-dimensions for Hamiltonian flows).

Let us consider the dynamical system in $\mathcal{R}^n$

$$\dot{x} = f(x), \tag{63}$$

where x, $f(x)$ are vectors in $\mathcal{R}^n$ and call $\varphi_t(x)$ the flow defined by Equation (70). We consider a *surface of section* Σ,

$$\Sigma \subset \mathcal{R}^n : \quad (n-1) - \mathrm{dim}$$

and we assume that the flow is transverse. This means that the velocity vector $f(x)$ of the flow is not tangent to the surface, i.e. $f(x) \cdot n(x) \neq 0$, where $n(x)$ is the tangent unit vector to the surface Σ (see Figure 11a).

The Poincaré map, Figure 11b, is the map

$$q \rightarrow p(q), \text{ where } p(q) = \varphi_\tau(q),$$

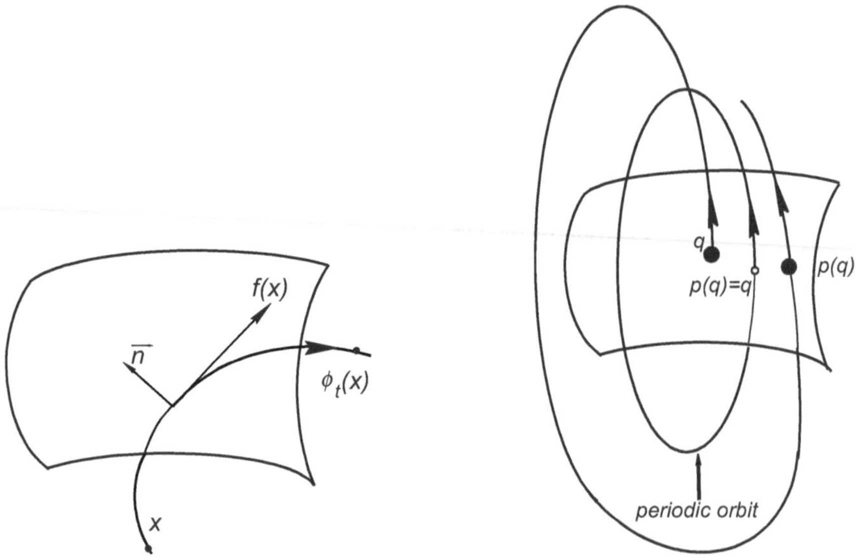

Figure 11. (a) The surface of section. (b) The Poincaré map on the surface of section

where τ is the time from the point of intersection q to the next point $p(q)$. We note that

- The consecutive points on the surface of section, $q, p(q), \ldots$ define accurately the state of the system.
- $p(q)$ is a continuous function of q.
- If $\bar{x}(t)$ is a T-periodic orbit (Figure 11b), the corresponding Poincaré map is a *fixed point*, $p(q) = q$, (maybe multiple).
- It can be proved that the stability of the invariant point is the same as the stability of the corresponding periodic orbit. The fixed point has the same set of eigenvalues, except the unit eigenvalue (which corresponds to the periodic orbit).

9. Poincaré maps in Hamiltonian systems

We consider the canonical equations

$$\dot{q} = \partial H/\partial p, \quad \dot{p} = -\partial H/\partial q, \tag{64}$$

where

$$q, p \subset \mathcal{R}^n$$

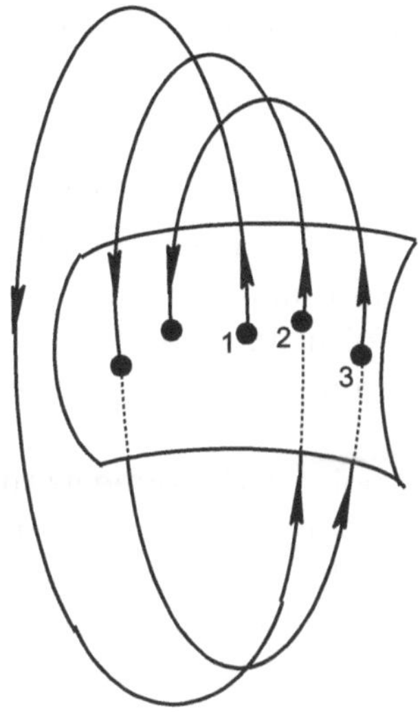

Figure 12. The consecutive points of intersection

and consider as surface of section Σ the $(2n-2)$-dimensional surface defined by

$$H = h, \quad f(q, p) = 0 \ \ \text{(for example } q_2 = 0). \tag{65}$$

The intersections of the continuous Hamiltonian flow in the $2n$-dimensional phase space, defined by Equation (71), with the surface of section defined by Equation (72), transforms the continuous flow to an equivalent discrete flow (map), on a $(2n-2)$-dimensional surface of section (see Figure 12).

9.1 Hamiltonian systems with two degrees of freedom

We will study in particular Hamiltonian systems with two degrees of freedom with Hamiltonian $H(q_1, q_2, p_1, p_2)$. We define

$$x_1 = q_1, \; x_2 = q_2, \; x_3 = p_1, \; x_4 = p_2,$$

and consider the surface of section

$$H(x_1, x_2, x_3, x_4) = h, \quad x_2 = 0, \ \text{with}\, x_4 > 0.$$

The map is on the two-dimensional space $x_1\, x_3$. The values of x_1, x_3 define exactly the state, because $x_2 = 0$ and x_4 is obtained from $H = h$, $x_4 > 0$.

The map is two-dimensional and can be expressed as

$$x_1 = g_1(x_1, x_3),$$

$$x_3 = g_2(x_1, x_3).$$

The map is symplectic. The fixed points of the map correspond to the periodic orbits of the Hamiltonian system.

The set of eigenvalues of a fixed point is the same as the set of the eigenvalues of the monodromy matrix of the periodic orbit, minus the two unit eigenvalues.

The stability of a fixed point is the same as the stability of the periodic orbit. Note that the shift along the perturbed orbit (due to the double eigenvalue), compared to the periodic orbit (see Figure 7), does not produce instability to the fixed point, so the stability of the fixed point of the Poincaré map is in fact *orbital stability.*

9.2 Invariant curves

The consecutive points of the map may lie on a smooth curve (ordered motion), or be scattered (chaotic motion).

Let us assume that another first integral of motion exists, independent of H =constant,

$$G(x_1, x_2, x_3, x_4) = c.$$

Then all the consecutive points of the map lie on smooth invariant curves. This can be proved as follows: Let (x_1, x_3) be a point of the map on the two-dimensional surface of section. We have $x_2 = 0$ and x_4 is expressed in terms of x_1, x_3, through the energy integral $H = h$, as

$$x_4 = x_4(x_1, x_2 = 0, x_3).$$

The points x_1, x_3 satisfy also the integral

$$G(x_1, x_2 = 0, x_3, x_4(x_1, x_2 = 0, x_3)) = c,$$

which is a relation of the form

$$F(x_1, x_3) = 0,$$

and this implies that the consecutive points (x_1, x_3) of the map lie on a smooth curve.

10. Near-integrable systems

Let us consider the Hamiltonian system

$$H = H_0(q_1, q_2, p_1, p_2) + \varepsilon H_1(q_1, q_2, p_1, p_2),$$

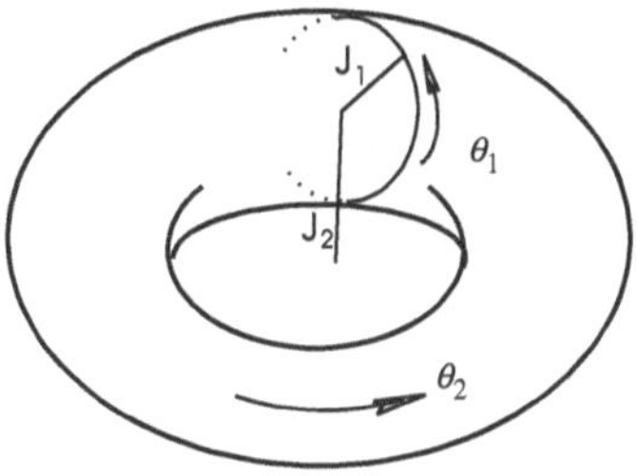

Figure 13. The 2-torus

where H_0 is integrable. We make a canonical transformation to action-angle variables of the integrable part,

$$J_1,\ J_2,\ \vartheta_1,\ \vartheta_2$$

and in these variables the Hamiltonian takes the form

$$H = H_0(J_1, J_2) + \varepsilon H_1(J_1, J_2, \vartheta_1, \vartheta_2).$$

The flow of the unperturbed Hamiltonian $H_0(J_1, J_2)$ is given by the equations

$$J_1 = J_{10},\ \ J_2 = J_{20},\ \ \dot{\vartheta}_1 = n_1,\ \ \dot{\vartheta}_2 = n_2,$$

where the frequencies n_1, n_2 are constant and are given by

$$n_1 = \partial H_0/\partial J_1,\ \ n_2 = \partial H_0/\partial J_2.$$

The flow takes place on a 2-torus, with constant radii J_1, J_2 and constant frequencies n_1, n_2 (Figure 13).

We define now a Poincaré map on the surface of section

$$H_0(J_1, J_2) = h,\ \ \vartheta_2 = 0 \mod 2\pi,$$

(or equivalently, $J_2 = J_{20}$ and $\vartheta_2 = 0$). The unperturbed mapping can be obtained analytically as

$$\begin{aligned} J_1 &\to J_1, \\ \vartheta_1 &\to \vartheta_1 + 2\pi\tfrac{n_1}{n_2}. \end{aligned} \tag{66}$$

This is a mapping in the variables J_1 ϑ_1 and depends on the parameter h (or J_2). At each iteration the angle ϑ_1 changes by

$$\Delta\vartheta_1 = 2\pi\frac{n_1}{n_2}.$$

The angle $\Delta\vartheta_1$ is called the *rotation angle*. It depends on the action J_1 only, for a given parameter h (or J_2). For this reason the map (73) is called a *twist map*. The map (73) is usually presented in the *Poincaré variables*

$$X = \sqrt{2J_1}\cos\vartheta_1, \quad Y = \sqrt{2J_1}\sin\vartheta_1.$$

All consecutive points of the map lie on smooth invariant curves that are circles with radii $\varrho = \sqrt{2J_1}$. The rotation angle $\Delta\vartheta_1$ as well as the ratio $\frac{n_1}{n_2}$ varies along the radius ϱ (see Figure 14a).

There are certain radii that are *resonant*: the ratio of the frequencies is rational,

$$\frac{n_1}{n_2} = \frac{s}{r}, \quad \text{or} \quad \Delta\vartheta_1 = 2\pi\frac{s}{r}.$$

All points on a resonant invariant curve (circle) are r-multiple fixed points: the point comes to the initial position after r rotations along the angle ϑ_2 on the 2-torus. ($\Delta\vartheta_1 = 2\pi s$ and $\Delta\vartheta_2 = 2\pi r$). It can be easily seen that the unperturbed mapping (73) can be obtained from the generating function

$$F = J_{1,n+1}\vartheta_{1,n} + G_0(J_{1,n+1}),$$

where G_0 is defined by

$$\partial G_0/\partial J_1 = 2\pi n_1/n_2,$$

by making use of the equations

$$J_{1,n} = \frac{\partial F}{\partial \vartheta_{1,n}}, \quad \vartheta_{1,n+1} = \frac{\partial F}{\partial J_{1,n+1}}.$$

Evidently, this mapping is symplectic.

We assume now that $\varepsilon \neq 0$ in the Hamiltonian H. Then, for sufficiently small ε, we have a perturbed twist map

$$\begin{aligned} J_1 &= J_1 + \varepsilon(...) \\ \vartheta_1 &= \vartheta_1 + 2\pi\frac{n_1}{n_2} + \varepsilon(...). \end{aligned}$$

The perturbed map is also symplectic and can be obtained from a generating function

$$F = J_{1,n+1}\vartheta_{1,n} + G_0(J_{1,n+1}) + \varepsilon G_1(J_{1,n+1}, \vartheta_{1,n}).$$

What happens to the unperturbed circular invariant curves when $\varepsilon > 0$?

- A non-resonant invariant curve survives as closed invariant curve, due to the KAM theorem (see Figure 14b).

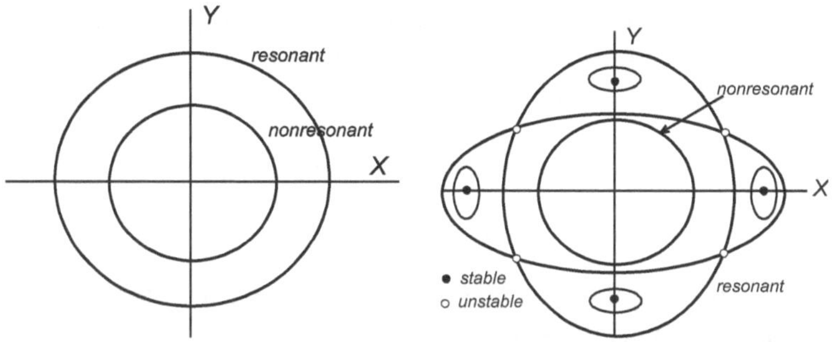

Figure 14. (a) The resonant and non resonant invariant curves of the integrable problem. (b) Only a finite number of fixed points survive on a resonant invariant curve after the perturbation is applied

- On a resonant invariant curve, out of the infinite set of r-multiple fixed points, only a finite (even) number survive, half stable and half unstable, as a consequence of the Poincaré-Birkhoff fixed point theorem (Arnold and Avez, 1968; Lichtenberg and Lieberman, 1983), as shown schematically in Figure 14b.

11. An Application to the Solar System

We shall apply the above described theory to the motion of a small body (asteroid, Kuiper belt object, satellite) moving around the Sun in a nearly Keplerian, elliptic, orbit, and perturbed by a major planet.

11.1 The unperturbed problem

We consider first a body of negligible mass moving around a body of finite mass m in an elliptic orbit. It can be proved (Murray and Dermott, 1999) that the action-angle variables of the two-body problem in the inertial frame, in the plane, are the Delaunay variables defined by

$$\hat{J}_1 = L, \quad \hat{\vartheta}_1 = M,$$

$$\hat{J}_2 = G, \quad \hat{\vartheta}_2 = \omega, \tag{67}$$

with Hamiltonian

$$\hat{H}_0 = -\frac{(Gm)^2}{2L^2}, \tag{68}$$

where M is the *mean anomaly*, ω the longitude of pericenter,

$$L = \sqrt{Gma}, \quad G = \sqrt{Gma(1-e^2)}, \tag{69}$$

a is the semimajor axis and e the eccentricity. We perform now a canonical change of variables to a rotating frame by the time dependent generating function

$$F_2 = J_1\hat{\vartheta}_1 + J_2(\hat{\vartheta}_1 + \hat{\vartheta}_2) - J_2\lambda',$$

$$\lambda' = n't + \omega',$$

where n' is the angular velocity of rotation of the rotating frame. The new action-angle variables are

$$\begin{aligned} J_1 &= L - G, \qquad & \vartheta_1 &= M = \lambda - \omega, \\ J_2 &= G, & \vartheta_2 &= \lambda - \lambda', \end{aligned}$$

where $\lambda = M + \omega$ is the *mean longitude*. The new Hamiltonian is

$$H_0 = -\frac{(Gm)^2}{2L^2} - n'G. \tag{70}$$

In terms of the elements of the orbit, the Hamiltonian (78) is expressed in the form

$$H_0 = -\frac{Gm}{2a} - n'\sqrt{Gma(1-e^2)}. \tag{71}$$

This is constant and is the energy integral

$$H_0 = h = \text{constant}.$$

According to Section 10, the motion takes place on a 2-torus, with radii the actions J_1 and J_2 and corresponding angles ϑ_1 and ϑ_2, as shown in Figure 13. The angular velocities n_1, n_2 are

$$n_1 = \frac{\partial H_0}{\partial J_1}, \quad n_2 = \frac{\partial H_0}{\partial J_2}, \tag{72}$$

and using Equation (77) and (78),

$$n_1 = n, \quad n_2 = n - n', \tag{73}$$

where $n = \sqrt{Gm}/a^{3/2}$ is the mean angular velocity of the elliptic orbit.

11.2 Poincaré map of the unperturbed problem in action-angle variables

As in the previous section, we define the map on the surface of section

$$J_2 = J_{20}, \;\; \vartheta_2 = 0, \;\; \mathrm{mod}\ 2\pi.$$

The mapping in the variables J_1, ϑ_1 is

$$\begin{aligned} J_1 \rightarrow &\quad J_1, \\ \vartheta_1 \rightarrow &\quad \vartheta_1 + 2\pi\tfrac{n}{n-n'}, \end{aligned}$$

and we present it in the Poincaré (canonical) variables $X = \sqrt{2J_1}\cos\vartheta_1$, $Y = \sqrt{2J_1}\sin\vartheta_1$. This map corresponds to the motion of the small body in a rotating frame with constant angular velocity n'. We note now that

- All invariant curves on the $X\,Y$ plane are circles with constant radii $\sqrt{2J_1}$, where $J_1 = \sqrt{Gma} - J_{20}$. It is clear that the radius of the invariant curve depends on a fixed value of the semimajor axis a.
- On a particular invariant curve, with radius $\sqrt{2J_1}$, there corresponds a certain eccentricity, obtained from $J_1 = \sqrt{Gma}(1 - \sqrt{1-e^2})$.
- As J_1 increases, the semimajor axis increases. Consequently the ratio n/n' varies, and passes through resonant values.
- The central fixed point $J_1 = 0$ corresponds to a circular orbit ($e = 0$).
- The resonant invariant curves correspond to the resonant elliptic periodic orbits, in the rotating frame.
- The non resonant invariant curves correspond to elliptic orbits in the inertial frame, which however are not periodic in the rotating frame.
- The angle ϑ_1 on the invariant curve defines the orientation ω. (At $t = 0$ and $\vartheta_2 = \lambda - \lambda' = 0$, and for $\lambda' = 0$ we have $\lambda = 0$, $\rightarrow$ $nt + \omega = 0$, $\rightarrow$ $M = -\omega$, $\rightarrow$ $\vartheta_1 = -\omega$.)

Resonant invariant curves

A resonant $n/n' = p/q$ elliptic periodic orbit is a multiple fixed point on the resonant invariant curve. The angle ϑ_1 changes during one iteration by

$$\Delta\vartheta_1 = 2\pi\frac{1}{1-\frac{n'}{n}} = 2\pi\frac{p}{p-q}.$$

The mapping after $p-q$ iterations is

$$\begin{aligned} \vartheta_1 \rightarrow \vartheta_1 + 2\pi\frac{1}{1-\frac{n'}{n}}(p-q) = \vartheta_1 + 2\pi p, \\ J_1 \rightarrow J_1. \end{aligned}$$

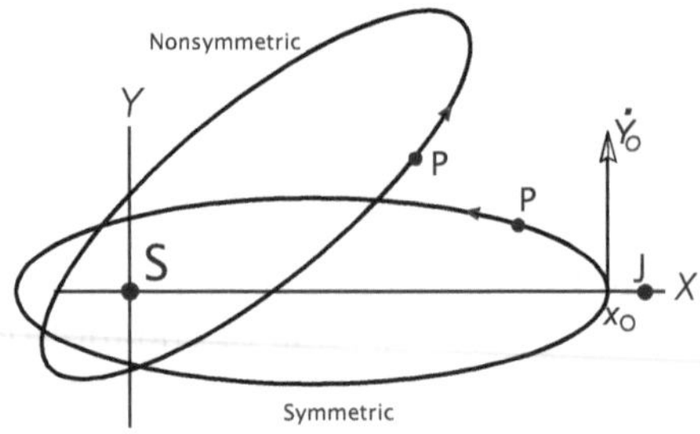

Figure 15. The resonant elliptic orbits for different orientations

All points on this invariant curve are fixed points. A linear analysis shows that the two eigenvalues of a fixed point are equal to 1. This means that the periodic elliptic orbits in the rotating frame have two pairs of unit eigenvalues.

We remark that all the fixed points on a resonant invariant curve of the Poincaré map correspond to elliptic motion of the small body, with the same semimajor axis a, such that n/n' is rational, and the same eccentricity e. They differ only in the orientation, which means that all these orbits have different values of ω, as shown in Figure (15).

12. Continuation of periodic orbits when a perturbation is applied

We perturb now the above two-body problem by adding to the model the gravitational attraction from a major planet (for example Jupiter), which we assume that revolves around the Sun in a circular orbit with constant angular velocity n'. The study of the periodic orbits will be made in the rotating frame that rotates with constant angular velocity n'. The new Hamiltonian has the form

$$H = H_0 + \varepsilon H_1,$$

where H_0 is the unperturbed integrable Hamiltonian (78) corresponding to the two-body problem. We want to study what happens to the circular, resonant and non resonant, invariant curves of the two-body problem when $\varepsilon > 0$.

According to the KAM theorem (Guckenheimer and Holmes, 1983), for sufficiently small ε, the non resonant invariant circles survive the perturbation as nearly circular invariant curves. These invariant curves represent nearly elliptic orbits of the small body that are not periodic both in the rotating frame and the inertial frame.

On a resonant circular invariant curve of the unperturbed problem each point is a fixed point of the Poincaré map, which in general is multiple. As soon as the perturbation is applied, out of the infinite set of (multiple in general) fixed points, only a finite number survive (usually only two), according to the Poinaré-Birkhoff fixed point theorem (Arnold and Avez, 1968, Lichtenberg and Lieberman, 1983). Half of them are stable and half unstable. This is shown in Figure 14b.

If the system has symmetries (as is the case with the restricted problem), usually the symmetric periodic orbits survive (but not always!). The resonant fixed points that survive correspond to monoparametric families of elliptic periodic orbits, in the rotating frame. These families bifurcate from the circular family, at the corresponding circular resonant orbits. From the above analysis we come to the conclusion that out of the infinite set of resonant elliptic periodic orbits of the two-body problem, with the same semimajor axes and the same eccentricities, but different orientations, as shown in Figure 15, only a finite number survive as periodic orbits in the rotating frame, and in most cases only two, usually, but not always, are symmetric.

References

Arnold, V.A. and Avez, A. (1968), *Ergotic Problems of Classical Mechanics*, W.A. Benjamin (New York).

Guckenheimer, J. and Holmes, Ph. (1983), *Nonlinear Oscillations, Dynamical Systems, and Bifurcations of Vector Fields*, Springer-Verlag (Berlin).

Jordan, D.W. and Smith, P. (1988), *Nonlinear Ordinary Differential Equations*, Clarendon Press (Oxford).

Lichtenberg, A.J. and Lieberman, M.A. (1983), *Regular and Stochastic Motion*, Springer-Verlag (Berlin).

Meyer, K.R. and Hall, G.R. (1992), *Introduction to Hamiltonian Dynamical Systems and the N-Body Problem*, Springer-Verlag (Berlin).

Murray, C.D. and Dermott, S.F. (1999), *Solar System Dyamics*, Cambridge Uiversity Press (Cambridge).

Yakubovich, V.A. and Starzhinskii, V.M. (1975), *Linear Differential Equations with Periodic Coefficients, vol. 1, 2*, Halsted Press.

PERIODIC SOLUTIONS IN GRAVITATIONAL PROBLEMS AND THE STRUCTURE OF SOLUTION SPACE

P.J. Message
University of Liverpool, L69 3BX
sx20@liverpool ac.uk

Abstract A brief review of some features of the system of periodic solutions, especially in the gravitational problem of three bodies, considering their part in the structure of the space of all solutions of the problem, and their relation to features of near-commensurability motion.

Keywords: Gravitational motion, periodic solutions, resonances, bifurcations

1. Introduction

In a system of bodies moving under their mutual gravitational attractions, if we know the motion to be periodic, then, to know the motion for all time, it is necessary to know it over just one period.

In fact, in the light of Poincaré's extensive studies into families of periodic solutions in the case of three bodies, he was led to conjecture that the periodic solutions are everywhere dense, in the sense that, given any motion in the system, a periodic solution may always be found keeping within any specified closeness to it for a chosen length of time. Schwarzschild gave the related, but different, conjecture, that, given any set of initial conditions, then there can be found, arbitrarily close in phase space, a set of initial conditions corresponding to a periodic solution (though the period may be very long, but will be finite). These conjectures embody the view that the structure of the families of periodic solutions form a sort of all-pervading "skeleton", around and amongst which are all the other solutions. We will go on to consider successive levels of bifurcation of families of periodic solutions (to families of

B.A. Steves et al. (eds.), Chaotic Worlds: from Order to Disorder in Gravitational N-Body Dynamical Systems, 81–101.

higher "genus"), continuing without limit, to illuminate this view. It is in the light of this that we must understand the significance of "wildness", or "chaos", which is now found to show itself in some solutions in many non-linear systems. Many of the topics forming part of this review have been dealt with in more detail in my presentations in previous summer schools, and in other publications, and appropriate references are given to these fuller accounts as we proceed.

2. The general, and restricted, three-body problem

2.1 The general three-body problem

We will mostly be considering a system in which three bodies, P_0, P_1, and P_2, move under their mutual gravitational attractions, and we will only be concerned with their relative motions. Therefore, in the plane problem, we have a system of just four degrees of freedom (two co-ordinates for each of the two relative positions), while in the three-dimensional problem, we have six degrees of freedom. There is always the integral of energy, since gravitational force is conservative, and there are the integrals of conservation of angular momentum (one component in the case of the plane problem, and three in the case of the three-dimensional problem; since we are concerned only with the relative motion, the integral of conservation of linear momentum does not appear). (See Poincaré (1892), sections 1 to 11.) The only exact solutions known, describable in closed form, are the Lagrangian collinear motions, and the Euler equilateral triangle solutions. In each of these, the figure defined by the three bodies remains similar at any time to that defined at any other time (i.e. they are *homothetic* solutions) and the relative motion in each case is, for each pair of bodies, in a conic section described according to Kepler's laws, as it would be for two-body motion.

2.2 The restricted problem of three bodies

Most numerical exploration of periodic solutions have in fact been carried out in the restricted problem of three bodies, the simplified version in which the mass, m_2, of the body P_2 is considered to be too small for it to influence the motion of the bodies P_0 and P_1, but remains fully influenced by the gravitational attraction of them. This system can be reached from the general three-body problem by a limiting process in which m_2 tends to zero, after the factor m_2 has been cancelled from the equation of motion for P_2. But it is more usual for its equations of motion to be derived from first principles. It possesses a single integral of motion, *Jacobi's integral*, which is usually derived from first principles from its equations of motion, but which can also be reached from the integrals of the general problem by the limiting process just mentioned, beginning with the appropriate linear combination of the integral of energy with the component of the angular momentum which is perpendicular to the plane of the (undisturbed keplerian) relative motion of P_0 and P_1, and, after subtraction of the parts relating to the motion of these two bodies, and then cancelling the factor m_2. (Carrying this process out on any other combination leads to the empty result $0 = 0$!) This restricted problem may be used as an approximate model for a number of actual situations in the Solar system, when studied in isolation from the rest of the system, for example, the case of a lone satellite moving around a planet (Jacobi's integral, provides a means of identifying cases where the satellite cannot leave the vicinity of the planet, using the fact that the square of the relative speed, in a rotating frame, must be non-negative, leading to the concept of *Hill stability*). See Szebehely (1967) for an extensive description of many features of this problem, including most of the approaches and methods which had been used by 1967, the date of its publication.

3. General Properties of periodic solutions, and motions near them. The linear equations of variation

Let $x = (x_1, x_2, \ldots, x_n)$ give the configuration, in phase space at any instant, of the system under consideration (e.g. the vector having as components the co-ordinates and components of momentum in the relative motion of the three, or more, bodies) and suppose then that the equations of motion of the full system are written

$$\dot{x} = X(x). \tag{1}$$

To begin the study of motions in the vicinity of a periodic solution, we use the *linear equations of variation*, which are the linear approximation to the equations of motion for the differences between the nearby motion and the periodic solution. Suppose that we are investigating motions in the vicinity of the periodic solution

$$x = \varphi(t). \tag{2}$$

which has period T, say. This solution must satisfy the equations (1), that is, we must have

$$\dot{\varphi}(t) = X\{\varphi(t)\}. \tag{3}$$

Then if

$$x = \varphi(t) + \delta x \tag{4}$$

is any neighbouring solution, it must satisfy

$$\begin{aligned} \dot{\varphi}(t) + \delta\dot{x} &= X\{\varphi(t) + \delta x\} \\ &= X\{\varphi(t)\} + \left(\frac{\partial X}{\partial x}\right)_{x=\varphi(t)} \delta x + O\{(\delta x)^2\} \end{aligned} \tag{5}$$

of which the linear approximation is

$$\delta\dot{x} = \left(\frac{\partial X}{\partial x}\right)_{x=\varphi(t)} \delta x. \tag{6}$$

These are the linear equations of variation. (See Poincaré (1892), section 53.) Since $\varphi(t)$ has period T, so then must $\left(\frac{\partial X}{\partial x}\right)_{x=\varphi(t)}$. So, if

$$\delta x(t) = \delta x^*(t) \tag{7}$$

is a solution of (6), then so must

$$\delta x(t) = \delta x^*(T + t) \tag{8}$$

be. We consider solutions δx^* having the property that

$$\delta x^*(t + T) = \lambda \delta x^*(t) \tag{9}$$

for some λ. Given a linearly independent set of solutions of (6), this gives an eigenvalue problem to find values of λ for which (9) is true. Suppose

$$x(t_0) = \varphi(t_0) + \beta \tag{10}$$

are starting conditions for a solution, on which

$$x(t_0 + T) = \varphi(t_0) + \beta + \psi. \tag{11}$$

Then the values of $\lambda - 1$ are the eigenvalues of the matrix of the $\frac{\partial \psi_i}{\partial \beta_j}$, evaluated along the solution (2). (See Poincaré (1892), section 60, and Message (1966a).)

Equivalently, the λ are the eigenvalues of the matrix of the $\frac{\partial x_i(t_0 + T)}{\partial x_j(t_0)}$. For each such λ, the value α satisfying

$$\lambda = \exp(\alpha T), \tag{12}$$

with imaginary part $\mathcal{I}(\alpha)$ in

$$-\pi/T < \mathcal{I}(\alpha) \leq \pi/T \tag{13}$$

is a *characteristic exponent* of the periodic solution (2).

Clearly $x(t) = \dot{\varphi}$ satisfies the linear equations (6) of variation, so, from (9), one of the λ must be equal to 1, and so the corresponding characteristic exponent α must be zero. So, in an autonomous system, there must always be at least one zero characteristic exponent. If the characteristic exponents $\alpha_1, \alpha_2, \ldots, \alpha_n$ are all different, the general solution of the linear equations of variation (6) may be written as

$$\delta x = \sum_{j=1}^{n} A_j \exp(\alpha_j t)\chi_j(t) \qquad (14)$$

where the A_j are constants of integration, and the functions $\chi_j(t)$ all have period T. If some of the characteristic exponents are equal, then the corresponding A_j is replaced by a polynomial in the time, of degree one less than the the number of characteristic exponents which are equal, the co-efficients being all constants of integration. (See Poincaré (1892), section 59, and Message (1966a).) If the system has integrals of the motion, then any periodic solution must have zero characteristic exponents, equal in number to one more than the number of independent integrals of motion. For the stability of the periodic solution (2), it is clear that all of the characteristic exponents α must have non-positive real parts. For a conservative system, able to be put into Hamiltonian form, which any n-body system moving under mutual gravitational attraction is, the characteristic exponents occur in equal and opposite pairs $\pm\alpha$. (See Poincaré (1892), sections 67 and 69, and Message (1966a).) So the stability of the periodic solution then requires that all of the characteristic exponents be pure imaginary, that is, of the form $\alpha = \pm\nu\sqrt{-1}$, say. Then the frequencies ν appear in the expressions for the δx, and represent the frequencies of the free librations about the periodic solution. See Message (1982a) for a discussion on how oblateness of the primary (P_0) affects the librations, leading to the concept of *proper eccentricities.*

4. The use of normal and tangential components in the linear equations of variation

4.1 The case of a single planet or satellite

For the purpose of determining the stability of a solution, we need only determine the non-zero characteristic exponents. To do this we may not need to solve the whole system of linear equations of variation. This is also true for the purpose of iterating initial conditions to improve the closeness with which a solution closes, so as to approach more closely an exactly periodic solution. For example, consider the case of a system of two degrees of freedom, as for example a single particle moving in the plane, or for the restricted problem of three bodies, in which the equations of motion may be written

$$\ddot{x} = f(x, y, \dot{x}, \dot{y}), \tag{15}$$

and

$$\ddot{y} = g(x, y, \dot{x}, \dot{y}). \tag{16}$$

We will write the linear equations of variation, using, as components of the displacement, the *tangential displacement*, p, given by

$$p = (\dot{x}\delta x + \dot{y}\delta y)/V \tag{17}$$

where $V = \sqrt{\dot{x}^2 + \dot{y}^2}$ is the speed of the particle in the orbit, and the *normal displacement*, q, given by

$$q = (-\dot{y}\delta x + \dot{x}\delta y)/V. \tag{18}$$

The second of these components, q, describes the displacement orthogonal to the motion in the orbit, and its equation of motion gives most of the information which we will need. Let us also suppose that there is a single first integral of the motion, say

$$F(x, y, \dot{x}, \dot{y}) = C, \tag{19}$$

which requires that

$$\dot{x}\frac{\partial F}{\partial x} + \dot{y}\frac{\partial F}{\partial y} + f\frac{\partial F}{\partial \dot{x}} + g\frac{\partial F}{\partial + \dot{y}} = 0 \tag{20}$$

is true for all $(x, y, \dot{x}, \dot{y})$. Then we find that the equations satisfied by q and p are

$$\ddot{q} + \Lambda(t)\dot{q} + \Theta(t)q = M(t)\delta C, \tag{21}$$

and

$$\frac{d}{dt}(p/V) = \delta C/\Psi(t) - \Phi(t)\,\dot{q}\,\mathcal{N}(t)\,q/V, \tag{22}$$

where Λ, Θ, M, Ψ, Φ, and $\mathcal{N}$ are given in terms of the expressions for $(x, y, \dot{x}, \dot{y})$ as functions of time on the orbit. (See Message (1982a).)

In the particular case of the restricted problem of three bodies in a plane, we have

$$f = 2\omega\dot{y} + \frac{\partial \Omega}{\partial x}, \tag{23}$$

and

$$g = -2\omega\dot{x} + \frac{\partial\Omega}{\partial y}, \tag{24}$$

and the first integral is Jacobi's integral,

$$\begin{aligned} F &= \dot{x}^2 + \dot{y}^2 - 2\Omega(x, y) \\ &= C. \end{aligned} \tag{25}$$

In this case $\Lambda = \Phi = 0$, $\Psi = 2V^2$, $M = \omega/V + \left(\dot{y}\frac{\partial\Omega}{\partial x} - \dot{x}\frac{\partial\Omega}{\partial y}\right)/V^3$, and then (21) becomes *Hill's equation*,

$$\ddot{q} + \Theta(t)q = M(t)\delta C. \tag{26}$$

The solution of this equation, with $\delta C = 0$, is sufficient to determine the non-zero characteristic exponents, and hence the linear stability, and also to improve the approximation to a closed orbit, when seeking initial values for a periodic orbit, by iteration.

4.2 The extension to two planets or satellites

The generalisation, to two bodies in a plane, moving under conservative forces, which also gives the equations for the relative motion in the general problem of three bodies in a plane, is

$$m_i\ddot{x} = -\frac{\partial\mathcal{V}}{\partial x_i},$$

and

$$m_i\ddot{y} = -\frac{\partial\mathcal{V}}{\partial y_i}, \tag{27}$$

for $i = 1, 2$, with the integral of energy,

$$\frac{1}{2}\sum_{i=1}^{2} m_i(\dot{x_i}^2 + \dot{y_i}^2) + \mathcal{V}(x_1, x_2, y_1, y_2) = C. \tag{28}$$

Suppose there is also an integral of angular momentum (which of course is so in the case of the general problem of three bodies),

$$\sum_{i=1}^{2} m_i(x_i\dot{y_i}^2 - y_i\dot{x_i}^2) = L, \tag{29}$$

which of course requires

$$\sum_{i=1}^{2}(x_i\frac{\partial\mathcal{V}}{\partial y_i} - y_i\frac{\partial\mathcal{V}}{\partial x_i}) = 0. \tag{30}$$

to be true for all $(x_1, y_1, \dot{x_1}, \dot{y_1}, x_2, y_2, \dot{x_2}, \dot{y_2})$.

Then, if we define tangential and normal components, p_i, and q_i, respectively, of the displacements of the two particles (using the equivalent equations to (17) and (18)) we find that we cannot separate the equations for the q_i to be completely independent of those for the p_i. In fact we will need a differential equation system of order five to determine all the non-zero characteristic exponents. In fact we include the equation for the quantity

$$s = p_1/V_1 - p_2/V_2 \tag{31}$$

where $V_i = \sqrt{\dot{x}_i^2 + \dot{y}_i^2}$, for $i = 1$ and 2.

Then the equations of motion are

$$\begin{aligned} m_1\ddot{q_1} + \Lambda_{11}(t)\dot{q_1} + \Theta_{11}(t)q_1 + \Lambda_{12}(t)\dot{q_2} + \Theta_{12}(t)q)2 + K(t)_1 s + \\ = M_1(t)(h_2\delta L - m_2V_2^2\delta C), \\ m_2\ddot{q_2} + \Lambda_{21}(t)\dot{q_1} + \Theta_{21}(t)q_1 + \Lambda_{22}(t)\dot{q_2} + \Theta_{22}(t)q)2 + K(t)_2 s + \\ = M_2(t)(h_1\delta L - m_1V_1^2\delta C), \end{aligned} \tag{32}$$

and

$$\begin{aligned} Z(t)\dot{s} - \Psi(t)s - W_1(t)\dot{q_1} + \tilde{W_1}(t)q_1 - W_2(t)\dot{q_2} + \tilde{W_2}(t)q_2 \\ = \Psi(t)\delta C - L(t)\delta L, \end{aligned} \tag{33}$$

where Λ_{ij}, Θ_{ij}, K_i, M_i, Z, Ψ, W_i, $\tilde{W_i}$, and ψ are given in terms of the expressions for $(x_i, y_i, \dot{x}_i, \dot{y}_i)$ as functions of time on the orbit.

See Message (1982a). In a similar way to the previous case, each of p_1/V_1 and p_2/V_2 is given by a single integration, given by an appropriate enlargement of the equation (22).

5. Solutions of second and higher genus

In the case of a stable periodic solution, if the frequencies of the librations, in the limit as the amplitude of the libration approaches zero, are all commensurable with the frequency $(2\pi/T)$ corresponding to the period T of the periodic solution itself, then there is a single overall period for the librating motion, usually a multiple $n > 1$ of the period of the original periodic solution. This leads to a bifurcation, from the original family of periodic solution, of a family of periodic solutions, whose limiting member, as the amplitude approaches zero, is the original solution covered n times. This new family may be described as being of the second *genus*, and there will be a bifurcation with a family, of this type, of periodic solutions, corresponding to each such commensurability. In the case of the restricted problem, there is only one frequency of libration, and so these bifurcations correspond to the rational numbers, and so are everywhere dense along the original family. In the case of the plane general three-body problem, there are two libration frequencies, so commensurability of the second frequency with commensurable pairs of the first frequency and the orbital frequency represent an everywhere dense set within an everywhere dense set, and so are also everywhere dense along the original family.

In the case of second-genus families which are stable, there will also be librations about each member periodic solution, and so, where there is commensurability of these librations with the period of the second genus solution, there will be a further bifurcation with a further family of periodic solutions, which we might call of the third genus, whose limiting member as its libration amplitude approaches zero is the member of the second-genus family from which it branches, covered a multiple number of times. Again, such bifurcations will be everywhere dense along the family of second genus solutions. This process of bifurcation of families of successively higher genus (and greater and greater multiplicity, and so of longer and longer period) can be expected to continue indefinitely. Thus we are led to expect an everywhere-dense set of initial conditions corresponding to periodic solutions, at least in that part of intitial-value space corresponding to stable periodic solutions.

6. Finding periodic solutions by analytical continuation

Poincaré (1892, sections 36-38) showed how to reach periodic solutions by analytical continuation from periodic motions in that limiting case in which the masses of two of the bodies are taken as zero, so that they have no mutual perturbations, finding solutions in the full perturbed motion in which the relative motion remains periodic.

Let us now consider the gravitational problem of three bodies, in such a way as to define the appropriate small parameter in terms of which to carry out the analytical continuation. Suppose the three bodies are P_0, P_1, and P_2, and that their masses are m_0, m_1, and m_2, respectively.

Put

$$\varepsilon\, m_1^* = m_0 m_1/(m_0 + m_1) \tag{34}$$

and

$$\varepsilon\, m_2^* = (m_0 + m_1) m_2/(m_0 + m_1 + m_2) \tag{35}$$

and choose ε as the greater of m_1/m_0 and m_2/m_0. Suppose the equations of motion are written

$$\begin{aligned} \dot{x} &= X(x, \varepsilon) \\ &= X_0(x) + \varepsilon X_1(x, \varepsilon) \end{aligned} \tag{36}$$

and, in the limiting case in which ε approaches zero,

$$\dot{x} = X(x_0), \tag{37}$$

which gives the motion of each of P_1 and P_2 as unperturbed motion under the attraction of P_0 (See, for example, Roy and Ovenden (1955); Message (1980); Message (1982b).) The analytical continuation to be carried out is then from zero to non-zero values of ε.

Poincaré's first sort of periodic solution arises by analytical continuation from circluar motions about the primary. These are not directly related to commensurability of orbital period, but we find that the condition for the analytical continuation to be possible in a simple way is that the periods be not in the ratio of two successive integers.

The second sort arises from motions in elliptic coplanar orbits, but in which the orbital periods are commensurable, so that the relative motion is periodic.

The third sort in non-coplanar ellipses, again with commensurable orbital periods, so that the relative motion is periodic.

6.1 General principles for the analytical continuation

Suppose that $x = x_0(t)$ is a motion with period T, in the perturbation-free limiting case in which ε is zero, so that

$$\begin{aligned} \dot{x_0} &= X(x_0, 0) \\ &= X_0(x_0), \end{aligned}$$

and

$$x_0(T + t) = x_0(t), \tag{38}$$

for all t.

We now seek a periodic solution in the full perturbed motion, that is, for ε not zero, and which co-incides with x_0 when ε is put equal to zero. We start with

$$x(t_0) = x_0(t_0) + \beta, \tag{39}$$

and suppose that this leads to

$$x(t_0 + T + \tau) \quad = \quad x_0(t_0 + T) + \beta + \psi. \tag{40}$$

Then this solution is periodic with period $T + \tau$ if

$$\begin{aligned} \psi &= (\psi_1, \psi_2, \ldots \psi_n) \\ &= 0, \end{aligned} \tag{41}$$

and we seek to satisfy this by choice of $\beta = (\beta_1, \beta_2, \ldots, \beta_n)$. (We will see very shortly why the period may need to differ from T.) Now ψ will be a function of β, τ, and ε, with the linear approximation

$$\psi_i \approx \sum_{j=1}^{n} \frac{\partial \psi_i}{\partial \beta_j} \beta_j + \frac{\partial \psi_i}{\partial \tau} \tau + \frac{\partial \psi_i}{\partial \varepsilon} \varepsilon \tag{42}$$

The general procedure to be followed is to put each of these equal to zero, and so solve for an approximation for the β_i, to first order in ε, and then proceed by successive approximation, using the complete expressions for the ψ_i, to obtain expressions for the β_i in powers of ε, to correspond to a periodic solution of the full perturbed motion. This could be done, taking $\tau = 0$, and so obtaining a solution of period T, provided that the Jacobian determinant

$$\left(\frac{\partial(\psi_1, \psi_2, \ldots, \psi_n)}{\partial(\beta_1, \beta_2, \ldots, \beta_n)} \right)_{x=x_0,\, \tau=\varepsilon=0} \tag{43}$$

were not zero.

But in fact it is not quite so straightforward! Note that $x = x(t + h)$ is also a solution of (36), for any value of h, corresponding to $\beta = x(t_0 + h) - x(t_0)$ and gives $\psi = 0$, so that the Jacobian determinant (43) is in fact zero. However, there are the n equations $\psi_i = 0$ to be solved by choice of the $n + 1$ quantities $\beta_1, \beta_2, \ldots, \beta_n$ and τ. So we may choose, say, $\beta_n = 0$, and still solve $\psi_i = 0$ for the n quantities $\beta_1, \beta_2, \ldots, \beta_{n-1}$ and τ, provided that

$$\left(\frac{\partial(\psi_1, \psi_2, \ldots, \psi_{n-1}, \psi_n)}{\partial(\beta_1, \beta_2, \ldots, \beta_{n-1}, \tau)} \right)_{x=x_0,\, \tau=\varepsilon=0} \tag{44}$$

is not zero.

In the systems we will be considering, there will be at least one integral of the motion. (The restricted problem has Jacobi's integral, and the general problem of three bodies has the integrals of energy and angular momentum.) If there is an integral

$$F(x, \varepsilon) \quad = \quad C, \tag{45}$$

so that there is the identity

$$\sum_{j=1}^{n} \frac{\partial F}{\partial x_j} X_j = 0, \tag{46}$$

it follows that

$$F(x_0(t_0) + \beta, \varepsilon) = F(x_0(t_0 + T + \tau) + \beta + \psi, \varepsilon), \tag{47}$$

which gives a relationship between the β, τ, ε, and the ψ. Therefore putting, say,

$$\psi_1 = \psi_2 = \cdots = \psi_{n-1} = 0, \tag{48}$$

must imply that $\psi_n = 0$, provided that $\frac{\partial F}{\partial x_n}$ is not zero, so that we need only solve the equations (48) for $\beta_1, \beta_2, \ldots, \beta_{n-2}$ and τ, which is possible provided that the Jacobian determinant

$$\left(\frac{\partial(\psi_1, \psi_2, \ldots, \psi_{n-2}, \psi_{n-1})}{\partial(\beta_1, \beta_2, \ldots, \beta_{n-2}, \tau)} \right)_{x=x_0,\, \tau=\varepsilon=0} \tag{49}$$

is not zero.

The existence of further integrals of motion would enable further reduction to be made in the number of equations which need to be solved to ensure a periodic solution. (For further details see Poincaré (1892), section 38). We will now return to the specific case of the general gravitational problem of three bodies, which of course possesses the integrals of angular momentum and energy.

6.2 Application to the General Problem of Three Bodies: periodic solutions of the First Sort

We are now considering three finite bodies, P_0, P_1, and P_2, of masses m_0, m_1, and m_2, respectively. The periodic solutions of the first sort are confined to a single plane. To reach these solutions Poincaré takes, for the solution

$(X = X_0(t))$ from which the analytical continuation begins, one in which P_1 and P_2 move about P_0 in circular orbits. The distances between the three bodies in pairs, and hence the relative configuration of them, can be expressed as functions of the quantities

$$\begin{aligned} x_1 &= a_1, \quad x_2 = a_2, \\ x_3 &= \lambda_1 - \lambda_2, \\ x_4 &= e_1 \cos \ell_1, \quad x_5 = e_1 \sin \ell_1, \\ x_6 &= e_2 \cos \ell_2, \quad x_7 = e_2 \sin \ell_2, \end{aligned} \tag{50}$$

where a_1 is the major semi-axis, and e_1 the eccentricty, of the orbit of P_1 about P_0, and ℓ_1 and λ_1 are the mean anomaly and mean longitude, respectively, in that orbit. The corresponding elements in the orbit of P_2 about P_0 are a_2, e_2, ℓ_2, and λ_2, respectively. In perturbation-free motion, with ε equal to zero, of course each a_i and e_i are constant, and each of ℓ_i and λ_i increase linearly with time at the rate

$$n_i = \sqrt{G(m_0 + m_i)a_i^{-3}}, \tag{51}$$

the mean motion in the unperturbed orbit, as given by Kepler's third law. After completion of a synodic period, $T = 2\pi/|n_1 - n_2|$, the mean elongation x_3 returns to its original value, and, since ℓ_i has increased by $n_i T$, we have

$$\begin{aligned} x_4(t_0 + T) &= x_4(t_0)\cos(n_1 T) - x_5(t_0)\sin(n_1 T), \\ x_5(t_0 + T) &= x_4(t_0)\sin(n_1 T) + x_5(t_0)\cos(n_1 T), \\ x_6(t_0 + T) &= x_6(t_0)\cos(n_2 T) - x_7(t_0)\sin(n_2 T), \end{aligned}$$

and

$$x_7(t_0 + T) = x_6(t_0)\sin(n_2 T) + x_7(t_0)\cos(n_2 T), \tag{52}$$

Since we are considering circular motion, with $e_1 = e_2 = 0$, the relative configuration recurs after a time interval T.

Now consider the perturbed situation, with ε different from zero, and consider a motion for which, at $t = t_0$,

$$\begin{aligned} x_1 &= a_{10} + \beta_1, \quad x_2 = a_{20} + \beta_2, \\ x_3 &= \lambda_{10} - \lambda_{20} + \beta_3, \end{aligned} \tag{53}$$

and, for $i = 4, 5, 6$, and 7,

$$x_i = x_i(t_0)_{\varepsilon=0} + \beta_i, \tag{54}$$

and for which, at $t = t_0 + T$,

$$\begin{aligned} x_1 &= a_{10} + \beta_1 + \psi_1, \quad x_2 = a_{20} + \beta_2 + \psi_2, \\ x_3 &= \lambda_{10} - \lambda_{20} + \beta_3 + \psi_3, \end{aligned}$$

and, for $i = 4, 5, 6$, and 7,

$$x_i = x_i(t_0 + T)_{\varepsilon=0} + \beta_i + \psi_i, \tag{55}$$

from which we derive, using equations (52),

$$\begin{aligned}\frac{\partial\psi_4}{\partial\beta_4} &= \cos(n_1T)-1, \ \frac{\partial\psi_4}{\partial\beta_5} = -\sin(n_1T),\\ \frac{\partial\psi_5}{\partial\beta_4} &= \sin(n_1T), \ \frac{\partial\psi_5}{\partial\beta_5} = \cos(n_1T)-1,\\ \frac{\partial\psi_6}{\partial\beta_6} &= \cos(n_2T)-1, \ \frac{\partial\psi_6}{\partial\beta_7} = -\sin(n_2T),\\ \frac{\partial\psi_7}{\partial\beta_6} &= \sin(n_2T), \ \frac{\partial\psi_7}{\partial\beta_7} = \cos(n_2T)-1. \qquad (56)\end{aligned}$$

Now we may use the integral of energy

$$F = C, \qquad (57)$$

and the integral of angular momentum to establish that

$$\psi_3 = \psi_4 = \psi_5 = \psi_6 = \psi_7 = 0, \qquad (58)$$

imply that also

$$\psi_1 = \psi_2 = 0. \qquad (59)$$

We may then solve (57) (that is, fix the energy), and (58) for β_1, β_2, β_4, β_5, β_6, and β_7, and then the Jacobian determinant, which must not vanish if the solution is to be possible, is

$$\Delta = \left(\frac{\partial(F,\psi_3,\psi_4,\psi_5,\psi_6,\psi_7)}{\partial(\beta_1,\beta_2,\beta_4,\beta_5,\beta_6,\beta_7)}\right)_{t=T,\, e_1=e_2=\varepsilon=0}. \qquad (60)$$

Most of its entries are zero, and we find that

$$\begin{aligned}\Delta &= \frac{\partial(F,\psi_3)}{\partial(\beta_1,\beta_2)}\frac{\partial(\psi_4,\psi_5)}{\partial(\beta_4,\beta_5)}\frac{\partial(\psi_6,\psi_7)}{\partial(\beta_6,\beta_7)}\\ &= -3n_1n_2\left(\frac{n_2a_2}{a_1}+\frac{n_1a_1}{a_2}\right)(\cos(n_1T)-1)(\cos(n_2T)-1)\end{aligned}$$

This is zero only if $n_1T = 2\pi k$, or $n_2T = 2\pi k'$, for integer k or k', that is, if $n_1/|n_1-n_2| = k$ or if $n_2/|n_1-n_2| = k'$, that is, if $n_1/n_2 = (k-1)/k$ or if $n_1/n_2 = (k'+1)/k'$.

In this way we are led to the family of periodic solutions of the general three-body problem, of Poincaré's *first sort*, except when the mean motions are in the ratio of two successive integers. With this exception (whose significance we will see later), then, the process leads, in the full (perturbed) system, to a family *(the first sort)* of periodic motions, which become circular motions in the limit as ε approaches zero. (See Poincaré (1892), sections 30-40.) For Poincaré's reference to Hill's variational orbit in the main lunar problem, as being an orbit of this type in a very similar problem, see Poincaré (1892), section 41.

6.3 Periodic solutions of the Second Sort

The solutions of the second sort (see Poincaré (1892), sections 42-47) are also planar motions, but now we start from orbits in the perturbation-free ($\varepsilon = 0$) case which have non-zero eccentricities. For the motion to be periodic, there must now be a common multiple of the periods of the two relative orbits. This means that there must be a commensurability between the two orbital periods $T_1 = 2\pi/n_1$ and $T_2 = 2\pi/n_2$, and so between the two mean orbital motions n_1 and n_2, that is, there must be integers p and q so that $(p+q)n_2 = pn_1$. After the elapse of a synodic period $T_s = 2\pi/(n_1 - n_2)$, the two bodies P_1 and P_2 return to the same relative configuration (e.g., conjunction, or opposition). After the elapse of an orbital period T_i, P_i returns to the same apse. Now $p(n_1 - n_2) = qn_2$, so that $qT_s = pT_2$, and likewise $qT_s = (p+q)T_1$, so that the entire relative configuration recurs after the elapse of the overall period $T = qT_s$.

Let us now follow Poincaré (1892), sections 42-46, in making use of a canonical formulation, but choose the co-ordinates

$$\ell_1, \ell_2, \gamma = \varpi_1 - \varpi_2, \text{ and } \chi = \varpi_1 + \varpi_2,$$

where ϖ_i is the longitude of the near apse of the orbit of P_i about P_0. The conjugate momenta are, respectively,

$$\begin{aligned} L_1 &= m_1^*\sqrt{\mu_1\, a_1}, \\ L_2 &= m_2^*\sqrt{\mu_2\, a_2}, \\ \Gamma &= \frac{1}{2}\left\{L_1\sqrt{1-e_1^2} - L_2\sqrt{1-e_2^2}\right\} \end{aligned}$$

and

$$X = \frac{1}{2}\left\{L_1\sqrt{1-e_1^2} + L_2\sqrt{1-e_2^2}\right\} \tag{61}$$

The relative positions of the three bodies may be expressed in terms of ℓ_1, ℓ_2, $\varpi_1 - \varpi_2$, a_1, a_2, e_1, and e_2, and so in terms of ℓ_1, ℓ_2, γ, L_1, L_2, Γ, and X. So the Hamiltonian
function is expressible in terms of these, and takes the form

$$\mathcal{H} = \mathcal{H}_0(L_1, L_2) - \varepsilon\mathcal{R}(\ell_1, \ell_2, \gamma, L_1, L_2, \Gamma, X). \tag{62}$$

So χ does not appear, that is, it is an ignorable co-ordinate, and its conjugate momentum, X, is a constant of the motion, in fact corresponding to the constancy of the total angular momentum.

We proceed now making use of Roy and Ovenden's mirror theorem (Roy and Ovenden (1955)), which refers to a *mirror configuration*, which is one in which every velocity is perpendcular to every position vector, and states that the motion subsequent to such a configuration is the mirror image of the motion prior to it. So the occurrence of two such configurations in the course of a solution is sufficient to establish that solution is periodic. In our present problem, such a configuration is given if each of ℓ_1, ℓ_2, and $\gamma = \varpi_1 - \varpi_2$, is either zero or π.

As in the previous section, we start by considering a solution of the unperturbed ($\varepsilon = 0$) motion which is periodic. Suppose that the values

$$L_i = L_{i0}$$

(for $i = 1, 2$),

$$\Gamma = \Gamma_0, \text{ and } X = X_0, \tag{63}$$

correspond to values $n_1 = n_{1,0}$ and $n_2 = n_{2,0}$ for which

$$(p+q)\, n_{2,0} = p\, n_{1,0} \tag{64}$$

for some integers p and q, and so corresponding to a periodic solution, of period T, say, in the unperturbed case.

Then, in the perturbed case, $\varepsilon \neq 0$, take, at $t = t_0$,

$$\ell_1 = \ell_2 = \gamma = 0, \tag{65}$$

and so giving a mirror configuration, and also

$$L_i = L_{i0} + \beta_i$$

(for $i = 1, 2$),

$$\Gamma = \Gamma_0 + \beta_3, \tag{66}$$

and

$$X = X_0 + \beta_4, \tag{67}$$

and suppose that this leads, at $t = t_0 + T + \tau$, to

$$\ell_i = n_{1,0} T + \psi_i$$

(for $i = 1, 2$), and

$$\gamma = \psi_3 \tag{68}$$

Then, for a second mirror configuration, and hence a periodic solution, it is sufficient that

$$\psi_1 = \psi_2 = \psi_3 = 0. \tag{69}$$

We may choose $\beta_3 = \beta_4 = 0$, thus fixing the scale and angular momentum of the solution, and solve the equations (69) for β_1, β_2, and τ. Now

$$\dot{\gamma} = -\varepsilon \frac{\partial \mathcal{R}}{\partial \Gamma} \tag{70}$$

so that

$$\psi_3 = -\varepsilon \int_{t_0}^{t_0+T+\tau} \frac{\partial \mathcal{R}}{\partial \Gamma} dt \tag{71}$$

which is zero for $\varepsilon = 0$, together with its derivatives with respect to β_1, β_2, and τ, so we must in fact solve $\psi_3/\varepsilon = 0$, if we are to hope for a non-zero Jacobian. Thus the set of equations to be solved is in fact

$$\psi_1 = \psi_2 = \psi_3/\varepsilon = 0. \tag{72}$$

and the Jacobian which must not be zero is

$$\Delta = \frac{\partial(\psi_1, \psi_2, \psi_3/\varepsilon)}{\partial(\beta_1, \beta_2, \tau)} \tag{73}$$

evaluated at

$$\beta_1 = \beta_2 = \beta_4 = \beta_5 = \tau = \varepsilon = 0 \tag{74}$$

It is found to be

$$\Delta = -3\, n_{1,0}\, n_{2,0}\, T^2 \frac{\partial^2 \mathcal{R}^*}{\partial \Gamma^2}$$

where

$$\mathcal{R}^* = \frac{1}{T} \int_{t_0}^{t_0+T} \mathcal{R} dt, \tag{75}$$

so we must have $\frac{\partial^2 \mathcal{R}^*}{\partial \Gamma^2}$ to be non-zero. We also need ψ_3/ε to be zero under the conditions (74), that is, we need also that $\frac{\partial \mathcal{R}^*}{\partial \Gamma_0}$ be zero. Thus we need to be able to find Γ_0 so that $\mathcal{R}^*(\Gamma_0)$ has a simple maximum or minimum.

Now e_1 and e_2 must be real, which can be seen from the equations (61) to require that

$$-X - L_1 \leq \Gamma \leq -X + L_1$$

and

$$X - L_2 \leq \Gamma \leq X + L_2 \tag{76}$$

so, according to the ordering of the sizes of L_1, L_2, and X, there will be linear combinations A and B of them so that Γ is actually confined by

$$A \leq \Gamma \leq B. \tag{77}$$

Put now

$$\Gamma_0 = \frac{1}{2}(A+B) + \frac{1}{2}(B-A)\cos\varphi, \tag{78}$$

so that $\mathcal{R}^*$ is expressible as as periodic function of φ. Its modulus is bounded, so it must have at least one maximum and one minimum. Now

$$\frac{\partial \mathcal{R}^*}{\partial \varphi} = -\frac{1}{2}(B-A)\frac{\partial \mathcal{R}^*}{\partial \Gamma_0}\sin\varphi \tag{79}$$

so, since we can show that neither occur when $\sin\varphi$ is zero, then, since Γ_0 and X_0 may be freely chosen (and thus, equivalently e_1 and e_2 may be), this leads to at least two families of periodic solutions with this kind of mirror configuration, and so of Poincaré's second sort.

For a more detailed discussion of the occurrence of symmetries in the more simple restricted problem case, see Message (1985). For an approach, using simple perturbation theory, to this type of solution, see Message (1999). Note that, if e_1 and e_2 were both zero, the motion in the perturbation-free case is then identical to one corresponding to the first sort of periodic solution. In fact there is a bifurcation of families of second sort solutions from the first sort for every pair of integers (p, q), except when $q = 1$, in which case there is a continuous transition from the first sort solution to the second sort. (See Message (1982a); Message (1966b).) For each pair (p, q) with $q \neq 1$, it is found that, in general, there are two families of periodic solutions of the second sort which branch from the family of those of the first sort. Of these two, one consists of stable solutions, and the other of unstable, at least for small enough values of ε. (In some cases in the restricted problem, as increasing values of the unperturbed eccentricity of P_2 are taken, a value is reached at which a bifurcation takes place with families of asymmetric solutions. At such a bifurcation, there is a transition between stability and instability along the family of symmetric second sort solutions. (See Message (1978); Message (1982a)). Now the ratios $(p+q)/p$ are everywhere dense, so bifurcations with families of periodic solutions of the second sort are everywhere dense on each family of periodic solutions of the first sort.

6.4 Periodic solutions of the Third Sort

The solutions of the third sort (see Poincaré (1892), sections 48-49) are not planar motions, and we start from orbits in the perturbation-free $\varepsilon = 0$ case which have non-zero orbital inclinations. As also for the second sort, for the motion to be periodic, there must be a common multiple of the periods of the two relative orbits, and, again, this means that there must be a commensurability between the two mean orbital motions n_1 and n_2 of the type $(p+q)n_1 = pn_2$. After the elapse of a synodic period $T_s = 2\pi/(n_1 - n_2)$, the two bodies P_1 and P_2 return to the same relative configuration (e.g., conjunction, or opposition). After the orbital period $T_i = 2\pi/n_i$, P_i returns to the same node and apse. As also for the second sort, the entire relative configuration recurs after the elapse of the overall period $T = qT_s$. For a demonstration of the existence of the corresponding solutions in the full perturbed motion, of this third sort, using Roy and Ovenden's mirror theorem (Roy and Ovenden (1955)), see Message (1982c). The case q = 1 shows special features, in that the solutions of the third sort are found to have non-zero eccentricities.

For a brief discussion of the effect on this type of solution of the oblateness of the primary, as seen in the case of Saturn's satellites Mimas and Tethys, see Message (1999).

7. Extension to systems of more than three bodies

Let us consider briefly the extension of some of these ideas to systems of more than three bodies. In the case of four bodies, we can extend the procedure to reach periodic solutions of the first sort, by analytical continuation from the case of motion, in the perturbation-free case, in which the three bodies P_1, P_2, and P_3 move in circular orbits about the primary body P_0. But, in order that these motions have a common period, there must be a commensurability between the differences of the mean motions of the sort

$$r(n_2 - n_3) = s(n_1 - n_2), \tag{80}$$

where r and s are integers. (The set of points in (n_1, n_2, n_3)-space for which this is true is of course everywhere dense.). The proof for the existence of periodic solutions in the full, perturbed problem, proceeds by extension to the four-body problem of the proof for the three-body given above. The case where it fails now being where the ratio n_1/n_2 is equal to the ratio of two integers differing by r, and where the ratio n_2/n_3 is equal to the ratio of two integers differing by s. The relation (80) shows that either of these implies the other. An example of motion close to this type of periodic solution is given by the relative motion of Jupiter and its three innermost great satellites, with $r = 2$, and $s = 1$.

But it is relevant to ask if the study of periodic solutions of this sort does constitute the relevant extension to four or more bodies of the periodic solutions of the three-body problem, in having the appropriate significant properties? Perhaps the natural counterparts, in four, or more, body motion, of the periodic solutions of the three-body problem, are the motions in which the mutual gravitational influence takes place without there being any free oscillation (whose absence is a property of the periodic solutions in the three-body problem), that is, motions in which the only frequencies present are the synodic frequencies of all of the pairs of orbits, and in which the *secular variations*, which are the free oscillations, all have zero amplitude. Then the fact of there being a relation of the type (80), or its counterparts in systems of more bodies, leading to the entire motion having a single period, would be more of an accidental feature of this type of motion. (See Message (1985); Message (1982b)).

References

Message, P.J. (1966a) Stability and Small Oscillations about equilibrium and Periodic Motions. In Rosser, J. Barkley, editor, *Space Mathematics, Part 1*, which is volume 5 of *Lectures in Applied Mathematics*, pages 77–99. American Mathematical Society.

Message, P.J. (1966b) On nearly-commensurable periods in the restricted problem of three bodies, with calculations of the long-period variations in the interior 2:1 case. In Contopoulos, G., editor, *The Proceedings of Symposium No. 25 of the International Astronomical Union*, pages 197–222.

Message, P.J. (1970) On Linear equations of variation in dynamical Problems. *Celestial Mechanics*, vol. 2, pages 360–367.

Message, P.J. and Taylor, D.B. (1978) On asymmetric periodic solutions of the plane restricted problem of three bodies, and bifurcations of families. In Szebehely, V., editor, *Dynamics of planets and satellites and theories of their motion (Proceedings of Symposium No. 41 of the International Astronomical Union)*, pages 319–323.

Message, P.J. (1980) On the existence of periodic solutions of Poincaré's second sort in the general problem of three bodies moving in a plane. *Celestial Mechanics*, vol. 21, pages 55–61.

Message, P.J. (1982a) Some aspects of motion in the general planar problem of three bodies; in particular in the vicinity of periodic solutions associated with near small-integer commensurabilities of orbital period. In Szebehely, V. editor, *Applications of Modern Dynamics to Celestial Mechanics and Astrodynamics (The Proceedings of the N.A.T.O. Advanced Study Institute, in Istituto Antonelli, Zuel, Italy, 1981)*, pages 77–101. Reidel.

Message, P.J. (1982b) Asymptotic series for planetary motion in periodic terms, in three dimensions. *Celestial Mechanics*, 26:25–39.

Message, P.J. (1982c) On the existence of periodic solutions of Poincaré's third sort in the general problem of three bodies in three dimensions. *Celestial Mechanics*, vol. 28, pages 107–118.

Message, P.J. (1985) Some results of resonance and periodic motions. In Szebehely, V., editor, *The Stability of the Solar System and its Minor Natural and Artificial Bodies*, pages 193–199. Reidel.

Message, P.J. (1999) Orbits of Saturn's satellites: some aspects of commensurabilities and periodic orbits. In Steves, B., and Archie Roy, A.E., editors, *The Dynamics of Small Bodies in the Solar System: A Major Key to Solar System Studies (The Proceedings of the N.A.T.O. Advanced Study Institute, in Acquafredda di Maratea, Calabria, Italy, 1997.)*, pages 207–225.

Poincaré, H. (1892) *Méthodes Nouvelles de la Mécanique Célèste.* Gauthier-Villars, Paris.

Roy, A.E., and Ovenden, M.W.(1955) On the Occurrence of Commensurable Mean Motions in the Solar System. *Monthly Notices of the Royal Astronomical Society*, vol. 115, pages 296–309.

Szebehely, V. (1967) *Theory of Orbits: The Restricted Problem of Three Bodies* Academic Press, New York.

GENERAL PROPERTIES OF THREE-BODY SYSTEMS WITH HILL-TYPE STABILITY

C. Marchal
General scientific direction ONERA, BP 72, 92322 Chatillon cedex, France
clbmarchal@wanadoo.fr

Abstract Three-body systems with Hill-type stability are the generalization to the general three-body problem of the Hill-stable orbits of the circular restricted three-body problem.

These systems have always a negative energy integral h and a large angular momentum c (in the axes of the center of masses), they are characterized by a product hc^2 smaller than or equal to that of the corresponding circular Euler motion with the same three masses. They have a "close binary" and a "third body" that can neither approach nor disrupt the close binary (well defined limit distances can be given in terms of the three masses and the initial conditions). However, and this is a major difference with the circular restricted three-body problem, the third body can sometimes escape to infinity.

A large majority of known triple stellar systems have the Hill-type stability, and so is the Sun-Jupiter-Saturn system (99.99% of the mass of the Solar System). The "close binary" is then the Sun and Jupiter, while Saturn is the third body.

The third body always rotate into the positive direction about the direction of angular momentum, and many general properties of its orbit can be obtained: small inclination, upper and lower limits of its angular momentum, lower limits of its approach to the close binary, etc. But much less results are known for the relative orbit of the close binary (inner orbit), besides its roundedness.

If the mutual inclination of the inner and outer orbits is small (or in the vicinity of 180°), they have generally only small and slow perturbations. But if that inclination is large, in the vicinity of 90°, the perturbations can become very large, especially for the eccentricity of the inner orbit and for the mutual inclination itself.

Beside the usual types of orbits: bounded, with a parabolic or hyperbolic escape of the third body, etc. we must notice the presence of the two types of oscillating orbits. The first type is infinitely rare: the motion of the third body has then an infinite number of larger and larger loops, the reunion of which is unbounded, but it always come back to small

B.A. Steves et al. (eds.), Chaotic Worlds: from Order to Disorder in Gravitational N-Body Dynamical Systems, 103–128.

distances. On the contrary the oscillating orbits of the second type fill in phase space a set of positive measure. These orbits are characterized by an infinite number of very close approaches of the two bodies of the close binary (their mutual distance has no positive lower bound and then their velocities are unbounded), even if strict collisions of point-masses remain infinitely rare.

Of course real bodies are not point-masses and thus oscillating orbits of the second type lead to collisions.

Let us now have a physical point of view. In our galaxy a majority of stars are binary stars. If then a weak binary star meets a strong binary star, an ordinary motion of exchange type can easily disrupt the weak binary and lead to the formation of a triple system with the strong binary. That new-born triple system has generally a Hill-type stability and if its motion is of the second oscillating type (which usually requires a large inclination) it will lead to a collision of the two stars of the binary... The probability of this phenomenon is of the order of the ratio of the inner period (that of the close binary) to the outer period (that of the third body).

The phenomenon of supernova requires an energy much larger than that of the usual collision of two stars, but the phenomenon of nova has an energy of similar order of magnitude. It is then likely that a proportion of novae appear in this indirect way, having, by far, a much larger probability than that of direct collisions of stars.

Keywords: Celestial Mechanics. 3 body-problem. Hill-type stability.

1. Introduction

The Hill stability of low energy orbits of the circular restricted three-body problem goes back to the nineteenth century and its extension to the general three-body problem appears in the seventies of the twentieth century simultaneously among several authors (for instance all Ref. before 1980 but Yoshida 1972 and Marchal 1978). Almost all known triple stellar systems have this type of stability that is characterized by a disconnection, in phase space, of the zone of possible motion into two or three parts: the three body system of interest remains forever in its original part and exchange motions are impossible. For this reason systems with Hill-type stability appears as much simpler than non-Hill stable systems even if they can undergo very large perturbations and have the full complexity of non integrable problems.

2. Usual notations

We will use the classical Jacobi decomposition of the three-body problem (Figure 1), with:

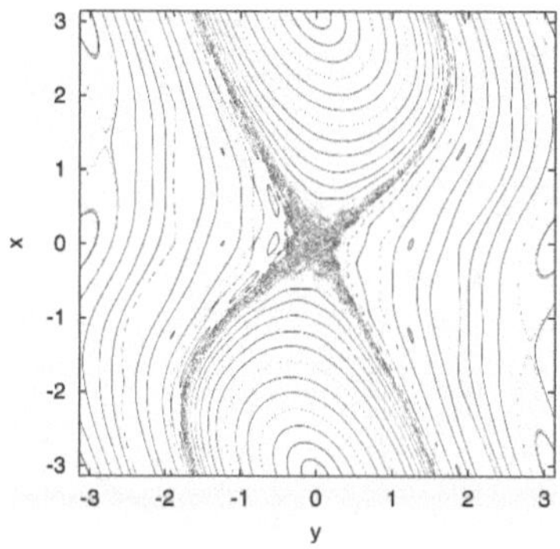

Figure 1. The Jacobi decomposition; $\vartheta = angle(\mathbf{R}, \mathbf{V})$

G = Cavendish constant (constant of the law of Newton) = 6.672 10^{-11} $m^3/kg.s^2$

m_1, m_2: the two masses of the "close binary" (with usually $m_1 \geq m_2$, but not in Figures 3-5)

m_3: "third mass" or "outer mass"

$M = m_1 + m_2 + m_3$ = total mass

$GM = \mu$ = gravitational constant of the system

$m = m_1 m_2/(m_1 + m_2)$ = "reduced mass of the binary"

$\mathrm{m} = m_3\ (m_1 + m_2)/M$ = "reduced mass of the outer system"

r= vector from m_1 to m_2; v = dr/dt = inner velocity vector

R = vector from I, center of mass of the binary to the mass m_3

V = dR/dt = outer velocity vector

$$\mathbf{c} = \text{angular momentum} = m\mathbf{r} \times \mathbf{v} + \mathrm{m}\mathbf{R} \times \mathbf{V} \tag{1}$$

$$h = \frac{mv^2 + \mathrm{m}V^2}{2} - G(\frac{m_1 m_2}{r} + \frac{m_1 m_3}{r_{13}} + \frac{m_2 m_3}{r_{23}}) \tag{2}$$

with, as usual, r_{ij} = distance between the masses m_i and m_j.

3. Some useful notations

$$\alpha = m_1/(m_1 + m_2) \quad ; \quad \beta = m_2/(m_1 + m_2) = 1 - \alpha \tag{3}$$

$$\mathbf{r}_{12} = \mathbf{r}; \quad \mathbf{r}_{13} = \mathbf{R} + \beta\mathbf{r}; \quad \mathbf{r}_{23} = \mathbf{R} - \alpha\mathbf{r} \tag{4}$$

a_i, e_i, i_i, Ω_i, ω_i M_i : orbital elements of the inner system (vectors r, v with the gravitational constant $G(m_1 + m_2)$ of the inner system).

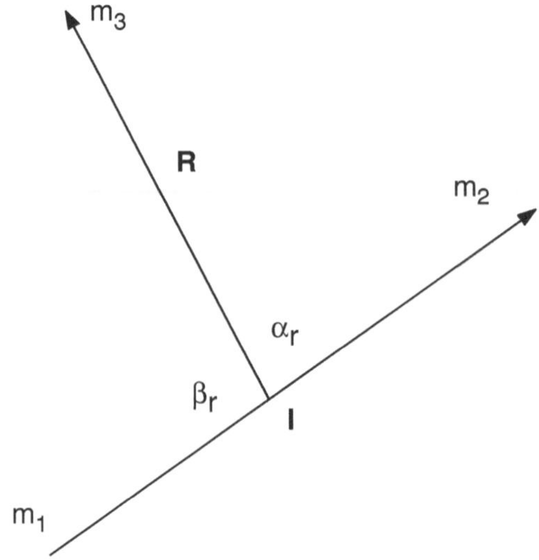

Figure 2. The mass ratios α and β

a_e, e_e, i_e, Ω_e, ω_e M_e: orbital elements of the outer or "exterior" system (vectors R and V with the gravitational constant GM equal to μ).

$$M^* = m_1 m_2 + m_1 m_3 + m_2 m_3 \tag{5}$$

$\varrho =$ "mean quadratic distance":

$$M^* \varrho^2 = m_1 m_2 r_{12}^2 + m_1 m_3 r_{13}^2 + m_2 m_3 r_{23}^2 \tag{6}$$

$\nu =$ mean harmonic distance

$$M^*/\nu = (m_1 m_2/r) + (m_1 m_3/r_{13}) + (m_2 m_3/r_{23}) \tag{7}$$

$$a = \text{generalized semi-major axis} = -GM^*/2h \tag{8}$$

$$p = \text{generalized semi-latus rectum} = Mc^2/GM^{*^2} \tag{9}$$

In a two body problem a and p are the usual semi-major axis and semi-latus rectum of the relative Keplerian orbit, and they are also these same elements of the relative orbit of the primaries in a restricted three-body problem.

4. A short digression: the fixed lengths a and p and the variable lengths ϱ and ν

The relations (1)-(9) can be easily generalized to the full n-body problem: M = Σ_k = total mass; GM = μ = gravitational constant.

$$
\begin{aligned}
M^* &= \sum_{1\le j<k\le n} m_j m_k; \\
M^* \varrho^2 &= \sum_{1\le j<k\le n} m_j m_k r_{jk}^2; \\
M^*/\nu &= \sum_{1\le j<k\le n} m_j m_k / r_{jk}
\end{aligned} \tag{10}
$$

The relations (8) and (9) defining the generalized semi-major axis a and the generalized semi-latus rectum p remain the same, with the integrals of motion c and h of the n-body problem (in the axes of the center of mass).

The main classical relations, with I being the moment of inertia in the axes of the center of mass and U the potential, gives a useful general insight of the n-body problem ($I = \Sigma\, m_j\, r_j^2 = M^*\varrho^2/M$; $U = G \sum_{1\le j<k\le n} m_j\, m_k \,/\, r_{jk} = GM^*/\nu$)

1. The quadratic mean of positive numbers is always larger than or equal to their harmonic mean (notice that the equivalent inequality, $IU^2 \geq G^2M^{*3}/M$, is generally ignored):

   $$\text{At all times:} \qquad \varrho \geq \nu \tag{11}$$

   The equality is obtained either in a two-body case (only one mutual distance) or when all mutual distances are equal (equilateral triangle, regular tetrahedron).

2. The Lagrange-Jacobi identity, $d^2I/dt^2 = 2U + 4h$, becomes:

   $$(\varrho^2)'' = 2\mu[(1/\nu) - 1/a] \tag{12}$$

   Since $\varrho = \nu$ in a two body case, the equation (12) gives directly the two-body evolution of the mutual distance; and because of the inequality (11) the two-body evolution appears as a limit of the general case.

3. The Sundman inequality, $8I\,(h + U) \geq 4c^2 + I'^2$, becomes:

   $$\varrho/\nu \geq j \qquad \text{with:} \qquad j = (p/2\varrho) + (\varrho/2a) + (\varrho\varrho'^2/2\mu) \tag{13}$$

We will call j the "Sundman function" since it is proportional to the original Sundman function $J = \frac{(4c^2+I'^2-8hI)}{\sqrt{I}} = 8jGM^{*3/2}M^{-1/2}$. With (12) its derivative is simple:

$$dj/dt = j' = (\varrho'/\varrho)[(\varrho/\nu) - j] \tag{14}$$

Notice that, with (13), j' and ϱ' have always the same sign; j and ϱ always vary in the same direction and this can give long term information. If for instance, at some time t_o, the distance ϱ_o is very small and ϱ'_o is positive, the non-decrease of j and the equation (13) show that ϱ' will remain positive and ϱ increasing at least as long as ϱ reaches the value ap/ϱ_o.

4. The Easton inequality, $U^2I + 2c^2h \geq 0$, (consequence of the Sundman inequality) is equivalent to:

$$\varrho^2/\nu^2 \geq p/a; \quad \text{with}: p/a = -2Mc^2h/G^2M^{*3} \tag{15}$$

We will call "Szebehely constant" the ratio p/a. It is proportional to the various Szebehely constants used and has a major interest: in the three and four-body problems the restrictions on configurations appear for p/a larger than one and that limit remains always very close to one for more-than-four-body problems. (Notice than in a two body problem of eccentricity e, the ratio p/a is (1 -e^2) and is thus always smaller than or equal to one).

These relations (11)-(15) give much information on the possible evolution of the mean quadratic distance ϱ (but less information on the evolution of the mean harmonic distance ν that has much more variations (Marchal (1990), pages 349-360)).

For instance let us consider first the limit and two-body case $\varrho = \nu$. For a negative energy h, we obtain an elliptic Keplerian motion with the period $T = 2\pi\sqrt{(a^3/\mu)}$. The successive pericenter and apocenter distances ϱ_m and ϱ_M verify:

$$\varrho_m + \varrho_M = 2a; \quad \varrho_m\varrho_M = ap \tag{16}$$

The interval of time between these two extrema is of course T/2, with a share given by the following usual Keplerian expressions:

$$\text{For the range} \quad \varrho_m \leq \varrho \leq a, \text{ a duration} \quad T_m = T(\frac{1}{4} - \frac{(a-\varrho_m)}{2\pi a}) \tag{17}$$

$$\text{For the range} \quad a \leq \varrho \leq \varrho_M, \quad \text{a duration} \quad T_M = T(\frac{1}{4} + \frac{(\varrho_M - a)}{2\pi a})$$

$$\text{We can verify that}: \quad T_m \leq T/4 \leq T_M, \quad \text{and}: \quad T_m + T_M = T/2.$$

If now we consider an arbitrary n-body motion, the limit character of the two-body motion appears in the following. Let us assume that the evolution $\varrho(t)$ has successively a minimum ϱ_m and a maximum ϱ_M, then (16) becomes:

$$\varrho_m + \varrho_M \geq 2a; \quad \varrho_m \varrho_M \geq ap; \quad \text{hence} \quad \varrho_M \geq sup\{a; \sqrt{(ap)}\} \qquad (18)$$

The durations T_m' and T_M' in the ranges $\varrho_m \leq \varrho \leq a$ and $a \leq \varrho \leq \varrho_M$ verify now, with the Keplerian durations T_m and T_M given in (17):

$$T'_m \leq T_m \leq T/4 \leq T_M \leq T'_M \qquad (19)$$

and there are many interesting general properties (Marchal (1990), pages 349-358] such as at any time (if $h < 0$):

$$\varrho(t) + \varrho[t + (T/2)] \geq 2a \qquad (20)$$

5. Interest of the Szebehely constant, i.e. the ratio p/a

We will see in Figures 3-5 how the restrictions on possible configurations appear as soon as the ratio p/a is larger than one, and this is perhaps a factor of stability since all known more-than-two-body systems have a ratio p/a larger than one.

	Value of the ratio p/a
Solar system	1.16
Jupiter and satellites	1.25
Saturn and satellites	1.043
Uranus and satellites	1.2
Multi stellar systems	from 1.15 to 7.2
The Galaxy	about 1.2

6. Hill's curves and curves of constant ratio ϱ/ν

The classical Hill's curves of the restricted three-body problem are given in Figure 3 for equal primaries. The curves of constant ratio ϱ/ν are given in Figures 4 and 5 in terms of the position of m_1 with respect to m_2 and m_3 for the following mass ratios: $2m_1 = m_2 = m_3$ (Figure 4) and $m_1 = m_2 = m_3$ (Figure 5).

The curve of constant ratio ϱ/ν represent the natural prolongation of Hill's curves with continuity between the corresponding Figures as Figures 3-5. The two minimums correspond to $\varrho/\nu = 1$ at the equilateral Lagrangian points and the three collinear saddle points are the usual Euler equilibrium points.

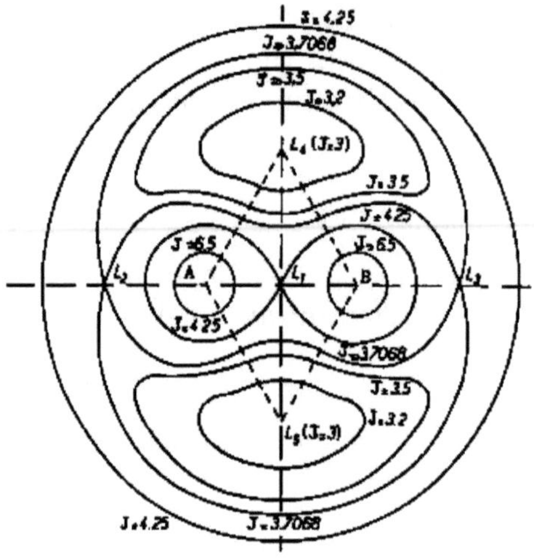

Figure 3. The classical Hill's curves for equal primaries. $AB = 1; J = (r_A^2 + r_B^2)/2 + 1/r_A + 1/r_B$

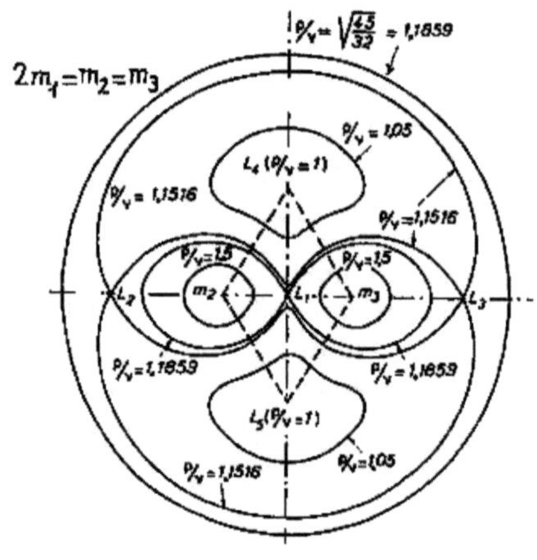

Figure 4. The ratio ϱ/ν for $m_2 = m_3 = 2m_1$ in terms of the position of m_1 with respect to m_2 and m_3 (Notice that the three collinear points L_1, L_2 and L_3 have the subscript of the mass that is between the two other masses)

7. The zones of possible motion

Let us consider the Easton inequality written in (16), that is: $\varrho^2/\nu^2 \geq p/a$.

The Figures 4 and 5 show that, as soon as the ratio p/a is larger than one, a forbidden zone appear about the Lagrangian points L_4 and

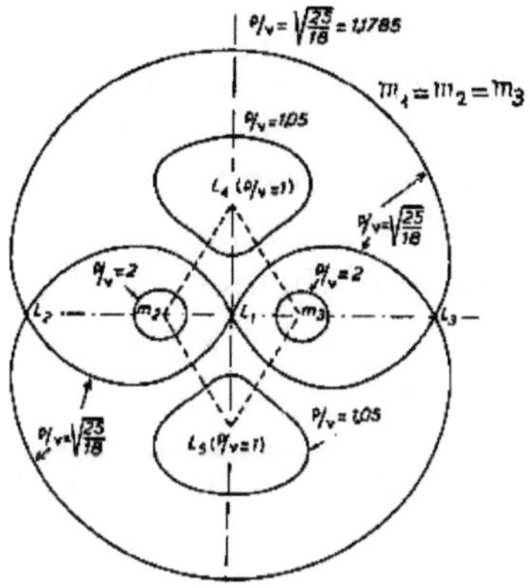

Figure 5. The ratio ϱ/ν for $m_2 = m_3 = m_1$ in terms of the position of $m_1$with respect to m_2 and m_3

L_5: the three-body system of interest cannot approach equilateral configurations. Furthermore if $\sqrt{p/a}$ is larger than the value of ϱ/ν at the saddle points the zone of possible motion becomes disconnected into two or three parts and we reach then the extension of Hill stability of the circular restricted three-body problem to the general three-body problem.

It is of course of interest to know the limit ratios $(p/a)_{Lj} = (\varrho^2/\nu^2 \geq p/a)L_j$, for j = 1, 2, 3 and for various mass ratios and, if as in Figures 4 and 5 we give to the point L_j the subscript of the mass that is between the two other masses, we obtain the following:

$$0 < m_A \leq m_B \leq m_C \qquad \text{implies:} \tag{21}$$
$$1 < (p/a)_{LC} \leq (p/a)_{LB} \leq (p/a)_{LA} \leq 343/243 = 1.4115.$$

We will see that the upper limit is obtained for $m_C = m_B = 1.6\ m_A$.

Notice that if the ratio p/a is equal to some $(p/a)_{Lj}$ with j = 1, 2, 3, 4 or 5, the point L_j only corresponds to the associated circular Eulerian or Lagrangian motion. Hence, with (21), the division of phase space is the following:

If $p/a < (p/a)_{LB}$, the portion of phase space corresponding to the given integrals of motion is connected.

If $(p/a)_{LB} \leq p/a < (p/a)_{LA}$, the portion of phase space corresponding to the given integrals of motion is disconnected into two parts: a part about m_A (Hill type stability, with the two largest masses m_B and m_C as the close binary and m_A as the third "outer" body) and a second part about m_B, L_A and m_C. Notice that, even if $p/a = p/a(L_B)$, no motion can cross L_B.

Finally if $(p/a)_{LA} \leq p/a$, the portion of phase space corresponding to the given integrals of motion is disconnected into three parts (in Figures 4 and 5, a part in the vicinity of m_2, a part in the vicinity of m_3 and a far away part). No motion can cross L_A, even if $p/a = p/a(L_A)$, and all motions have then the Hill type stability.

If we consider the system of Figures 1 and 2 and if we assume $m_1 \geq m_2$, as we will do henceforth, the two conditions for a Hill type stability of the close binary m_1, m_2 are:

$$p/a \geq (p/a)_{L2}; \quad r/R \leq (r/R)_{L2} \tag{22}$$

(let us recall that the Szebehely constant p/a is equal to $-2Mc^2h/G^2M^{*3}$).

If these two inequalities are at their limit, the three-body motion is a circular Eulerian motion with m_2 between m_1 and m_3 (or perhaps with m_1 between m_2 and m_3, if we have also $m_1 = m_2$).

The limit value $(p/a)_{L2}$ can easily be obtained in the following way: it is the minimum of the ratio ϱ^2/ν^2 for collinear configurations with m_2 between m_1 and m_3:

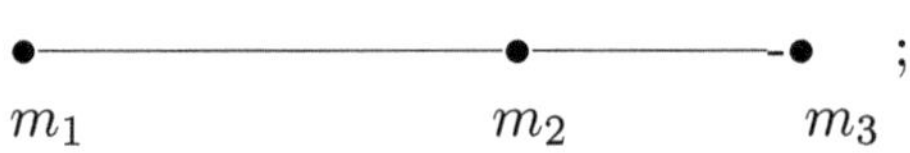

; $r_{23}/r_{12} = x$

$(p/a)_{L2}$ = inf of $(\varrho^2\nu^2)$ for x > 0 = inf, for x > 0, of the quantity:

$$[m_1m_2 + m_1m_3(1+x)^2 + m_2m_3x^2][m_1m_2 + \frac{m_1m_3}{(1+x)} + \frac{m_2m_3}{x}]^2 M^{*(-3)} \tag{23}$$

Of course that minimum is obtained when x is a solution of the classical Euler equation:

$$\begin{aligned}(m_1+m_2)x^5 + (3m_1+2m_2)x^4 + (3m_1+m_2)x^3 \\ = (m_2+3m_3)x^2 + (2m_2+3m_3)x + m_2 + m_3\end{aligned} \tag{24}$$

and the corresponding ratio $(r/R)_{L2}$ is given by:

$$(r/R)_{L2} = \frac{1}{(x+\alpha)}; \quad \text{with } \alpha = m_1/(m_1+m_2), \quad \text{as given in (3)} \tag{25}$$

That limit ratio is always smaller than unity, and thus we have ever $r < R$ in motions with Hill type stability.

Notice that the case x = 1 in (24) gives already a simple upper bound of $p/a(L_2)$, that is: $p/a(L_2) \leq (1+6u)(1-u)^2$, with $u = \frac{m_1m_3}{2M^*} \leq 1/2$.

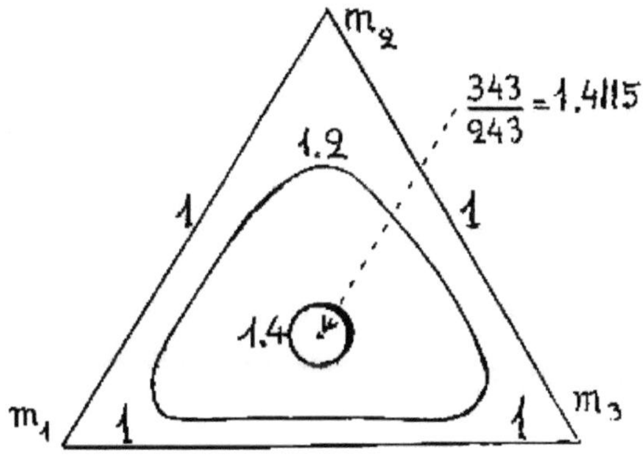

Figure 6. The ratio $(p/a)_{L_2}$ in terms of the mass ratios.

That upper bound is reached when $m_1 = m_3$ and gives the maximum of $(p/a)_{L2} = 343/243$ for $u = 2/9$, that is for $m_1 = m_3 = 1.6m_2$.

Also notice that the classical Euler condition (24), for a collinear central configuration with m_2 between m_1 and m_3, gives *the position of the center of mass in terms of the positions of the three bodies (and independently of the three masses themselves).* Indeed if r_j, with j = 1, 2, 3 are the three radius vectors from the center of mass, that Euler condition (24) is equivalent to the following beautiful symmetrical condition:

$$(\mathbf{r}_1/r_{23}^2) + (\mathbf{r}_3/r_{12}^2) = (\mathbf{r}_2/r_{13}^2) \tag{26}$$

8. Limitations related to the integrals of motion

Let us recall the two integrals of motion (1) and (2):

$$\mathbf{c} = \text{angular momentum} = m\mathbf{r} \times \mathbf{v} + \mathrm{m}\mathbf{R} \times \mathbf{V} \tag{27}$$

$$h = \frac{mv^2 + \mathrm{m}V^2}{2} - G(\frac{m_1 m_2}{r} + \frac{m_1 m_3}{r_{13}} + \frac{m_2 m_3}{r_{23}}) \tag{28}$$

These two integrals gives of course a lot of classical information, the most interesting are the following.

8.1 Conditions for the vectors r and R

The plane of the three bodies is the r, R plane and, for given r and R the velocity components normal to that plane are given by (27):

$$\mathbf{c}.\mathbf{r} = \mathrm{m}(\mathbf{r} \times \mathbf{R}).\mathbf{V}; \qquad \mathbf{c}.\mathbf{R} = -m(\mathbf{r} \times \mathbf{R}).\mathbf{v} \tag{29}$$

These two normal components give the in-plane part of the angular momentum c and it remains to find some in-plane components of v and V that satisfy (27) and (28). This classical problem is easy and solutions are possible if and only if:

$$(mv_p^2 + \mathrm{m}V_p^2).(mr^2 + \mathrm{m}R^2) \geq c_n^2 \tag{30}$$

where v_p and V_p are the in-plane part of v and V, while $\mathrm{c}_n$is the normal component of c.

With (28) and (29) this condition (30) leads to the following necessary condition for the vectors r and R:

$$(mr^2 + \mathrm{m}R^2)\{[2h + 2G\Sigma(m_i m_j/r_{ij})](\mathbf{r} \times \mathbf{R})^2 - \mathbf{c}.\mathbf{r})^2/\mathrm{m} - \mathbf{c}.\mathbf{R})^2/m\} \geq [\mathbf{c}.(\mathbf{r} \times \mathbf{R})]^2 \tag{31}$$

Notice that:

A) For a motion with the Hill type stability we must add the second condition (22):

$$r/R \leq (r/R)_{L2} \tag{32}$$

B) If the two vectors r and R are collinear the analysis (29)-(31) is no more valid, but a similar analysis leads to the following conditions:

$$\mathbf{c}.\mathbf{r} = \mathbf{c}.\mathbf{R} = 0; \quad (mr^2 + \mathrm{m}R^2)\{2h + 2G\Sigma(m_i m_j/r_{ij})\} \geq c^2 \tag{33}$$

8.2 Conditions for the lengths r and R

Two lengths r and R are given and satisfy (32) in a three-body problem with Hill type stability. Are there some suitable orientations that satisfy (31) or (33)? The optimal orientations are with collinear r and R, in the same direction, and both normal to c.

With (33) the corresponding condition is then:

$$2(mr^2 + \mathrm{m}R^2)\{h + G[(m_1 m_2/r) + m_1 m_3/(R + \beta r) + m_2 m_3/(R - \alpha r)]\} \geq c^2 \tag{34}$$

In the R-r plane, the zone of possible motion is along the R-axis from zero to infinity with an upper limit for the ratio r/R and another upper limit for the length r (see the Figure 7 for the cases of three equal masses).

Let us call k the maximum of the ratio r/R. It is obtained for $\nu = a$ and $\varrho = \sqrt{ap}$, which leads to easy computations; k is the only root given by:

The zone of possible motion of the point (R, r) for three equal masses and for three values of the ratio p/a. (The value 25/18 corresponds to the limit of Hill type stability). The points A, B, C (r/R maximum) and A', B', C' (r maximum) :

	A	B	C	A'	B'	C'
p/a	25/18	1.5	2	25/18	1.5	2
R/a	5/4	1.357 700	1.707 229	2.189 268	2.714 479	4.151 993
r/a	5/6	0.690 361	0.584 360	0.886 521	0.854 047	0.788 594
k(= sup(r/R)	2/3	0.501 825	0.342 286	*On all three curves* R→∞ *implies* r/a → 2/3		

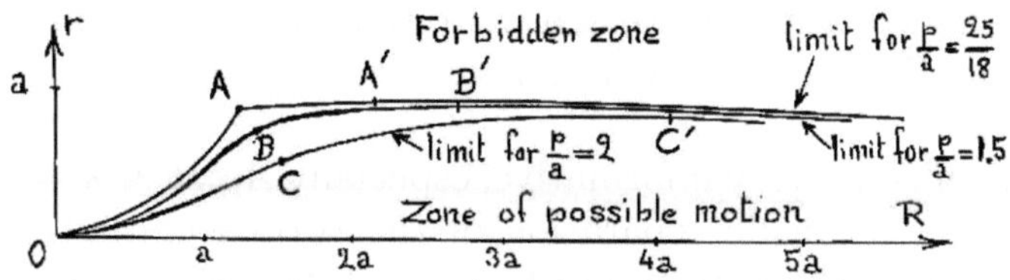

Let us add that with: $A_i = a_i(1 + e_i)$ = *apocenter distance of the inner orbit*
$P_e = a_e(1 - e_e)$ = *pericenter distance of the outer orbit*
the points (R, A_i) *and* (P_e, A_i) *also belong to the same zone of possible motion.*

Figure 7. The zone of possible motion for the point (R, r).

$$0 < k = sup(r/R) \leq (r/R)_{L2}$$

$$p/a = \frac{M(\mathrm{m} + mk^2)}{M^{*3}}[(m_1m_2/k) + m_1m_3/(1+\beta k) + m_2m_3/(1-\alpha k)]^2 \qquad (35)$$

8.3 Conditions for the inner apocenter distance A_i and the outer pericenter distance P_e

Let us consider a three-body system with Hill type stability. The corresponding R, r point is of course in the zone of possible motion of a suitable figure as Figure 7 (conditions (32) and (34)) and the corresponding radius vectors R and r and velocity vectors V and v satisfy the equations (27) and (28) of the two integrals of motion. However notice that, in order to satisfy the condition (34) it is sufficient to find two velocity vectors V' and v' such that:

$$|m\mathbf{r} \times \mathbf{v}'| + |\mathrm{m}\mathbf{R} \times \mathbf{V}'| \geq |\mathbf{c}| \qquad (36)$$

$$[(mv'^2 + \mathrm{m}V'^2)/2] - [(m_1m_2/r) + \frac{m_1m_3}{(R+\beta r)} + \frac{m_2m_3}{(R-\alpha r)}] \leq h \qquad (37)$$

and these two conditions are satisfied, with the velocity vectors of the osculating orbits, as soon as the inner distance is larger than or equal to r and the outer radius is smaller than or equal to R.

In this analysis the most interesting points of the two osculating orbits are of course the apocenter of the inner orbit and the pericenter of the outer orbit. The corresponding distances are $A_i = a_i(1+e_i)$ = apocenter distance of the inner orbit, and $P_e = a_e(1 - e_e)$ = pericenter distance of the outer orbit. On the other hand, by continuity, the condition (32) remains satisfied and thus we can write that the zone of possible motion of figures as Figure 7 also contains at all times the points (R, A_i) and (P_e, A_i).

These results are very informative, especially $A_i/P_e \leq k = sup(r/R) < 1$, that shows the wide separation of the two osculating orbits. They are of course much more interesting than the similar classical results (Marchal (1990), p. 334) as at all times $a_i < a$, or, almost symmetrically, $p_e > p$ if $m_3/M \leq 0.75$ and $p_e > 0.98p$ if $m_3/M > 0.75$ where $p_e = a_e(1 - e_e^2)$ is the semi-latus rectum of the outer orbit.

8.4 Conditions for angular momenta $m.r \times v$ and m.R × V

With (1) we already know that the sum of the inner angular momentum m.r×v and the outer angular momentum m.R×V is the total angular momentum c.

The limits of these two partial angular momenta are obtained when the vectors r and R are collinear, in the same direction, normal to c and to the velocities v and V.

Let us choose, as usual, a set of axes with the angular momentum c in the z-direction, and let us put the vectors r and R into the x-direction. The velocity vectors v and V will be then:

$$\mathbf{v} = (0, y', z'); \quad \mathbf{V} = (0, Y', Z') \tag{38}$$

with:

$$\mathrm{m}.R \times V = (0, -\mathrm{m}RZ', \mathrm{m}RY') = (0, Y, Z) \tag{39}$$

$$m.r \times v = (0, -mrz', mry') = c - \mathrm{m}.R \times V = (0, -Y, c - Z) \tag{40}$$

$$h = [(mv^2 + \mathrm{m}V^2)/2] - G[(m_1m_2/r) + \frac{m_1m_3}{(R+\beta r)} + \frac{m_2m_3}{(R-\alpha r)}] \tag{41}$$

that is, after elimination of Y', Z', y', and z':

$$\mathrm{m}R^2.\{2h + 2G[\frac{(m_1m_2}{r}) + \frac{m_1m_3}{(R+\beta r)} + \frac{m_2m_3}{(R-\alpha r)}]$$
$$-[\frac{c^2}{(mr^2 + \mathrm{m}R^2)}]\} = \{Y^2 + [ZZ_q]^2\}.\frac{(mr^2 + \mathrm{m}R^2)}{mr^2} \tag{42}$$

with:

$$Z_q = \mathrm{cm}R^2/(mr^2 + \mathrm{m}R^2) \tag{43}$$

For a given value of the ratio q = r/R, the parameter Z_q is given and the widest possibilities in (42), i.e. the largest value of the left member, are obtained in the case when:

$$2h = -G[\frac{(m_1 m_2}{r}) + \frac{m_1 m_3}{(R+\beta r)} + \frac{m_2 m_3}{(R-\alpha r)}] \tag{44}$$

that is $\nu = a$, with the notations (7), (8). It remains thus to study (38)-(44) with the only remaining parameter, the parameter q equal to r/R and thus smaller than or equal to k, the maximum of r/R, given in (35).

We have chosen the two vectors r and R in the x-direction, but of course all directions of the Oxy plane are possible. Hence, for a given q, the possible values of the outer angular momentum m.R $\times$V fill a sphere of center (0, 0, Z_q), with Z_q = cm /(mq^2 + m) and of radius R given by:

$$R^2 = c^2 m\mathrm{m}q^2 Q/pM *^3 \, (\mathrm{m} + mq^2)^2$$

with:

$$Q = aM(\mathrm{m} + mq^2)[\frac{m_1 m_2}{q} + \frac{m_1 m_3}{1+\beta q} + \frac{m_2 m_3}{1-\alpha q}]^2 - pM^{*3} \tag{45}$$

The equation (35) shows that R = 0, when q = k. The domain of the outer angular momentum is given by all these spheres for q in the range $0 \leq$ q $\leq$ k (see Figure 8 for three equal masses).

The ratio R/Z_q is simple:

$$\begin{aligned} R^2/Z_q^2 \quad &= mq^2Q/p\mathrm{m}M^{*3} = (m/\mathrm{m})\{[aM(\mathrm{m}+mq^2)/pM^{*3}] \\ &[m_1m_2 + m_1m_3q/(1+\beta q) + m_2m_3q/(1-\alpha q)]^2 - q^2\} \end{aligned} \tag{46}$$

The maximum of this ratio for q in the range (0, k) is always between zero and one, and gives the square of the sinus of the maximum possible inclination of the outer orbit with respect to the invarible plane (normal to c). Indeed the angle between the two momenta c and m.R $\times$V is also the inclination i_e of the outer orbit, and sup $sini_e$ = sup (R/Z_q).

Of course the maximum of the right member of (46) is not easy to obtain exactly, but a simple upper bound can be obtained as follows: When |m.r$\times$v| is maximum this momentum has the direction of the momentum m.R $\times$V, which implies |m.r$\times$v| + |m.R $\times$V|= c.

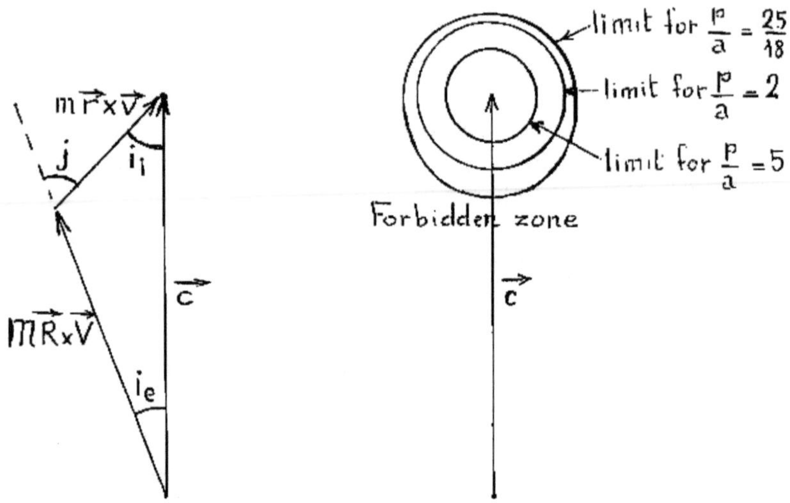

Figure 8. The limits of the possible zone of the outer momenta m.$R \times V$ for three equal masses and three different values of the Szebehely ratio p/a. The corresponding largest inclination i_{eM} and the upper bound $\alpha_m = Arcsin\ s/(1+s)$ are successively for $p/a = (25/18)$; 2 and 5: $i_{eM} = 13°\ 11'$; $10°\ 27'$ and $6°\ 15'$; $\alpha_M = 19°\ 28'$; $15°\ 18'$ and $14°\ 50'$

On the other hand, from $A_i/P_e \leq k = sup(r/R) < 1$, we deduce:
$|\mathrm{m.r} \times \mathrm{v}| \leq \mathrm{m}\ .\ [\ \mathrm{G\ A}_i\ (\mathrm{m}_1 + \mathrm{m}_2)]^{1/2}$; $|\mathrm{m.R} \times \mathrm{V}| \geq \mathrm{m}.\ [\ \mathrm{G\ P}_e\ \mathrm{M}\ ]^{1/2}$

$$|m.\mathbf{r} \times \mathbf{v}|/\mathrm{m}.R \times V| \leq m.[k(m_1+m_2)/M]^{1/2}/\mathrm{m} = s \qquad (47)$$

Hence, with that ratio s:
$|\mathrm{m.r} \times \mathrm{v}| \leq sc/(1+s)$

$$\sup \sin i_e \quad \leq s/(1+s) \qquad (48)$$

These upper and lower bounds are rather weak (see Figure 8), nevertheless they give some essential elements: both inner and outer angular momenta remain bounded, while the z-component of m.R ×V remains forever above c/(1 + s) . The outer body always rotates about the z-axis in the positive direction with a limited and generally small inclination, while the inner orbit may have any inclination and also any eccentricity in the range from 0 to 1.

8.5 Other conditions related to the integral of motions

The remaining informations related to the integral of motion have fewer interest.

For instance let us consider two lengths r and R satisfying the necessary conditions (32) and (34). What is then the range of possible latitudes φ of the vector R with respect to the invariable plane, the plane normal to the angular momentum c? That range is of course symmetrical with respect to $\varphi = 0$; the absolute value of the possible latitudes has a maximum φ_M, that can be obtained with (31) and is given by:

$$sin^2\varphi_M = mr^2[2(mr^2 + \mathrm{m}R^2)(h+U) - c^2]/\mathrm{m}R^2c^2 \tag{49}$$

the potential U being somewhere in the following small range:

$$\begin{aligned} G[(m_1m_2/r) + (m_1m_3 + m_2m_3)/(R^2 + \alpha\beta r^2)^{1/2}] &\leq U \\ \leq G[(m_1m_2/r) + m_1m_3/(R+\beta r) + m_2m_3/(R-\alpha r)] \end{aligned} \tag{50}$$

For instance if R is very large, the potential U is $\mathrm{G}m_1m_2/\mathrm{r}$, and the largest latitude $\varphi_M\infty$ is then obtained for $r = -Gm_1m_2/2h$, that is $U = -2h$ which gives:

$$\varphi_M\infty = \arcsin\{(m_1m_2/M^*)^3(a/p)[M/(m_1+m_2)]\}^{1/2} \tag{51}$$

For R in the vicinity of the generalized semi-latus rectum p the maximum φ_M of the latitude can be a little larger than $\varphi_M\infty$, but when R is very small the maximum latitude is equivalent to $(R/p).\{Mm_3m_1^3m_2^3\}^{1/2}/M^{*2}$ and goes to zero with R.

9. Information given by a test of escape and by d^2R/dt^2

Let us consider first the classical Yoshida test of escape (Yoshida 1972). That test uses a fixed and suitable distance D and considers the evolution of the distance R during an interval (t_o, t_f) of time in which the relation $\mathrm{r} \leq \mathrm{D}$ is always verified:

Let us put:

$$\Sigma = (dR/dt)^2 - 2GM\{[\alpha/(R+\beta D)] + [\beta(R-\alpha D)]\} \tag{52}$$

If at the time t_o:

$$r < D; \quad R > \alpha D; \quad dR/dt \geq 0 \quad \text{and } \Sigma \tag{53}$$

then the function Σ will remain non negative and even non decreasing all along the interval (t_o, t_f) in which r remains smaller than or equal to D.

That test can easily be used as a test of escape, with $t_f = \infty$, if the equality $r = D$ is impossible in the future, for instance if at t_o:

$$h \leq (m\Sigma/2) - (Gm_1m_2/D) \tag{54}$$

Let us try now to improve that test as much as possible.

It is impossible to substitute a function $S = V^2 - 2GM\{[\alpha/(R + \beta D)] + [\beta(R - \alpha D)]\}$ to the function Σ of (52), because that function would decrease in many cases, but we can try something as:

$$S = V^2 - 2GM\{[\alpha/(R + \beta D)] + [\beta(R - \alpha D)] + f(\vartheta).g(R)\} \tag{55}$$

the angle ϑ being the angle (R, V) given in Figure 1.

The analysis of the case $D << R$ leads to $g(R) \sim 3\alpha\beta D^2 / R^3$, and to:

$$2sin\vartheta.(df/d\vartheta) = [4 + 5cos^2\vartheta]^{1/2} - (3 + 12f)cos\vartheta \tag{56}$$

that leads to the complex but also rigorous expression:

$$f(\vartheta) = [F(\vartheta)/sin^6\vartheta] - 0.25 \tag{57}$$

with:

$$\begin{aligned} F(\vartheta) = \{&[-39cos\vartheta + 70cos^3\vartheta - 25cos^5\vartheta][4 + 5cos^2\vartheta]^{1/2}/300\} \\ &-0.06 - 0.296\sqrt{5}Ln\{[4 + 5cos^2\vartheta]^{1/2} + \sqrt{5}cos\vartheta]/[3 + \sqrt{5}]\} \end{aligned} \tag{58}$$

$f(\vartheta)$ is an increasing function of ϑ that is equivalent to $\vartheta^2/24$ for small ϑ and that reaches the following values:

$$f(0) = 0; f(30°) = 0.012703; f(60°) = 0.072161; f(90°) = 0.327005 \tag{59}$$

For the general case let us keep the same function $f(\vartheta)$ and, as in the above Yoshida test, let us try to obtain a non decreasing function S in a domain as large as possible.

An excellent solution for the function g(R) is:

$$g(R) = [\alpha R/(R + \beta D)^2] + [\beta R/(R - \alpha D)^2] - (1/R) \tag{60}$$

We can verify that g(R) is usually small and is equivalent to $3\alpha\beta D^2/R^3$ when $D << R$.

The new test is an efficient improvement of the Yoshida test (52)-(54). We can almost write that the third body escapes as soon as it exceeds the parabolic escape velocity given by $V^2 = 2GM/R$, but notice that the following conditions of application similar to (53) are a little less broad because of the necessity to ensure $dS/dt \geq 0$, and also $d^2R/dt^2 > 0$ when $dR/dt = 0$.

Three of the four conditions of (53) remain almost the same:

$$r < D; \qquad dR/dt \geq 0 \quad \text{and} \quad S \geq 0 \tag{61}$$

but the condition $R > \alpha D$ becomes only:

$$R \geq 1.171D \quad \text{and /or} \quad 4.2(R - \alpha D) \geq R + \beta D(62) \tag{62}$$

And the test is: "With the function S defined in (55)-(60), if at the time t_o the conditions (61)-(62) are satisfied they will remain satisfied, with a non-decreasing function S, all along the interval (t_o, t_f) in which r remains smaller than D".

We can even add that all along that interval of time the derivative $d^2(R^3)/dt^2$ remains positive.

Let us also add the two following propositions: A) This test is valid even for three-body systems without the Hill type stability, even if the energy integral h is positive. B) A condition similar to (54) allows to extend the test to $t_f = \infty$ and to give a test of escape. It is sufficient that:

$$h \leq (\mathrm{m}S/2) - (Gm_1m_2/D) \tag{63}$$

and the variations of the major axis $2a_i$ of the inner orbit will forever remain bounded:

$$2a_i < 2Gm_1m_2/(\mathrm{m}S - 2h) \leq D \tag{64}$$

The escape is then a "hyperbolic-elliptic" final evolution.

Because of (63) the fixed length D is very often chosen equal to $(-Gm_1m_2/h)$, but this choice is of course not at all required (especially if h is positive or zero!).

Let us consider now the second derivative d^2R/dt^2 of the length R. The vector R has a second derivative given by Newton's law:

$$\mathbf{d^2R/dt^2} = -GM[(\alpha\mathbf{r}_{13}/r_{13}^3) + (\beta\mathbf{r}_{23}/r_{23}^3)] \tag{65}$$

and d^2R/dt^2 is given by:

$$d^2R/dt^2 = (V^2 sin^2\vartheta/R) + (\mathbf{R.d^2R/dt^2})/R \tag{66}$$

The product m$RV sin\vartheta$ is equal to the modulus of the outer angular momentum $|m.R \times \mathbf{V}|$, hence this angular momentum being bounded (see Figure 8), the first right term of (66) is always less than some K/R^3. However the last term is negative and, for large R, is equivalent to $(-GM/R^2)$. These two properties imply that for sufficiently large R the second derivative d^2R/dt^2 is negative (and below some $[K/R^3] - GM/R^2]$). All pericenters of the outer orbit occur at a distance smaller than GM/K and pictures such as Figure 9 can be drawn.

These pictures are related to the possible motions of the vector R = (X,Y,Z), with Z in the c direction.They have of course a symmetry of revolution about the Z-axis and we find the following zones: A) Two zones of large latitudes, above and below, in which the vector R cannot enter. B) A small annulus corresponding to OAB. If R enter in that annulus, the test of escape (55)-(64) shows that it escapes immediately to infinity (for both t $\rightarrow\rightarrow \pm\infty$) the motion is a "simple fly-by" with only one pericenter and no apocenter. C) The annulus ABCD contains

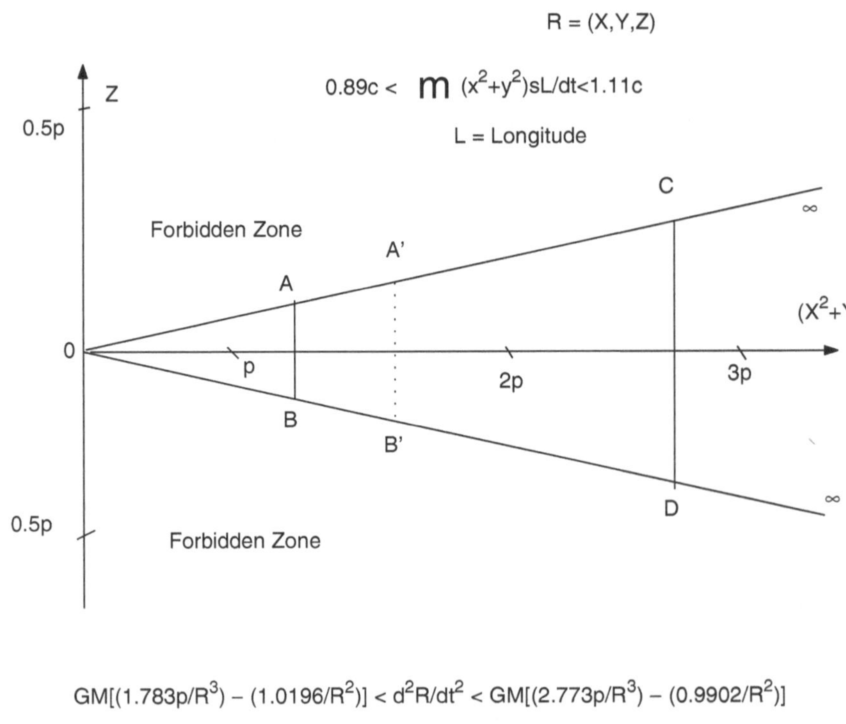

Figure 9. Discussion of the possible motions of the outer body for three equal masses and for a ratio p/a *equal to 5. Zone OAB: the conditions (55)-(63) of the test of escape are satisfied (simple "fly by" with only one pericenter). Zone OA'B':* $d^2R/dt^2 > 0$, *no apocenter. Zone* $\infty DC\infty$:$d^2R/dt^2 < 0$, *no pericenter*

all pericenters of the other motions, that are usually either bounded motions or motions with a "temporary capture" and with an hyperbolic-elliptic evolution at both ends through the fourth zone, the outer annulus ∞CD∞.

There are also some infinitely rare motions: with parabolic-elliptic original and/or final evolution, with an escape in the past and a capture in the future – or inversely, or even with an *oscillating motion of the first type* i.e with an unbounded motion composed of an infinite number of larger and larger loops in the ∞CD∞ annulus, but with all pericenters in the second annulus ABCD.

Let us recall that all these motions rotate in the positive direction about the Z-axis (the Z-component of the outer momentum is bounded both above and below by positive bounds), and we will have a rather good qualitative idea of the possible motions of the outer body.

10. The wide perturbations of the inner orbit

While the outer orbit has a relatively simple and stable motion with slow and generally small perturbations, the inner orbit can have very large long-term perturbations especially if its inclination is large. Its only element that remains almost always with long-term close bounds is its major axis. For instance in the three-body Sun-Jupiter-Saturn system, the bounds of the Jupiter semi-major axis, even for very long-term motions, are 4.472 AU and 5.216 AU (notice that the upper bound is very close to the present value 5.202 AU; this bound is only given by the integrals of motion, while the lower bound take also account of the theoretical possibilities of escape of Saturn).

An example of the wide perturbations of the inner orbit for large mutual inclination has been computed by John Hadjidemetriou and is presented in Ref. 11, page 103-106. The initial conditions at t = 0 are the following:

$$G = 1; m_1 = m_2 = m_3 = 1$$

$$a_i = 0.5; \quad e_i = 0.2; \quad i_i = \ 83^\circ\ 39'\ 9"; \quad \Omega_i = \omega_i = 0^\circ; M_i = 180^\circ \quad (67)$$

$$a_e = 3; \quad e_e = 0; \quad i_e = 14^\circ\ 5'\ 24"; \quad \Omega_e = 180^\circ; \quad \omega_e + M_e = \ 0^\circ$$

$$\text{j} = \text{mutual inclination} \ = i_i + i_e = 97^\circ 44' 33''; \quad (cosj = -\sqrt{0.01815}) \quad (68)$$

The initial period of the inner orbit is $T_i = \pi/2$ and that of the outer orbit is $T_e = 6\pi$.

The motion has been integrated until the time t = 170 and, during that interval of time, the variations of the elements a_i, a_e, e_e are small:

$0.494 < a_i < 0.501$; $2.94 < a_e < 3.04$; $0 \leq e_e < 0.02$. On the contrary the inner eccentricity e_i begins by a large and regular increase: starting from 0.2 at t = 0 it reaches 0.999 303 at the inner pericenter t = 161.380, after 103.5 revolutions of the binary, and the distance of the two inner bodies is then only 0.000346. The eccentricity e_i will then decrease and come back to the vicinity of 0.2 for t twice larger.

This phenomenon is very general when the inclinations i_i and j are far from 0° and 180°, but the plane motions with Hill type stability undergo only small and slow perturbations.

The first order study (Marchal, 1978) shows, beside the two integrals of motion, the existence of some "quasi-integrals" the secular part of which has only extremely slow variations: the two semi-major axes a_i and a_e the outer eccentricity e_e and the following quantity W:

$$W = (1 - e_i^2).(1 + \sin^2 j) + 5e_i^2 \sin^2 j \sin^2 \omega_i; \quad 0 \leq W \leq 5 \qquad (69)$$

If the ratio a_i/p_e is small, a good rough idea of the long-term evolution can be obtained with these integrals and quasi-integrals and with:

$$de_i/dt = Gm_3[15e_i \sin^2 j \sin 2\omega_i \sqrt{(1 - e_i^2)} + \varepsilon]/8n_i b_e^3 \qquad (70)$$

with, as usual:

n_i = mean angular motion of the inner orbit = $[G(m_1 + m_2) / a_i^3]^{1/2}$

$$b_e = \text{semi-minor axis of the outer orbit } = a_e\sqrt{1 - e^2} \qquad (71)$$

and with:

$$\varepsilon = O[(a_i/p_e) \sup\{(m_1 - m_2)/(m_1 + m_2); [Ma_i/(m_1 + m_2)p_e]^{1/2}\}] \quad (72)$$

We are thus led to the classification of the types of motion for three-body systems with Hill type stability.

There are many types that represent a set of measure zero in phase space: orbits with an original and/or final parabolic escape, orbits with an asymptotic motion to some periodic or quasi-periodic solution, orbits with an *oscillating motion of the first type* that we have already met in Section 9, orbits open in the past and bounded in the future (complete capture) or inversely, motions leading to a collision of the two point-masses of the binary, etc. and there are also three main types, three types that represent sets of positive measure in phase space:

1.) The "fly-by" either simple (only one pericenter of the outer orbit) or with a temporary capture. The evolution is hyperbolic-elliptic at both ends and the third body arrives from infinity and goes back to infinity.

2.) The bounded motions. Almost all of them belong to an Arnold torus of quasi-periodic motions.

3.) The *oscillating orbits of the second type.* They also belong to Arnold tori of quasi-periodic motions, but with a major difference with respect to the bounded motions: they have an infinite number of very close approaches, as close as desired, of the two bodies of the binary and, even if the radius-vectors remains forever bounded, the velocities are unbounded.

This type of orbit is of course related to the large perturbations of the inner eccentricity presented with the example (67). It requires theoretical and practical analyses.

10.1 Theoretical analysis

1.) Motions that do not belong to an Arnold torus of quasi-periodic motions are "chaotic". However for the given values of the integrals of motion the phase space is open and the Arnold tori have an insufficient dimensionality to enclose a part of phase space. On the other hand the measure of the vicinity of collisions is finite (for finite integrals), hence, if the "conjecture of Arnold diffusion" (Marchal, 1990, page 509) is true, almost all chaotic motions belong to the fly-by type. They are "temporary captures" even if such captures can sometimes be extremely long.

2.) A Hamiltonian analysis can help to understand the structure of the set of oscillating orbits of the second type.

Let us use the usual Delaunay's orbital elements L_i, G_i, l_i, g_i, L_e, G_e, l_e, g_e. with:

$L_i = mn_i a_i^2$; where n_i is the inner mean angular motion defined in (71) (73)

$$G_i = L_i\sqrt{1 - e_i^2} = |m.\mathbf{r} \times \mathbf{v}| \tag{74}$$

$$l_i = M_i = \text{inner mean anomaly} \tag{75}$$

$$g_i = \omega_i = \text{inner argument of pericenter} \tag{76}$$

$$L_e = \mathrm{m} n_e a_e^2;\ \text{with } n_e = \text{ outer mean angular motion } = [GM/a_e^3]^{1/2} \tag{77}$$

$$G_e = L_e\sqrt{1 - e_e^2} = |\mathrm{m}.\mathbf{r} \times \mathbf{v}| \tag{78}$$

$$l_e = M_e = \text{ outer (or “exterior”)mean anomaly} \tag{79}$$

$$g_e = \omega_e = \text{ outer argument of pericenter} \tag{80}$$

With these eight Delaunay's elements, and with the z-axis in the direction of the angular momentum c , the four remaining Delaunay's elements H_i, h_i, H_e, h_e are ignorable ($H_i = G_i \cos i_i = [c^2+G_i^2-G_e^2]/2c$; $H_e = G_e \cos i_e = [c^2 + G_e^2 - G_i^2]/2c$; $h_i = \Omega_i$; $h_e = \Omega_e$; $h_e = h_i+180°$).

For given values of the integrals of motion h and c, the Hamiltonian H is equal to the energy integral h and can be expressed in terms of the total angular momentum c and the above eight first Delaunay elements.

$$h = H = H(L_i, G_i, l_i, g_i, L_e, G_e, l_e, g_e, c) \tag{81}$$

This elimination of nodes leads to the usual Hamiltonian equations:

$$\partial H/\partial L_i = dl_i/dt; \partial H/\partial l_i = -L_i/dt; \text{ etc. and } \partial H/\partial c = dh_i/dt = dh_e/dt \tag{82}$$

For given values h and c, we have thus an eight-dimensional phase space of the eight Delaunay's parameters, and the motion remains in the seven-dimensional subspace given by the condition H = h. In that subspace we find infinitely many four-dimensional Arnold tori and a four-dimensional manifold of collisions.

Many Arnold tori intersect the collision manifold along a one-dimensional submanifold and have thus a two-dimensional submanifold of solutions leading to a strict collision. The remainder of these Arnold tori has oscillating orbits of the second type and the measure analysis shows that the corresponding subset has a positive measure (Féjoz,1999, p. 76).

10.2 Practical analysis

In real three-body systems the bodies are not point masses and the oscillating orbits of the second type lead to collisions. Is that phenomenon frequent?

We need first an evaluation of the importance of this type of orbit.

The first-order analysis (Marchal, 1978, page 762) gives the following, with G_{es}being the secular part of G_e.

The domain of oscillating orbits of the second type corresponds to the non-resonant orbits (Arnold tori) that satisfy:

$$|c - G_{es}| \leq [Gmm_3 a_i^2 (5 - W)(3 + 4e_e)/8n_e b_e^2] + \text{ upper order terms} \tag{83}$$

n_e, b_e and W being given in (77), (71) and (69).

Of course that first result is insufficient, and no triple system with a long past can be of the oscillating type of the second type: it would have already led to a collision. But this type of motion is possible in a new born triple system.

How can a new triple system be born? Consider our galaxy, with many single stars but with a majority of Binary systems. When a weak binary meets a strong binary, an ordinary motion of exchange type can easily disrupt the weak binary and form a new born triple system.

What is the probability P of an oscillating motion of the second type for this new born triple system? We can analyse the following conditions: the orbital elements a_i, e_i, a_e, e_e are given (the semi-major axes and the eccentricities) while the angular parameters are arbitrary, with an isotropic repartition of probability.

In these conditions the angular momenta G_i and G_e are also given and the probability P is: If $G_i > 2G_e$ (rare case): $P = 0$

If $G_i < 2G_e$:

$$\begin{aligned} P \approx\ & m_3 n_e (3 + 4e_e) \\ & \times [4G_e^2(6 - e_i^2) + G_i^2(2 + 3e_i^2)]/64 M n_i G_e^2 (1 - e_e^2)^{3/2} (1 - e_i^2)^{1/2} \end{aligned} \tag{84}$$

That probability P is always larger than $9m_3T_i\ /8M\ T_e$ where $(T_i/\ T_e)$ equal to (n_e/n_i) is the ratio of orbital periods.

This indirect way of collision gives a final probability much larger than the probability of direct collision of stars and this allows the possibility that a part of the phenomena of nova is a consequence of star collisions. Of course the explosions of supernova (about 10^{46} joules) cannot be explained by star collisions (about 10^{42} or 10^{43}joules only), but the energy of usual nova explosions has a similar order of magnitude.

Another physical possibility related to the existence of oscillating orbits of the second type is the formation of a large number of very strong binary stars, binary stars almost at contact. Indeed during the close approaches the tidal forces are very strong and the orbital energy of the binary undergoes great losses. This hypothesis can be tested: what proportion of binaries almost at contact belong to a triple system?

11. Conclusions

In spite of a greater simplicity than the general three-body problem, and especially in spite of the absence of motions of exchange type, the three-body systems with Hill type stability have a wide variety of solutions and, when the mutual inclination is large, they undergo large perturbations that can sometimes lead to the collision of the two bodies of the binary.

Nevertheless many information of several types can be obtained. The perturbations of the outer body have well defined limitations, sometimes very narrow. The three main types of motion, fly-by, bounded and *oscillating of the second type* – i.e. with infinitely many close approaches of the binary, approaches as close as desired, – these three main types are very interconnected in phase space. If the conjecure of Arnold diffusion is true, the fly-by type is everywhere dense, even if it has only an extremely small local measure in the regions far from the "escape velocity" and appears there as very long "temporary captures".

These results have far-reaching consequences on the long-term stability of triple stellar systems. The danger of dislocation – the third body escape to infinity – is generally small, and systems with plane motion and Hill type stability have only small and slow perturbations. However the perturbations become very large if the mutual inclination is large and close approaches can easily occurs with their consequences – large tidal effects and loss of energy, danger of collision, etc. These conclusions can be extended to multi stellar systems.

The complexity of all these results give a small taste of the far larger complexity of the general three-body problem with possible escape of any of the three bodies. It would be especially of great interest to look for motions of exchange type and to determine how close they can approach to the Hill type stability.

References

Bozis G. (1976). Zero velocity surfaces for the general planar three-body problem. *Astrophysics and Space Sciences* 43, p. 355-368.

Chen Xiang-Yan, Sun Yi-Sui and Luo Din-Jun (1978). *Acta Astronomica Sinica* 19, p. 119.

Easton R. (1971). Some topology of the three-body problem. *Journal of differential equations* 10, p. 371-377.

El Mabsout B. (1973). Espace des phases dans le problème plan des trois corps. *Comptes-Rendus à l'Académie des Sciences.* A 276, p. 495, also A 278, p 459, 1974.

Féjoz J. (1999). Dynamique séculaire globale dans le problème plan des trois corps et application à l'existence de mouvements quasi-périodiques. *Thèse de Doctorat de l'Université Paris* 13.

Golubev V. G. (1968). *Soviet Phys. Dokl.* 13, p. 273.

Marchal C. (1971). Qualitative study of a n-body system: a new condition of complete scattering. *Astronomy and Astrophysics*, 10, 2, p. 278-289.

Marchal C. (1990). *The three-body problem.* Elsevier Science Publisher B.V.

Marchal C. (1978). Collision of stars by oscillating orbits of the second kind. *Acta Astronautica*, Vol 5, p. 745-764.

Marchal C. and Saari D.G. (1975). Hill's regions for the general three-body problem. *Celestial Mechanics* 12, p. 115.

Smale S. (1970). Topology and Mechanics – II *Inventiones Math.* 11, p. 45-64.

Tung Chin-Chu (1974). *Scienta Sinica* 17, No. 3.

Yoshida J. (1972). *Publication of Astronomical Society of Japan*, 24, 3, p. 391-408.

Zare K. (1976). *Celestial Mechanics* 14, p. 73-83, also 16, p35, 1977.

II

TOOLS FOR INVESTIGATION OF CHAOS AND ITS SOURCES

THE FINE STRUCTURE OF HAMILTONIAN SYSTEMS REVEALED USING THE FAST LYAPUNOV INDICATOR

Claude Froeschlé
Observatoire de Nice, Bv. de l'Observatoire, B.P. 4229, 06304 Nice cedex 4, France

Elena Lega
Observatoire de Nice, Bv. de l'Observatoire, B.P. 4229, 06304 Nice cedex 4, France

1. Introduction

Since the pioneering work of Hénon (1964) the search for chaos indicators, especially for Hamiltonian flows, has known several developments. For systems with two degrees of freedom the basic tool is the visual treatment through the Poincaré section in which dispersed points are hints of chaos. Although this approach has been successfully extended to systems with three degrees of freedom, i.e. by analyzing points on a four dimensional space through stereoscopic views, slice cuttings (Froeschlé 1970,a, Froeschlé 1972), etc., it is clear that it becomes almost impossible for systems with more than three degrees of freedom.

Since the very beginning non graphical methods for the detection of chaos where implemented, taking into account the exponential divergence of nearby chaotic orbits (Froeschlé 1970,b). The introduction, in a comprehensive way, of the Lyapunov Characteristic Exponents (LCEs hereafter) and a method for computing all of them (Benettin et al. 1980) made a major breakthrough for the characterization of chaos. Actually, the largest Lyapunov Exponent was already computed earlier (Froeschlé 1970,b) but was called indicator of stochasticity since the works of the Russian mathematician where not known by the author (Froeschlé 1984).

Whatever the method used for computing the LCEs, we are always faced with the problem that such quantities are defined by a limit of

B.A. Steves et al. (eds.), Chaotic Worlds: from Order to Disorder in Gravitational N-Body Dynamical Systems, 131–165.

time going to infinity. More precisely, we compute the Lyapunov Characteristic Indicators (LCIs hereafter) which are the truncated values of the LCEs for a time T. Moreover, in order to study the dynamical behavior of a given portion of the phase space, it is necessary to compute at least the largest Lyapunov Indicator for a large number of orbits. But this could take a huge amount of time in particular for weakly chaotic orbits.

According to Nekhoroshev (1977) and to Morbidelli and Giorgilli (1995), the old and crucial question of stability of a dynamical system turns out to be related to the structure and density of invariant tori which foliate the phase space. For instance the puzzle of the 2/1 gap of the asteroidal belt distribution was explained showing that the corresponding region of the phase space is a weak chaotic one (Nesvorny and Ferraz-Mello 1997).

For such kind of studies it is interesting to define new methods of analysis which should be as sensitive to weak chaos as the LCIs but cheaper in computational time.

Different methods have been developed either for a rapid computation of the LCIs (Cincotta and Simó 2000) or for detecting the structure of the phase space (chaotic zones, weak chaos, regular resonant motion, invariant tori). Especially for this last purpose we quote the frequency map analysis (Laskar 1990, Laskar et al. 1992, Laskar 1993, Lega and Froeschlé 1996), the sup-map method (Laskar 1994, Froeschlé and Lega 1996), and more recently the fast Lyapunov indicator (hereafter FLI, Froeschlé et al. 1997, Froeschlé et al. 2000) and the Relative Lyapunov Indicator (Sandor et al. 2000). The definitions and comparisons between different methods including a preliminary version of the FLI have been discussed in Froeschlé and Lega (1998, 1999).

In this work we focus the attention on the FLI, because it is easy to implement, cheap in computational time and very sensitive for the detection of weak chaos and particularly for distinguishing between regular resonant orbits and regular non resonant ones.

The paper is organized as follows: in Section 2 and 3 we define the Fast Lyapunov Indicator and give some examples on the 2 dimensional standard map and on a Hamiltonian model. The special case of periodic orbits will be detailed in 4 and thanks to a model of linear elliptic rotation we will be able to recover the structure of the phase space in the vicinity of a noble torus. The use of the FLI for detecting the transition between the stable Nekhoroshev regime to the diffusive Chirikov's one will be recalled in Section 5. In 6 and 7 we will make use of the FLI results for the detection of the Arnold's diffusion.

2. The Fast Lyapunov Indicator revisited

When computing the Lyapunov Characteristics Indicators the attention is focused on the length of time necessary to get a reliable value of their limit, but very little importance has been given to the first part of the computation. Actually, this part was considered as a kind of transitory regime depending, among other factors, on the choice of an initial vector of the tangent manifold.

Already in 1997, Froeschlé et al. have remarked that the intermediate value of the largest LCI (which was called fast Lyapunov Indicator: FLI), taken at equal times for chaotic, even weakly chaotic, and ordered motion, allows to distinguish between them. It turns out, that the FLI allows also to distinguish among ordered motions of different origins, like resonant and non resonant motion (Froeschlé and Lega 2000, Guzzo et al. 2002). We remark that in both cases the LCI tends to zero when t goes to infinity.

2.1 Definition of the FLI

Given a mapping M from R^n to R^n, an initial condition $\vec{x}(0) \in R^n$, and an initial vector $\vec{v}(0) \in R^n$ of norm one, let us define the FLI function $FLI(\vec{x}(0), \vec{v}(0), T)$, T belonging to R^+, as:

$$FLI(\vec{x}(0), \vec{v}(0), T) = \sup_{0<t\leq T} \log ||\vec{v}(t)||, \tag{1}$$

where $\vec{v}(t)$ is given by the system:

$$\begin{cases} \vec{x}(t+1) & = & M\vec{x}(t) \\ \vec{v}(t+1) & = & \frac{\partial M}{\partial \vec{x}}(\vec{x}(t))\, \vec{v}(t). \end{cases} \tag{2}$$

The same definition holds for a continuous flow, where given a set of differential equations:

$$\frac{d}{dt}\vec{X} \quad = \quad \vec{F}(\vec{X}), \quad \vec{X} = (x_1, x_2....x_n) \tag{3}$$

for some suitable regular function $\vec{F} \in R^n$, the evolution $\vec{v}(t)$ of any vector $\vec{v}(0) \in R^n$ is obtained by integrating the linear variational equations:

$$\frac{d\vec{v}}{dt} = (\frac{\partial \vec{F}}{\partial \vec{X}})\vec{v} \quad . \tag{4}$$

2.2 Model problems

The standard map. We consider as a model problem the two dimensional standard map (Froeschlé 1970b, Chirikov 1979, Lichtenberg and

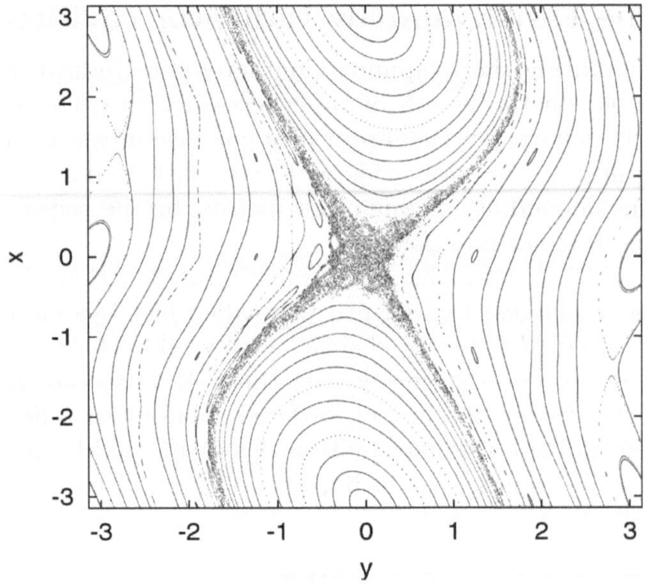

Figure 1. A set of orbits of the standard map (5) for $\varepsilon = 0.7$.

Lieberman 1983):

$$M = \begin{cases} x(t+1) = x(t) + \varepsilon \sin(x(t) + y(t)) & (\text{mod} 2\pi) \\ y(t+1) = y(t) + x(t) & (\text{mod} 2\pi) \end{cases} \quad (5)$$

Figure 1 displays orbits of the standard map of equation 5 for $\varepsilon = 0.7$.

For this value of the perturbing parameter a lot of orbits are still invariant tori. Some resonant curves are displayed surrounding some elliptic points and a chaotic, though well confined, zone is generated by the existence of the hyperbolic point at the origin.

Figure 2,a shows the variation of the FLI with time for 4 different kinds of orbits. The upper curve, with initial conditions $(10^{-4}, 0)$ in the chaotic zone just described, shows an exponential variation of the FLI with time. The upper value of 20 is a computational threshold that allows to avoid floating overflow.

The second curve corresponds to a weak chaotic orbit of initial conditions $(0.4839, 0)$ while the third one corresponds to a regular invariant torus of initial conditions $(2.5, 0)$ and the lowest one to a resonant curve of initial conditions $(1.2, 0)$.

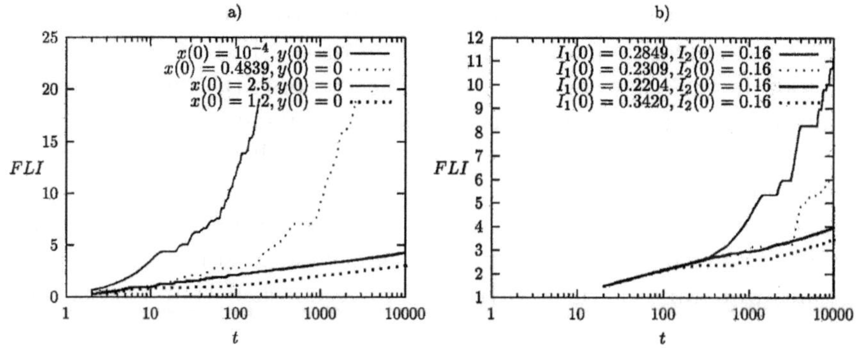

Figure 2. Variation of the Fast Lyapunov Indicator with time. a) Orbits of the standard map of equation (5) with $\varepsilon = 0.7$. The upper curve is for a chaotic orbit with initial conditions $x(0) = 10^{-4}, y(0) = 0$, the second one is for a weak chaotic orbit with $x(0) = 0.4839, y(0) = 0$, the third one is for a non resonant orbit with $x(0) = 2.5, y(0) = 0$ and the lowest one is for a resonant orbit with $x(0) = 1.2, y(0) = 0$. b) Orbits of Hamiltonian (6) with $\varepsilon = 0.004$. The upper curve is for a chaotic orbit with initial conditions $I_1(0) = 0.2849$, the second one is for a weak chaotic orbit with $I_1(0) = 0.2309$, the third one is for a non resonant orbit with $I_1(0) = 0.2204$ and the lowest one is for a resonant orbit with $I_1(0) = 0.342$. The other initial conditions are $I_2(0) = 0.16$, $I_3(0) = 1$, $\varphi_1(0) = \varphi_2(0) = \varphi_3(0) = 0$.

The Hamiltonian model. We consider, as we did in previous papers (Froeschlé et al. 2000 , Guzzo et al. 2002 , Lega et al. 2003), the following Hamiltonian system:

$$H = \frac{I_1^2}{2} + \frac{I_2^2}{2} + I_3 + \varepsilon \frac{1}{\cos(\varphi_1) + \cos(\varphi_2) + \cos(\varphi_3) + 4} \tag{6}$$

where $I_1, I_2, I_3 \in R$ and $\varphi_1, \varphi_2, \varphi_3 \in T$ are canonically conjugate and ε is a small parameter. Figure 2,b shows the variation of the FLI with time for 4 orbits of Hamiltonian (6) with $\varepsilon = 0.004$. As well as for the discrete case, the method quickly allows to reveal chaotic motion ($I_1(0) = 0.2849$), weakly chaotic motion ($I_1(0) = 0.2309$) and to distinguish between tori ($I_1(0) = 0.2204$) and regular resonant orbits ($I_1(0) = 0.2204$).

2.3 Computation of the FLI-map

The discrete case. We have computed the FLI for a set of 1000 initial conditions, of the mapping of equation 5, regularly spaced on the

x-axis in the interval $[0, \pi]$. For each orbit $y(0) = 0$. Figure 3,a shows what we call the FLI-map, i.e. the value of the FLI after T iterations ($T = 1000$ in this case) against $x(0)$. The vectors $\vec{v}(0)$ have always been taken in the direction of the x-axis. A lot of orbits appear to have a value of FLI very close to $\log T = 3$. Actually, they all appear to be regular invariant tori.

Values slightly greater than $\log T$ indicate either very thin chaotic layers or invariant tori close to very thin chaotic zones. By making a zoom around some of these orbits we have checked that this is indeed the case.

The orbits having an FLI value lower than $\log T$ correspond to chains of islands.

The continuous case. In Figure 3,b the FLI map is represented for a set of 1000 initial actions $I_1(0)$ regularly spaced in the interval $[0.2, 0.5]$ for the Hamiltonian of equation 6 with $\varepsilon = 0.004$. The other initial conditions are: $I_2(0) = 0.16$, $I_3(0) = 1$ and $\varphi_1(0) = \varphi_2(0) = \varphi_3(0) = 0$. The initial vectors have components: $\vec{v}(0) = (1, 0.5(\sqrt{3} - 1), 1, 1, 1, 1)$. The three different kind of dynamics: regular tori (values of the FLI$\simeq$ $\log T$, regular resonant orbits (having FLI lower than $\log T$) and chaotic motions (FLI greater than $\log T$) are differenciated as well as for the discrete case.

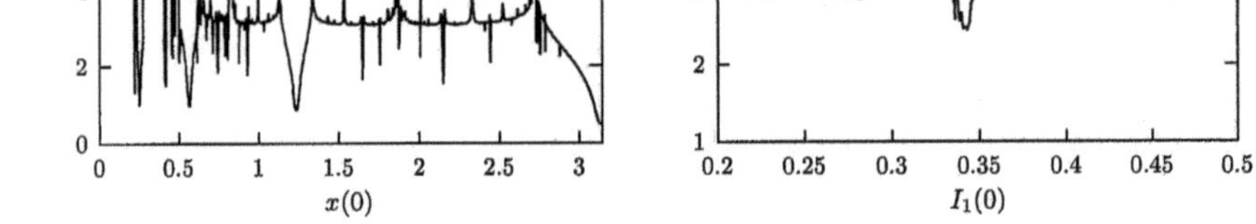

Figure 3. Variation of the FLI as a function of the initial action for a set of 1000 regularly spaced orbits. a) The case of the mapping for $\varepsilon = 0.7$. The initial actions are in the interval $[0, \pi]$ while $y(0) = 0$. The initial vectors $\vec{v}(0)$ have components $(1, 0)$, i.e. they are almost perpendicular to the corresponding invariant curves. b) The Hamiltonian case of equation 6 for $\varepsilon = 0.004$. The initial actions $I_1(0)$ are in the interval $[0.2, 0.5]$ while the other initial conditions are: $I_2(0) = 0.16$, $I_3(0) = 1$ and $\varphi_1(0) = \varphi_2(0) = \varphi_3(0) = 0$. The initial vectors have components: $\vec{v}(0) = (1, 0.5(\sqrt{3} - 1), 1, 1, 1, 1)$.

2.4 The FLI-map: a powerful tool to discriminate among regular resonant and non resonant orbits

Let us recall that the largest Lyapunov Exponent is defined as the limit for t going to infinity of $\ln ||\vec{v}(t)||/t$, where $\vec{v}(t)$ is obtained integrating (4), and $\vec{v}(0)$ is a generic vector $\vec{v}(0) \in R^n$ not belonging to some lower dimensional linear space. If the mapping is a symplectic one or, in the continuous case, if equation (3) is Hamiltonian and if the motion is regular (except for some peculiar hyperbolic structures, such as whiskered tori) then the largest Lyapunov Exponent is zero, otherwise it is positive. This property has been largely used to discriminate between chaotic and ordered motion. However, among regular motions the Lyapunov exponent does not distinguish between KAM tori and resonant islands. In contrast, the FLI, as shown in Figure 2 distinguishes between resonant and non resonant regular motions. An analytic explanation of the above result for the Hamiltonian case can be found in (Guzzo et al. 2002) while for mappings a discussion is provided in (Froeschlé and Lega 2000). Therefore, without entering in the details, in the present paper we will only point out that the FLI growths linearly with time for both kinds of regular orbits but with a different coefficient which turns out to be lower for resonant regular motion than for tori. Moreover, such a coefficient is the same for all tori while it changes smoothly for resonant regular orbits. Such a peculiar behaviour of the FLI allows us to obtain very accurate charts of the geometry of resonances.

3. The geometry of resonances revealed by the FLIs

In previous papers (Froeschlé et al. 2000, Guzzo et al. 2002, Lega et al. 2002) we used the FLI to describe the geometry of the resonances, integrating orbits of the Hamiltonian system of equation 6 and of the following 4-dimensional symplectic map:

$$T_1 = \left\{ \begin{array}{lcll} x(j+1) & = & x(j) + \varepsilon_1 \sin(x(j) + y(j)) + & \\ & & b\{\sin[x(j) + y(j) + z(j) + t(j)] + & \\ & & \sin[x(j) + y(j) - z(j) - t(j)]\} & (\mathrm{mod} 2\pi) \\ y(j+1) & = & y(j) + x(j) & (\mathrm{mod} 2\pi) \\ z(j+1) & = & z(j) + \varepsilon_2 \sin(z(j) + t(j)) + & \\ & & b\{\sin[x(j) + y(j) + z(j) + t(j)] - & \\ & & \sin[x(j) + y(j) - z(j) - t(j)]\} & (\mathrm{mod} 2\pi) \\ t(j+1) & = & z(j) + t(j) & (\mathrm{mod} 2\pi) \end{array} \right. \qquad (7)$$

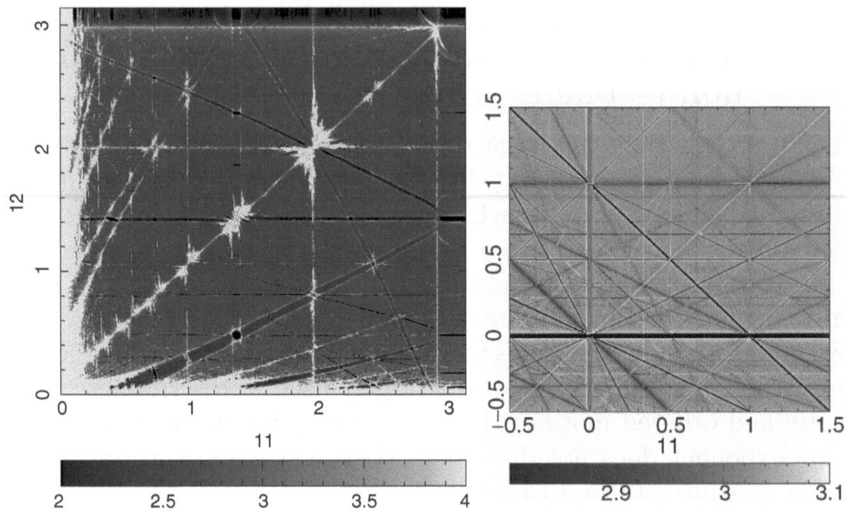

Figure 4. left) Geography of resonances in the plane $x - z$ for the mapping T_1 with $\varepsilon_1 = 0.4$, $\varepsilon_2 = 0.3$ and $b = 0.001$. The computation has been done for a set of 500×500 initial conditions regularly spaced on x, z. The other initial conditions are $y = 0$, $t = 0$, and the initial vectors are $\vec{v} = (1, 0.5(\sqrt{3}-1), 1, 1)$. The grey scale range from black ($FLI < 2$) to white ($FLI > 4$). right) Same as (left) for the Hamiltonian system (equation 6) with $\varepsilon = 0.001$. The grid is made of 500×500 initial conditions regularly spaced on I_1, I_2. The other initial conditions are $I_3 = 1$, $\varphi_1 = \varphi_2 = \varphi_3 = 0$, and the initial vectors are $\vec{v} = (1, 0.5(\sqrt{3} - 1), 1, 1, 1, 1)$. The grey scale range from black ($FLI < 2.8$) to white ($FLI > 3.1$).

Figure 4, (left) shows, at $T = 1000$, the FLI for a grid of 500×500 initial conditions regularly spaced on x, z. The parameters of the mappping are $\varepsilon_1 = 0.3$, $\varepsilon_2 = 0.4$ and $b = 0.001$. The other initial conditions are $y = 0$, $t = 0$, and the initial vectors are $\vec{v} = (1, 0.5(\sqrt{3} - 1), 1, 1)$. The FLI is reported with a grey scale: the black lines correspond to regular resonant motions while the white lines represent both chaotic resonant motions or the vicinity of a separatrix. The orbits having an FLI value of about $\log(T)$ constitute the background of KAM tori. Let us remark that for Hamiltonian (6), when $\varepsilon = 0$ the frequencies are strictly equal to the actions, therefore the FLI charts are nothing but frequency charts, which display a very well detailed Arnold's web.

As far as the mapping is concerned, the same occurs when considering the variables $x(j)$ and $z(j)$ which again are the frequencies of the unperturbed mapping T_1 ($\varepsilon_1 = \varepsilon_2 = b = 0$). The FLI chart in the plane $x - z$ shows clearly the Arnold's web. When comparing Figure 4 (left) to Figure 4 (right), obtained at $T = 1000$ for the Hamiltonian (equation 6) with $\varepsilon = 0.001$, less resonant lines appear in the case of the mapping. In fact the perturbing function of the mapping has only two harmonics at order b while for the Hamiltonian we have chosen a perturbing function having a full Fourier spectrum at order ε.

4. Periodic orbits and the structure of the neighborhood of noble tori

4.1 The particular behavior of the FLI for periodic orbits

Let us consider the elliptic point $(1.239, 0)$ of the standard map (equation 5) for $\varepsilon = 0.7$, which corresponds to a periodic orbit of order 4. We have computed the evolution of the FLI with time for ten orbits regularly spaced on the x–axis in the interval $[1.239, 1.339]$ starting from the periodic point.

We observe (Figure 5) the expected linear increase of the FLI with time for all the orbits except the periodic one. For such orbit the FLI after a transitory linear increase becomes constant. Our aim is to explain such a behavior but before we study the relation between the FLI values and the order of periodic orbits. At this purpose we have computed the FLI as a function of time for a particular set of periodic orbits, the Fibonacci sequence.

Let us recall that the Fibonacci sequence is the set of the successive rational ratios p_k/q_k, $k = 1, ...\infty$, obtained when developing the golden

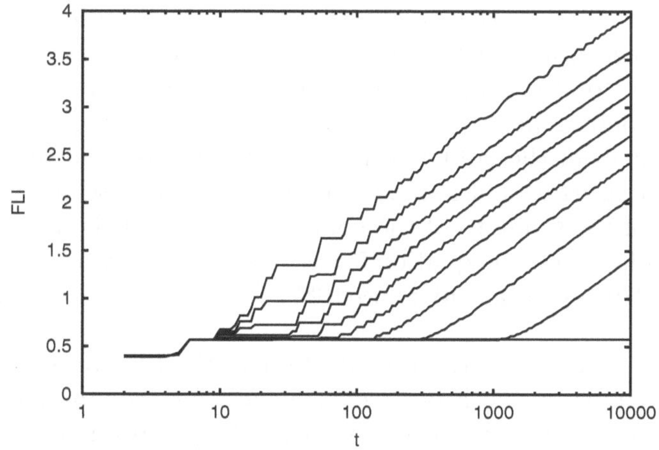

Figure 5. Variation of the FLI with time for 10 orbits of the standard map with $\varepsilon = 0.7$. The initial conditions are regularly spaced in the interval $1.239 < x(0) < 1.339$, $y(0) = 0$.

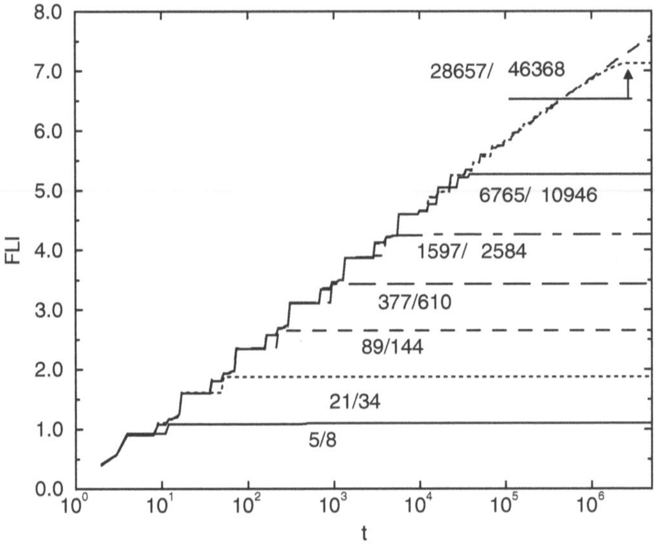

Figure 6. Variation with time of the FLI for a set of periodic orbits belonging to the Fibonacci sequence for the standard map with $\varepsilon = 0.9715$. The rational ratios p_k/q_k, $k = 1, ...7$, are written in the figure near the corresponding FLI curve. The FLI curve which grows linearly with time all over the interval $0 < t < 5\,10^6$ iterations is obtained for the golden torus.

number through the continued fraction process and therefore is the sequence which best approximates[1] the golden number.

As explained in Lega and Froeschlé (1996), because of the smallness of the denominators the amplitude of the perturbation associated to the Fibonacci terms is large. This makes the sequence a good candidate with a view to studying regular resonant motion, and in particular of the corresponding periodic orbits, up to a large order q_k. Such terms are also further from the golden number than any other set $\tilde{p}_k/\tilde{q}_k$ from its corresponding irrational number (see Lega and Froeschlé 1996). This is the origin of the strength of the golden torus, i.e. of the last KAM torus of the standard map of equation 5 which is broken down for $\varepsilon = 0.971635$ (Greene 1979, Olivera and Simó 1987, Laskar et al. 1992 and Mackay 1993).

Figure 6 shows the variation with time of the FLI for a set of periodic orbits belonging to the Fibonacci sequence. for the standard map with $\varepsilon = 0.9715$. For this value of the perturbing parameter the majority

[1]We say that p/q is a best approximation of x if $|qx - p| < |q'x - p'|$ for all p', q' such that $0 < q' \leq q$ and $p'/q' \neq p/q$.

of tori have disappeared and the phase space shows a rich structure of chains of islands and chaotic orbits. As obtained in Figure 5, the FLI of periodic orbits after a transitory linear increase (which is very short for the periodic orbit of order 4 of Figure 5) becomes constant, i.e. the norm of $\vec{v}(t)$ reaches some maximum value.

Let us consider the trivial case of the two dimensional standard map with $\varepsilon = 0$. In this case the norm of the vector $\vec{v}$ is equal to

$$||(\vec{v}(t)|| = \sqrt{t^2 v_x(0)^2 + v_y(0)^2} \quad (8)$$

independently on the initial conditions, i.e. for a given initial vector of components $\vec{v}(0) = (v_x(0), v_y(0))$ the FLI is the same for both a periodic and a quasi-periodic orbit. The difference with the perturbed case is that in this last case a periodic orbit is surrounded by quasi-elliptic regular resonant orbits. We think that the reason why the FLI becomes constant for the periodic orbits in the perturbed case must be found exactly on the fact that the orbits close to the periodic ones are well represented by an elliptic rotation. Therefore, we will introduce below a simple area-preserving mapping of linear elliptic rotation.

In Figure 6 the FLI curve which grows linearly with time all over the interval $0 < t < 5\,10^6$ iterations is obtained for the golden torus.

It is interesting to observe that, during the transitory linear increase, the FLI values obtained for the whole set of Fibonacci terms coincide with those of the golden torus. In fact, the ellipses surrounding the periodic orbits are more and more elongated for increasing order q_k, they become almost tangent to the near invariant torus, while the chaotic zone surrounding the islands shrinks to zero. Surprisingly, this argument based on continuity seems to hold also for lower order periodic orbits, although the chaotic zone surrounding the libration islands is large and probably a very small number of tori remains.

We can observe in Figure 6 that both the duration of the transitory linear increase of the FLI and its maximum and constant value seem to be related to the order q_k of the periodic orbit.

In what follows we introduce a simple model in order to explain the observed features: the linear increase of the FLI, the constancy after a transient interval of time and the dependence on the order q_k.

4.2 A simple model based on elliptic rotation

A small libration island close to the periodic orbit can be interpreted, through linearization, by an ellipse of ratio $\gamma = a/b$ between the semi-major axis a and the semi-minor axis b. Since there is no differential rotation in the case of periodic orbits, we consider a model without differential rotation. Let us define the following area-preserving mapping:

$$N = \begin{cases} x(t+1) &= x(t)\cos(\alpha) - \gamma y(t)\sin(\alpha) \\ y(t+1) &= \frac{1}{\gamma}x(t)\sin(\alpha) + y(t)\cos(\alpha) \end{cases} \tag{9}$$

where α is the mean rotation angle of the radius vector of the ellipse. In fact, since this mapping is area preserving, for a given number of iterations the density of the points will be higher for the regions further from the center and the angle α turns out to be a mean rotation angle as it is for the islands of the standard mapping.

Through elementary computations it appears that the elliptic mapping can be written as:

$$N = \begin{cases} x(t+1) &= x(0)\cos(t\alpha) - \gamma y(0)\sin(t\alpha) \\ y(t+1) &= \frac{1}{\gamma}x(0)\sin(t\alpha) + y(0)\cos(t\alpha) \end{cases} \tag{10}$$

and the tangent mapping is the same, since the tangent mapping of a linear mapping is the mapping itself:

$$\begin{cases} v_x(t) &= v_x(0)\cos(t\alpha) - \gamma v_y(0)\sin(t\alpha) \\ v_y(t) &= \frac{1}{\gamma}v_x(0)\sin(t\alpha) + v_y(0)\cos(t\alpha) \end{cases} \tag{11}$$

When $\gamma >> 1$, as it is in our study of the Fibonacci sequence, then the FLI, i.e. the supremum of the norm of v behaves like $\gamma v_y(0)$, i.e. like a since $v_y(0) \simeq b$. We remark that if $v_x(0) = 0$ then $v_y(0) = b$. A question remains about the transitory regime in which the FLI grows linearly with time. Actually, in order to reach its maximum value, the vector $\vec{v}(t)$ has to "visit" the ellipse from the the semi-minor to the semi-major axis, i.e. has to rotate at least an angle of 90 degrees. We were able to reproduce very well the evolution of the FLI with time of the seven Fibonacci periodic orbits (Lega and Froeschlé 2000), confirming the validity of the simple model for explaining the behavior of the FLI for periodic orbits.

4.3 Applications to the topology of the neighborhood of noble tori: an illustration of the Morbidelli-Giorgilli theorem

A theorem demonstrated by Morbidelli and Giorgilli (1995 a,b), using both the results of KAM and Nekhoroshev theorem, gives to KAM tori a special condition of chief "tori". More precisely, the theorem says that a torus found by the KAM construction, i.e. having a large diophantine constant is surrounded by a set of tori with low diophantine constant, let's say of "slave" tori, and that the diffusion in this region is super exponentially slow.

Using the frequency map analysis we visualized (Lega and Froeschlé 1996) the predicted result on the topology of the neighborhood of noble tori for values of the perturbing parameter well above the ones allowed by the mathematical demonstration. Moreover, we have measured the size of the complementary set of tori, showing that the size of islands and chaotic zones decreases exponentially when the distance to the noble torus goes to zero.

The same results can be obtained with the FLI, and in particular the size of the islands can be obtained in an easier and quicker way by the simple computation of the FLI for the periodic orbits. Figure 7 shows the FLI map as a function of the logarithmic distance to the golden torus. Figure 7,a has been obtained computing the FLI for only $t = 1000$ iterations. Far from the golden torus only chaos and islands appear, while for a distance lower than 10^{-5} from the golden torus the phase space seems completely filled by tori. Of course weak chaos and small islands can exist in this region and it is necessary to explore the set of orbits for a longer time. Figure 7,b is an enlargement of the box plotted in Figure 7,a obtained computing the FLI for $t = 10\,000$ iterations. We obtain a picture quite similar to the previous one, with the region filled by tori which shrinks to distances lower that $10^{-6.5}$. By the same procedure we compute the FLI on a time $t = 10^5$. Again we obtain a similar picture (Figure 7,c) like in a sort of renormalization process. The last enlargement, obtained for $t = 5\,10^6$ iterations shows a very different regime: the phase space is almost completely filled by invariant tori and only a very small island and a small chaotic region are detected. Up to a distance $d = 10^{-7.6}$ the "slave" tori protect the golden "chief" torus.

Measure of the size of the islands. In the light of the results obtained with the simple model, we can test if by computing the FLI for periodic orbits we can recover the geometric parameter $\gamma = a/b$.

Qualitative aspect

Figure 8 shows the relation between the FLI and the order q of the periodic orbits. The FLI behaves like $\log(q)$ up to the order $q = 610$, i.e. for the first four periodic orbits. Then a sort of change of regime occurs, i.e. the FLI increases exponentially with the order. In the hypothesis that the semi-major axis a of the ellipses behaves like π/q and that the FLI values are of the order of $\log(a/b)$, then the exponential increase

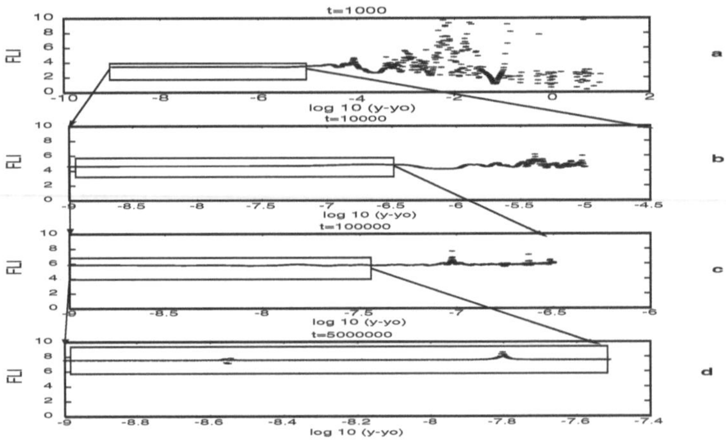

Figure 7. FLI-map as a function of the logarithmic distance to the golden torus for the standard map with $\varepsilon = 0.9715$ From top to bottom: enlargement of the box of the previous figure analyzed on a longer time as indicated on each figure.

corresponds to an exponential decrease of the size b of the islands as predicted by the Morbidelli and Giorgilli theorem. In fact the results obtained with the FLI agree, at least qualitatively (Figure 8), with those obtained when b was computed through the frequency analysis in Lega and Froeschlé 1996 and a is considered equal to π/q. We have also computed the ratio a/b using the linearization of the mapping M^q, i.e. by a direct computation and use of the elliptic rotation, as explained in appendix A and in Locatelli et al. 2000. Hereafter, we will call this method the linearized mapping approximation (LMA).

Again, we have a qualitative good agreement with the other two methods, but in the case of the LMA we expect also to have a quantitative agreement. Actually this method is based on the linear approximation as well as the model of elliptic rotation introduced for explaining the FLI for periodic orbits.

Quantitative aspect

When computing the FLI we consider the supremum of the norm of the vector v on a chain of islands which have different geometries, i.e. different values of a/b. Moreover, as shown in the previous section about the elliptic model, the FLI gives $\gamma v_y(0)$. In order to have a quantitative agreement with the LMA we need to compute a local FLI, i.e. the supremum of the norm of the mapping M^q for the same initial conditions used for the LMA. Then we have to correct the value $\gamma v_y(0)$, i.e. we need to find γ independently on the initial vector $v(0)$.

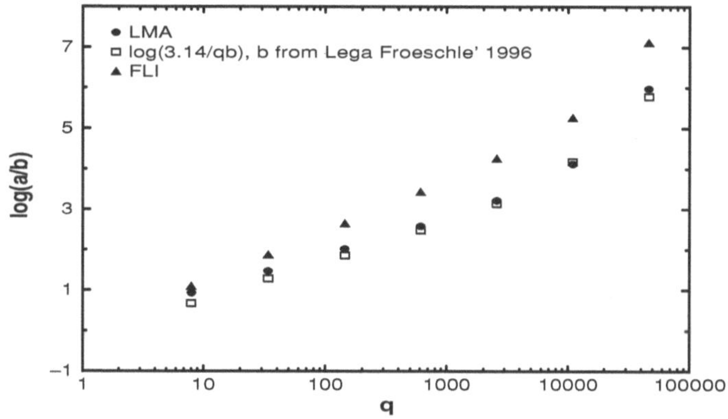

Figure 8. Variation of the quantity $\log(a/b)$ versus the order for the set of Fibonacci orbits of the standard map with $\varepsilon = 0.9715$ analyzed in Figure 6. The points labeled with a triangle stand for the $\log(a/b)$ obtained by the FLI computation. The points labeled by a square correspond to the $\log(a/b)$, with $a = \pi/q$ and b measured with the frequency-map analysis by Lega and Froeschlé 1996. The points labeled by a circle correspond to the LMA (explained in appendix A and in Locatelli et al. 2000).

Going back to the elliptical model, let us remark that the minimum value of $||v(t)||$ is approximatively $v_y(0)$ and that $v_y(0)$ is a good approximation of b in the case $\gamma >> 1$. Therefore, in order to have a good estimate of $\log(a/b)$ we computed the quantity $\tilde{\gamma} = \log(\sup ||v(t)||) - \log(\inf ||v(t)||)$. As shown in Figure 9,a in the case of the elliptic model we obtain $\tilde{\gamma} = \gamma$ even starting with a vector almost perpendicular to the y direction. In the case of the Fibonacci sequence (Figure 9,b,c), the test is made on the periodic orbit of order $q = 144$. The local FLI changes for different choices of $v(0)$, while the quantity $\tilde{\gamma}$ is independent on the orientation of the initial vector. Let us remark that our usual choice $v(0) = (1,0)$ for the Fibonacci sequence. with initial conditions selected on the $y(0) = 0$ axes is a quite good approximation of the direction of the semi-minor axes b of the ellipses surrounding the periodic orbits and therefore $\tilde{\gamma} \simeq \log(\sup ||v(t)||)$.

Figure 10 shows that the values of $\tilde{\gamma}$ computed for the Fibonacci sequence are in a very good agreement with the values obtained with the LMA method.

Concerning the estimate of $log(\gamma)$ through the method of frequency analysis we have a good quantitative agreement (Figure 10) although we introduced the approximation $a = \pi/q$. Moreover, let us remark that we did not compute exactly the semi-minor axes b but a conjugated diameter with respect to the x direction. It appears from the good quantitative

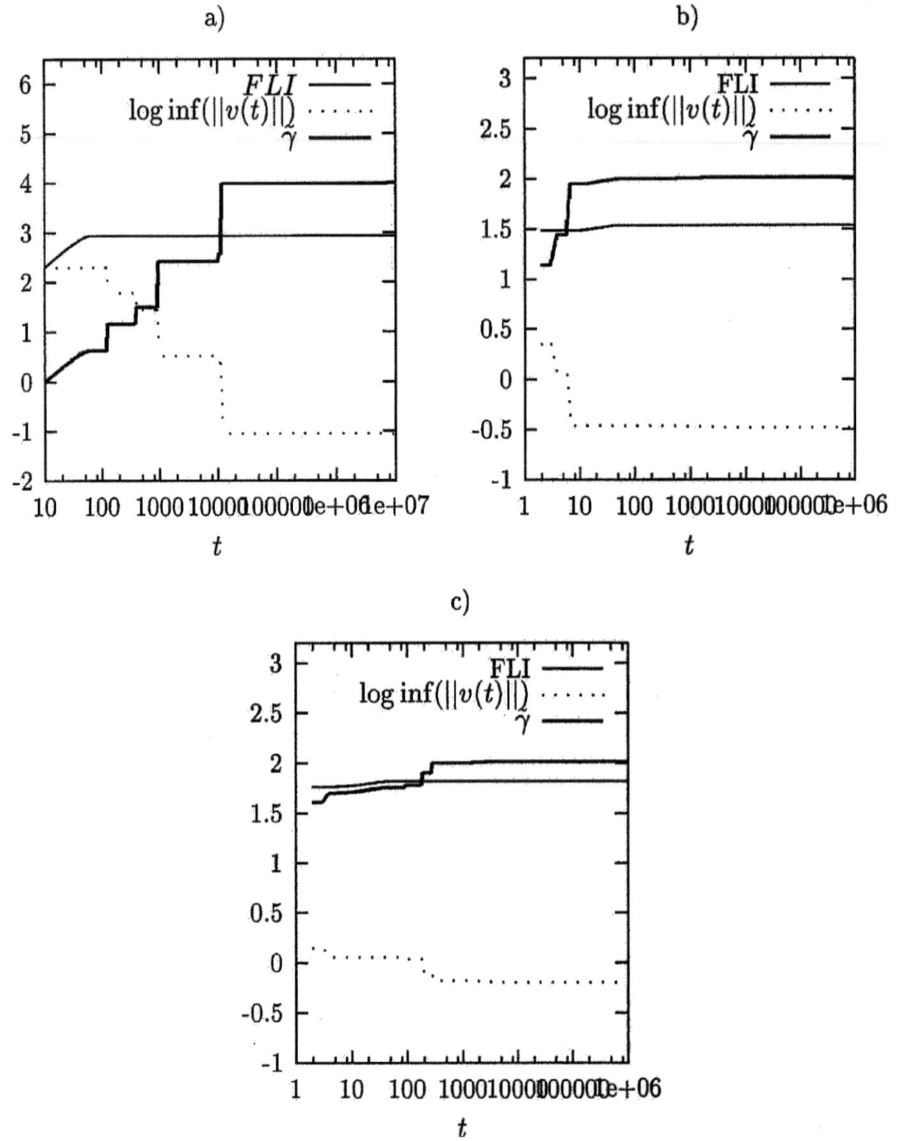

Figure 9. Variation with time of the local FLI, of $\log(\inf ||v(t)||)$ and of $\log(\tilde{\gamma})$. a) For the elliptic model with $\gamma = 10000$ and an initial vector $v(0) = (\cos\beta, \sin\beta)$ with $\beta = 5$ degrees. The FLI reaches exactly the value $\log(\gamma v_y(0)) = 2.94$, while $\log(\tilde{\gamma})$ after $t = 10000$ gives exactly $\log\gamma$. b),c) For the Fibonacci periodic orbit of order $q = 144$ with initial condition $x(0) = 4.1769873213140279$, $y(0) = 0$. The initial vector is $v(0) = (\cos\beta, \sin\beta)$ with $\beta = 88$ degrees in b) and $\beta = 0$ degrees in c).

agreement of results and from the previous study on the Fibonacci of order $q = 144$ (Figure 9b,c) that the x direction is a good approximation of the semi minor axis direction.

5. From Nekhoroshev to Chirikov

As we said in the introduction, the crucial question of stability of a dynamical system is related to the structure and density of invariant tori which foliate the phase space. This is in fact the geometrical representation of the Nekhoroshev theorem (1977). We recall that in a quasi-integrable system with Hamiltonian:

$$H_\varepsilon(I, \varphi) = h(I) + \varepsilon f(I, \varphi), \tag{12}$$

non degenerate and convex, an eventual instability of the actions obeys the following exponential law:

$$\|I(t) - I(0)\| \leq I_0\, \varepsilon^a \qquad \text{for} \quad |t| \leq t_0\, e^{(\varepsilon_0/\varepsilon)^b}, \tag{13}$$

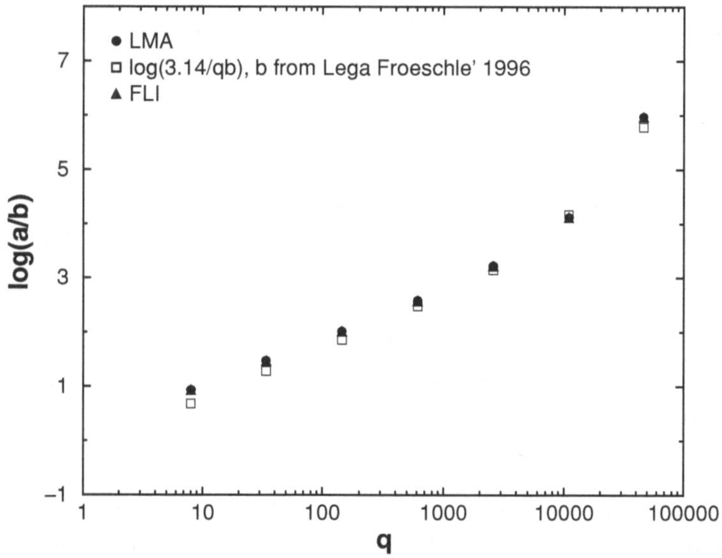

Figure 10. Same as Figure 8 but the FLIs values are replaced by the computation of the quantity $\tilde{\gamma}$, i.e. of a local corrected FLI as explained in the text.

where $I_0, t_0, \varepsilon_0, a, b$ are suitable positive constants. We will say that a system is in a Nekhoroshev regime when the diffusion of the action is exponentially slow or equivalently when the phase space is dense of tori.

For large values of the perturbation parameter, this description fails and diffusion becomes a rapid phenomenon thanks to the well known Chirikov (1960) criterium of overlapping of resonances.The transition from the Nekhoroshev's to the Chirikov's regime is a very delicate subject since it is a smooth process involving firstly the superposition of small resonances and then going from a slow Nekhoroshev diffusion to a slow Chirikov one. Therefore, such a slow diffusion is not easily recognized using even very long numerical integrations looking at the variations of the actions. It is very difficult to distinguish between the two regimes with purely analytic tools although many improvements in this direction have been obtained (Celletti and Chierchia 1995, Celletti and Chierchia 1997, Celletti et al. 2000, Locatelli and Giorgilli 2000). Therefore we made use of numerical tools to investigate the problem in an indirect way, i.e. by detecting the structures of the phase space, i.e. the geometry of resonances (Froeschlé et al. 2000, Froeschlé et al. 2001, Guzzo et al. 2002)

6. Arnold's diffusion

The possibility of the drift of the actions along a resonance was shown by Arnold (1964) using an ad hoc quasi-integrable Hamiltonian model. Up to now similar results have been proved for classes of specific quasi-integrable systems (see for example Bessi et al. 2001), but no general result exists yet. Moreover, as we recalled in the previous section, this kind of diffusion, related to the Nekhoroshev regime, is exponentially slow and therefore very difficult to detect numerically.

We have tried to measure it (Lega et al. 2003) taking profit of the previous results, namely:

1 we used the Hamiltonian of equation 6 which has a full Fourier spectrum, i.e. which has all harmonics already at order ε and a symplectic 4 dimensional mapping.

2 we worked within a given low order resonance. In fact, as explained in Nekhoroshev (1977), Morbidelli and Giorgilli (1995, a,b), the speed of an eventual diffusion decreases as the order of the resonance increases. We recall that for a resonance defined as a neighborhood of the manifold: $\sum_i k_i \omega_i(I) = 0$, with some integers $(k_1, \ldots, k_n) \in Z^n \backslash 0$ and $\omega_j = \partial H_0 / \partial I_j$ its order is: $|k| = \sum_i |k_i|$.

3 we chose initial data in the hyperbolic border of the resonance taking profit of the detailed representation of the geometry of resonances given by the computation of the FLI. The evolution of the orbits is also reported on the FLI pictures, where we can appreciate even small drifts along the chosen resonance because the section I_1, I_2 (x-z for the mapping) eliminates the large projection effects which could mask them.

4 thanks to previous results on the detection of the transition between Nekhoroshev and Chirikov (Guzzo et al. 2002) we chose the perturbing parameter sufficiently large for having measurable diffusion but still sufficiently small to be in the Nekhoroshev regime.

7. Diffusion along resonances

7.1 Qualitative aspects

Both for the Hamiltonian system (equation 6) and for the 4 dimensional symplectic mapping (equation 7) we have computed the FLI charts for different values of the perturbing parameter and we have selected a low order resonance. In the case of the Hamiltonian system we have considered the unperturbed resonance $I_1 = 2I_2$, while for the mapping we have chosen $z = -x/2$.

For Hamiltonian (6) the critical value for the transition between the Nekhoroshev and the Chirikov regime was found in the interval $0.03 < \varepsilon_0 < 0.032$ (Guzzo et al. 2002) . Starting from an upper bound of $\varepsilon = 0.022$, we have looked for diffusive orbits in the Nekhoroshev regime.

Figure 11 shows the FLI charts of the actions space around $I_1 = 0.3$, $I_2 = 0.15$ for different values of ε. In these pictures the region between the two white lines is the resonance associated to $I_1 - 2I_2 = 0$, and the two white lines correspond to its hyperbolic border where diffusion is confined. These charts provide us the possibility of choosing initial conditions in the hyperbolic border. Taking the double section $\sigma = |\varphi_1| + |\varphi_2| \leq 0.05$, $\varphi_3 = 0$ of the solution, we can follow its evolution on the FLI chart. Let us remark, that in such a way we minimize all projection effects and fast quasi-periodic movements. What remains is a very slow drift along the border of the resonance.

We have taken a set of 100 initial conditions in the small interval $0.303 \leq I_1 \leq 0.304$, $0.143 \leq I_2 \leq 0.144$ corresponding to orbits of the FLI charts having FLI values larger than $1.2\log(T)$, i.e. corresponding to chaotic orbits at the border of the resonance, far from the more stable crossing with other resonances. We have plotted on the FLI charts all the points in the double section described above. We remark that such points will appear on both side of the resonance (in fact the two white lines are connected by a hyperbolic region in the six dimensional phase space).

For $\varepsilon = 0.02$, Figure 11(top, left) we have plotted the evolution of the 100 orbits up to $t = 9\,10^3$ and up to $t = 5\,10^5$ (top, right). The phenomenon is already visible, although the higher order resonances intersecting the main one become evident, and consequently the region of diffusion extends a little also in the direction transversal to the resonance.

When decreasing the perturbation, $\varepsilon = 0.007$, the diffusion along the resonant line appears very clearly. In Figure 11 (middle) we have plotted again the intersections of the 100 orbits with the double section $\sigma = |\varphi_1| + |\varphi_2| \leq 0.05$, $\varphi_3 = 0$ on the FLI chart after respectively $t = 1.5\,10^6$ (middle, left) and $t = 2.3\,10^7$ (middle, right). The phenomenon still appears very clearly even for $\varepsilon = 0.0025$, Figure 11 (bottom), i.e. an order of magnitude lower than the threshold of transition between the two regimes, only the speed of the fastest orbits has decreased from about 10^{-9}, for $\varepsilon = 0.007$ to10^{-11} for $\varepsilon = 0.0025$.

We have observed the phenomenon down to $\varepsilon = 0.001$. According to Nekhoroshev theorem, the actions are stable for a time which is exponentially long with the inverse of the perturbation parameter. For such a low value of ε, we have detected a very slow diffusion after $t = 2\,10^9$, while for $\varepsilon = 0.002$ we had to wait "only" for a time of $t = 2\,10^8$: we are evidently in the presence of an exponential law.

Concerning the mapping, we had obtained a local threshold b^* of transition in the interval $0.03 < b^* < 0.05$. As for the Hamiltonian case, we have integrated a set of 100 chaotic initial conditions, satisfying the condition $FLI(T) > 1.2\log(T)$, taken in the resonant line $z = -x/2$ far from the crossing of resonances. The plot of the intersections with the double section $|y| + |t| < 0.005$ on the FLI charts allows to detect and to follow the slow diffusion along the chosen resonant line. Close to the threshold b^*, i.e. for $b = 0.02$ Figure 12 (top), we observe a quite rapid diffusion, but, as for the Hamiltonian case, when decreasing b, (for example, for $b = 0.0075$ Figure 12 (middle)) the diffusion becomes much slower: the integrations times are $j = 10^8$ (middle,left) and $j = 10^9$ (middle,right). Let us remark that, although we do not have a Nekhoroshev theorem for mappings, we found all the features of the Hamiltonian case. For $b = 0.004$ Figure 12 (bottom) we started to observe a very slow diffusion after 10^9 iterations and we could go down to $b = 0.0025$ observing diffusion after $t = 4\,10^9$ iterations obtaining a

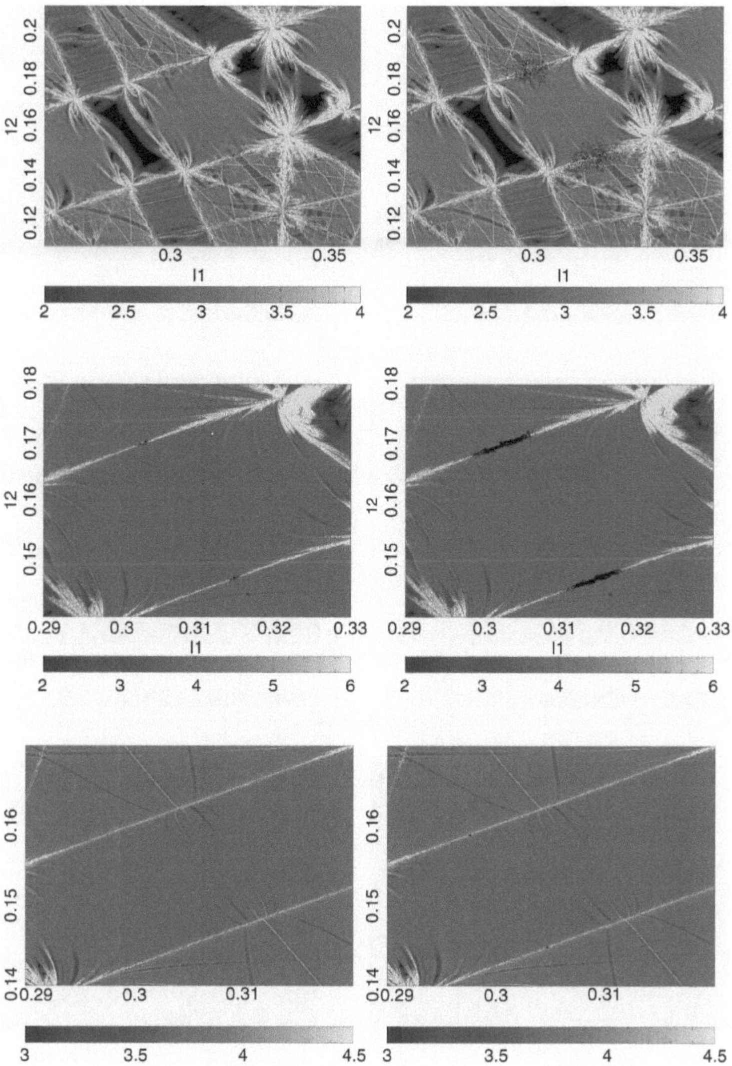

Figure 11. Diffusion along the resonant line $I_1 = 2I_2$ of the Hamiltonian of equation 6 for $\varepsilon = 0.02$ (top), $\varepsilon = 0.007$ (middle), $\varepsilon = 0.0025$ (bottom) of a set of 100 initial conditions taken in the hyperbolic border of the resonance in the interval $0.303 < I_1 < 0.304$ and $0.143 < I_2 < 0.144$. The black points are the intersections of the orbits on the double section $|\varphi_1| + |\varphi_2| \leq 0.05$, $\varphi_3 = 0$. The integration times are respectively: $t = 9\,10^3$ (top, left), $t = 5\,10^5$ (top, right), $1.5\,10^6$ (middle, left), $2,3\,10^7$ (middle, right), $2\,10^8$ (bottom, left), $2\,10^9$ (bottom, right). The color scale range from black to white.

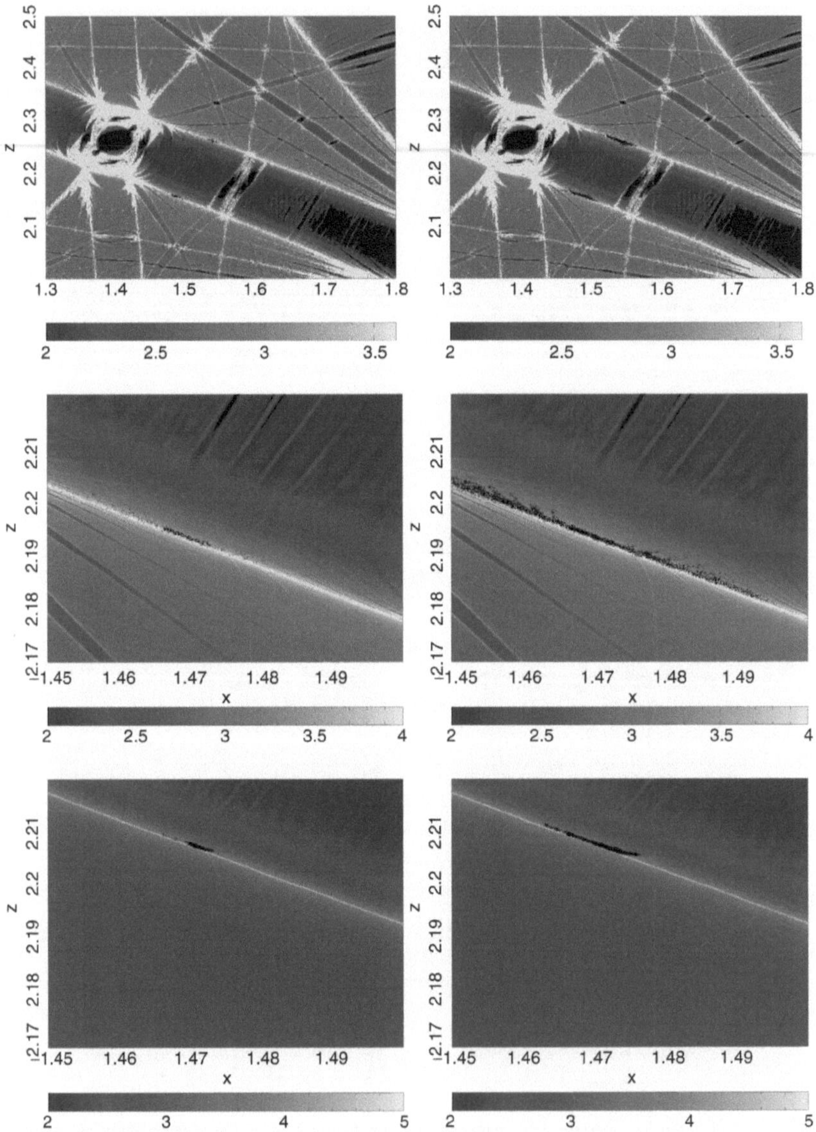

Figure 12. Diffusion along the resonant line $z = -x/2$ of the mapping T_1 for $b = 0.016$ (top), $b = 0.0075$ (middle), $b = 0.004$ (bottom) for a set of 100 initial conditions taken in the chaotic border of the resonance. The black points are the intersections of the orbits on the double section $|y| + |t| \leq 0.005$. The number of iterations on the set of orbits are respectively: $j = 10^6$ (top, left), $j = 10^7$ (top, right), 10^8 (middle, left), 10^9 (middle, right), 10^9 (bottom, left), $4\,10^9$ (bottom, right). The color scale ranges from black to white.

very good qualitative agreement with the Hamiltonian case, i.e. with the predictions of Nekhoroshev theorem.

7.2 Quantitative aspects

In order to measure the diffusion coefficient we have considered the phenomenon as if it was a Brownian motion, since we do not have yet an analytic model for a diffusion like the one we observed. We look therefore for a linear increase with time of the mean square distance from the initial conditions. We are aware that this is a very crude assumption and that for systems like the symplectic map diffusion can be anomalous. Such a phenomenon, to our knowledge, has been revealed for the 2 dimensional standard map for very high values of the perturbation parameter where no invariant curves remain, while here we are very close to the integrable case. This is a very interesting problem that we are studying, but which goes beyond the purpose of the present paper, mostly because we are constrained by computational limitations: we can't take a very large number of initial conditions since we need very long integration times.

Notwithstanding these difficulties, we observed indeed an averaged linear increasing with time of the mean squared distance from the initial conditions. Moreover, in order to reduce the contributions due to fast motion, we have only taken into account the points on the double section.

More precisely, denoting with $x_i(0)$ and $z_i(0)$, $i = 1, ..., N$ the initial conditions of a set of N orbits, with $x_i(j)$ and $z_i(j)$ the corresponding

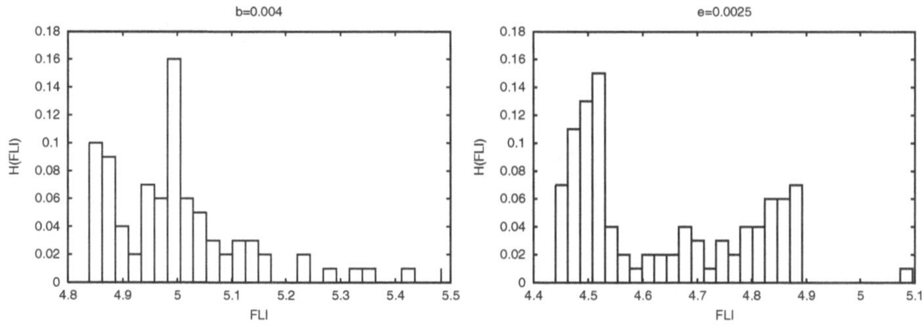

Figure 13. a,b: Histogram of the FLI values for the 100 initial conditions selected on the chaotic border of a resonance.

values at time j, and choosing a fraction T of the total integration time, we considered the quantity:

$$S(nT) = \frac{1}{M_n} \sum_{i:(|y_i(j)|+|t_i(j)|)<0.005} [(z_i(j)-2x_i(j))-(z_i(0)-2x_i(0))]^2 \quad (14)$$

where M_n is the number of points on the double section for j in the interval $(n-1)T \leq j \leq nT$. We observe (Figure 14) a linear increase with time of S and we estimate the diffusion coefficient D as the slope of the regression line. Let us remark that the drift of the action starts at $j \geq 4\,10^8$. As before, results are in agreement with the Nekhoroshev theorem which states that actions are stable on a time at least exponentially long with the inverse of the perturbation parameter b.

Following Zaslavsky and Edelman (2000, 2001), we can have diffusion, more precisely anomalous diffusion, driven by orbits with zero Lyapunov exponent. Therefore, we have computed the Lyapunov exponents of some initial conditions both for the discrete and the continuous case. More precisely, we computed the Lyapunov indicators, i.e. the truncated values

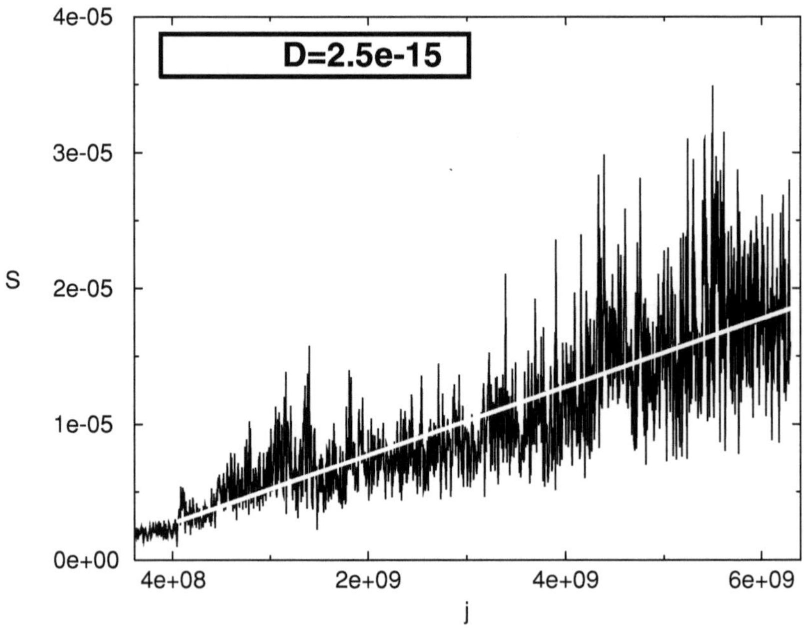

Figure 14. Evolution of the quantity $\bar{S}(nT)$ (see text) with $T = 5\,10^6$, for 100 orbits of the mapping T_1 with $b = 0.004$. The slope of the regression line gives the diffusion coefficient $D = 2.5\,10^{-15}$.

of the Lyapunov exponents which are defined by a limit for time going to infinity.

Figure 15 shows the evolution with time of the two positive Lyapunov indicators of three orbits of the standard map for $b = 0.004$ and for the Hamiltonian of equation 6 for $\varepsilon = 0.0025$. The orbits have been chosen considering the FLI distribution (shown in Figure 13) of the 100 initial conditions: the first one has the lowest FLI (4.85), the second has an FLI in the middle of the distribution (4.97) while the third has the largest FLI value (5.48). We recall that the FLI chart for $b = 0.004$ was computed on $T = 5000$ iterations and that the orbits are considered chaotic when they have an FLI value larger than $1.2 \log T$, i.e. $FLI > 4.43$ in the case considered. The first indicator is positive for the orbits considered (Figure 15), ensuring that we are really following chaotic motions. The sensitivity of the FLI method allows us to select chaotic initial conditions on a time that, for the case $FLI = 4.85$, is three orders of magnitude lower than the time needed for the Lyapunov indicators to stop to decrease and to reach a positive value.

The same investigation in the case of the Hamiltonian flow (Figure 17) shows exactly the same qualitative situation. Not only we are confident that we are not facing anomalous diffusion but also that the orbits correspond really to chaotic motion and not to regular tori which could have FLI larger than $\log T$ because of the proximity to a separatrix.

The estimates of D versus $1/b$ are reported in Figure 18 in a logarithmic scale. Clearly, data are not well fitted with a linear regression, which would correspond to a power law $D(b) = C(1/b)^m$. We have defined three sets of data, performing a local regression for each of them, and found three different slopes. The first set contains the values of D for $1/b \leq 62.5$, the second for $66.6 \leq 1/b \leq 133.3$ and the third for $1/b \geq 166.6$, and the corresponding slopes are respectively $m_1 = -2$, $m_2 = -6.5$ and $m_3 = -12$. This is sufficient to exclude a global power law, and the changes of slope are in agreement with the expected exponential decrease of D, although an exponential fit of the form $D(b) = C' \exp(-\kappa/b)^\alpha$ needs a larger interval of measure in b and this is outside our computational possibilities.

When comparing these results to those obtained for the Hamiltonian (6) (Figure 19) we obtain a surprising quantitative agreement; also for this case we could select three different sets of data well fitted by power laws with different values of the corresponding slopes (Lega et al. 2003). Unfortunately, even in this case we could not estimate an exponential fit, which remains up to now outside our computational possibilities.

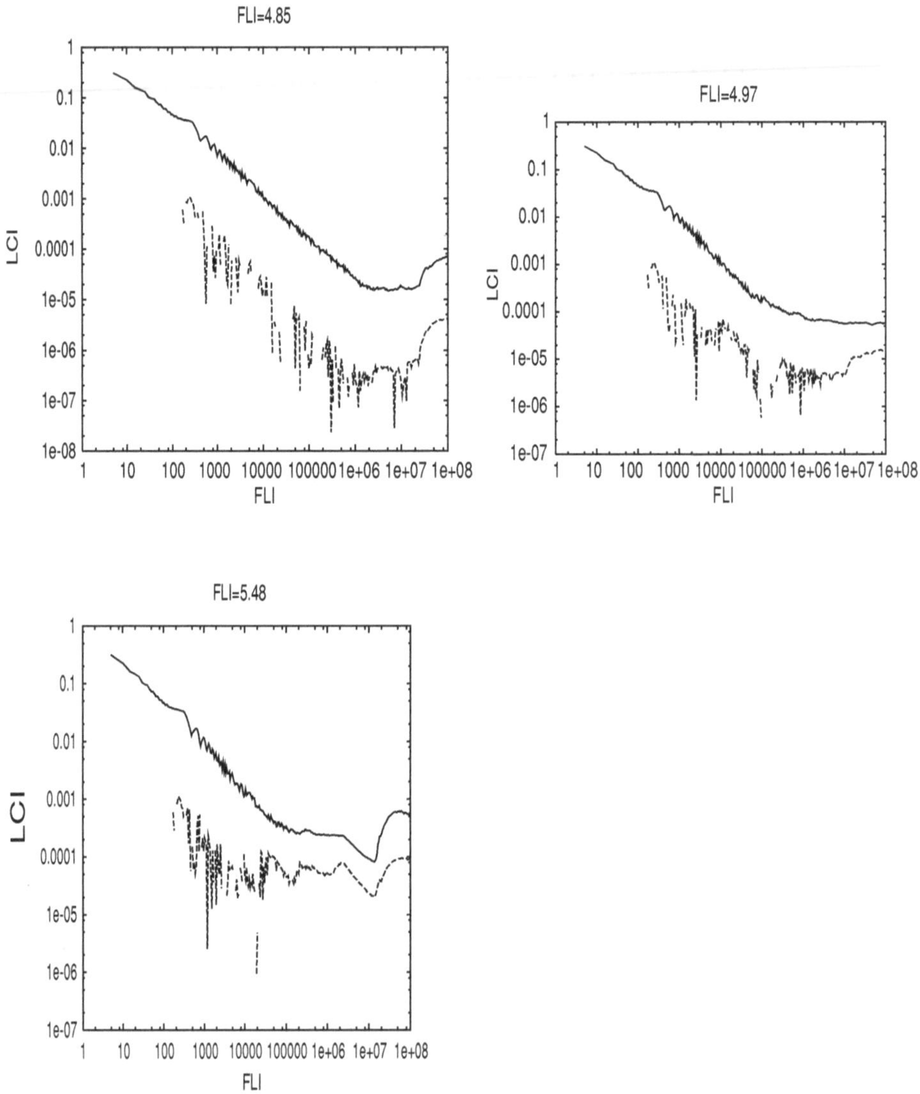

Figure 15. Evolution of the two positive Lyapunov indicators with time for three orbits of the standard map (a,b,c) T_1 with $b = 0.004$.

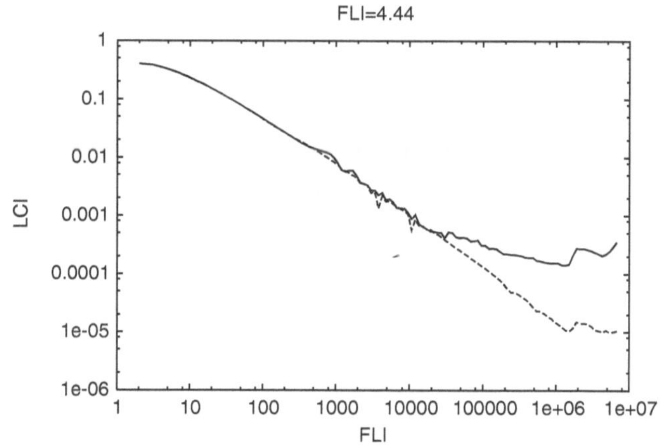

Figure 16. Hamiltonian of equation 6. Initial conditions: $I_1 = 0.3046217311$, $I_2 = 0.1449260426$, $I_3 = 1$; $\varphi_1 = \varphi_2 = \varphi_3 = 0$.

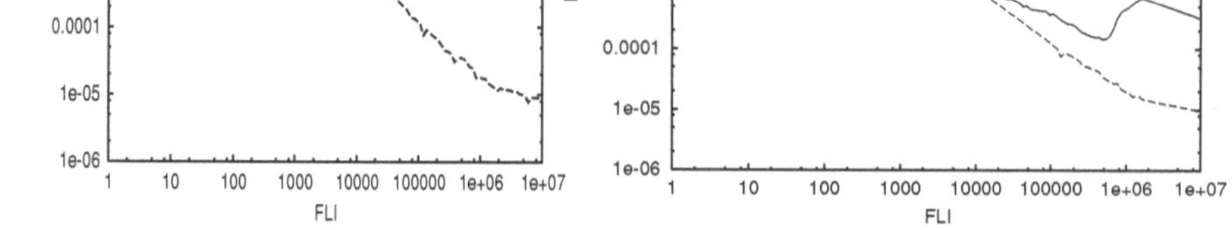

Figure 17. Evolution of the two positive Lyapunov indicators with time for three orbits (a,b,c) of the Hamiltonian of equation 6 with $\varepsilon = 0.0025$.

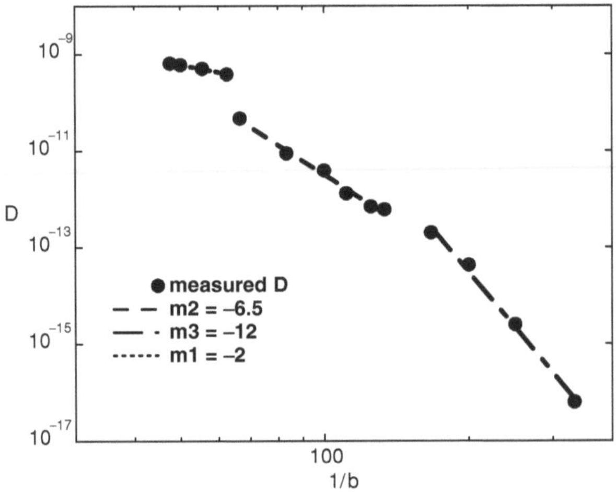

Figure 18. Measure of the diffusion coefficient as a function of $1/b$.The change of the slope of the three power law fits is in agreement with the expected exponential decrease of D.

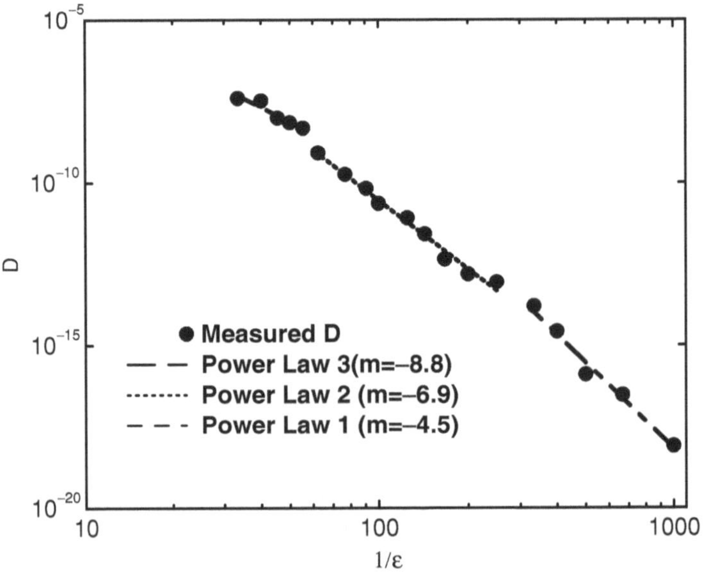

Figure 19. Measure of the diffusion coefficient as a function of $1/\varepsilon$. The change of the slope of the three power law fits is in agreement with the expected exponential decrease of D.

Appendix: Appendix A: Computation of the geometric parameters of the ellipses

In order to calculate the ratio δ between the semi-minor and the semi-major axes of the ellipses surrounding the fixed points of a periodic orbit we consider a two dimensional area preserving mapping and a point $(\bar{x}, \bar{y})$ of a periodic orbit of frequency P/Q. Let us write the Jacobian of the Qth iteration of the map as follows:

$$J = \begin{pmatrix} a & b \\ c & d \end{pmatrix} \tag{A.1}$$

A.1. A first method

Given a simmetrix matrix A of elements:

$$A = \begin{pmatrix} \alpha & \beta \\ \beta & \gamma \end{pmatrix} \tag{A.2}$$

the points x_i and x_{i+1} close to the periodic point $(\bar{x}, \bar{y})$ satisfy the relations:

$$\begin{cases} x_{i+1} & = & Jx_i \\ x_{i+1}^t A x_{i+1} & = & x_i^t A x_i \end{cases} \tag{A.3}$$

By solving the system of equations (A.3) we obtain the relation between the quadratic form A and the matrix J:

$$J^t A = AJ^{-1}, \tag{A.4}$$

$$\begin{pmatrix} a\alpha + c\beta & a\beta + c\gamma \\ b\alpha + d\beta & b\beta + d\gamma \end{pmatrix} = \begin{pmatrix} \alpha d - \beta c & -\alpha b + a\beta \\ \beta d - \gamma c & -\beta b + a\gamma \end{pmatrix} \tag{A.5}$$

The coefficients of the quadratic form describing the neighbourhood of a periodic point are therefore:

$$\begin{cases} \alpha & = & 2c^2 \\ \beta & = & c(d-a) \\ \gamma & = & -2bc\,. \end{cases} \tag{A.6}$$

We remark that the previous expression holds for both an elliptic and an hyperbolic periodic point. More precisely, we are able to write the equation of either the ellipse or the hyperbola centered in the periodic point $(\bar{x}, \bar{y})$ looking for the eigenvalues λ_1 and λ_2 of the matrix A.We will obtain:

$$\frac{x^2}{\sigma_+^2} \pm \frac{y^2}{\sigma_-^2} = 1 \tag{A.7}$$

with $1/\sigma_+^2 = \lambda_1$ and $1/\sigma_-^2 = \lambda_2$. In the case of an ellipse σ_+ and σ_- are nothing but the semi-major and the semi-minor axes of the ellipse. In the hyperbolic case the quantity of interest is the angle φ between the asymtotes, which satisfies: $\delta \equiv \tan\frac{\varphi}{2} = \sigma_-/\sigma_+$. Now, by solving:

$$\lambda^2 - 2(c-b)\lambda - (4bc + (d-a)^2) = 0. \tag{A.8}$$

we obtain:

$$\delta^2 \equiv \frac{\lambda_2}{\lambda_1} = \frac{(c-b) + \sqrt{(c+b)^2 + (d-a)^2}}{(c-b) - \sqrt{(c+b)^2 + (d-a)^2}} \tag{A.9}$$

Taking into account that the det $J = (ad - bc) = 1$ then:

$$\delta = \frac{\sqrt{|4 - (a+d)^2|}}{|(c-b) - \sqrt{(c+b)^2 + (d-a)^2}|} \tag{A.10}$$

A.2. A second method

We can obtain the previous result by solving the eigenvalue equation of the matrix J:

$$\lambda^2 - (a+d)\lambda + 1 = 0 \tag{A.11}$$

1 **The elliptic case**
In the hypothesis: $|TrJ| < 2$, then the eigenvalues are on the unitary circle:

$$\lambda = e^{-i\vartheta}, \ \lambda^* = e^{i\vartheta} \tag{A.12}$$

where $*$ stands for the complex conjugate and:

$$\cos\vartheta = \frac{a+d}{2}, \quad \sin\vartheta = \sqrt{1 - \frac{(a+d)^2}{4}} \tag{A.13}$$

Considering an eigenvector $\mathbf{u}$, $\mathbf{u} \in \mathbf{C^2}$ then $\mathbf{u}$ and $\mathbf{u}^*$ satisfy:

$$J\mathbf{u} = \lambda\mathbf{u} \quad J^*\mathbf{u}^* = \lambda^*\mathbf{u}^* \tag{A.14}$$

Now considering the real and the imaginary part of vector $\mathbf{v}$:

$$\mathbf{v} = \mathcal{R}\mathbf{u} = \frac{\mathbf{u}+\mathbf{u}*}{\mathbf{2}} \quad \mathbf{w} = \mathcal{I}\mathbf{u} = \frac{\mathbf{u}-\mathbf{u}*}{\mathbf{2}} \tag{A.15}$$

we find:

$$\begin{cases} J\mathbf{v} & = & \cos\vartheta\mathbf{v} + \sin\vartheta\mathbf{w} \\ J\mathbf{w} & = & -\sin\vartheta\mathbf{v} + \cos\vartheta\mathbf{w} \end{cases} \tag{A.16}$$

The application of the matrix J on a generic vector on the basis $(\mathbf{v}, \mathbf{w})$, let's take for instance $\xi\mathbf{v} + \eta\mathbf{w}$, turns out to be a rotation of an angle ϑ:

$$J(\xi\mathbf{v} + \eta\mathbf{w}) = (\xi\cos\vartheta - \eta\sin\vartheta)\mathbf{v} + (\xi\sin\vartheta + \eta\cos\vartheta)\mathbf{w}. \tag{A.17}$$

If we write the coordinates of the transformed point as:

$$\xi_1 = \xi \cos\vartheta - \eta \sin\vartheta, \quad \eta_1 = \xi \sin\vartheta + \eta \cos\vartheta \tag{A.18}$$

then:

$$\xi_1^2 + \eta_1^2 = \xi^2 + \eta^2, \tag{A.19}$$

which gives the invariant curves on the **v**, **w** basis. The last step consists in going back to the initial system of coordinates x, y. We call (v_x, v_y) and (w_x, w_y) the coordinates of respectively **v** and **w** on the initial **i**, **j** basis. Then we get:

$$\xi \mathbf{v} + \eta \mathbf{w} = (v_x \xi + w_x \eta)\mathbf{i} + (v_y \xi + w_y \eta)\mathbf{j} \tag{A.20}$$

i.e. the change from coordinates (ξ, η), in the basis $(\mathbf{v}, \mathbf{w})$ to coordinates (x, y) in the basis $(\mathbf{i}, \mathbf{j})$ is:

$$\begin{pmatrix} x \\ y \end{pmatrix} = \begin{pmatrix} v_x & w_x \\ v_y & w_y \end{pmatrix} \begin{pmatrix} \xi \\ \eta \end{pmatrix} = P \begin{pmatrix} \xi \\ \eta \end{pmatrix} \tag{A.21}$$

and by inversion we find:

$$\begin{pmatrix} \xi \\ \eta \end{pmatrix} = \frac{1}{D} \begin{pmatrix} w_y & -w_x \\ -v_y & v_x \end{pmatrix} \begin{pmatrix} x \\ y \end{pmatrix} = P^{-1} \begin{pmatrix} x \\ y \end{pmatrix} \tag{A.22}$$

where:

$$D = \det \begin{pmatrix} v_x & w_x \\ v_y & w_y \end{pmatrix} = \det P \tag{A.23}$$

It remains now to compute the eigenvectors. For it we need to solve:

$$\begin{cases} (a - e^{-i\vartheta})u_x + bu_y & = & 0 \\ cu_x + (d - e^{-i\vartheta})u_y & = & 0 \end{cases} \tag{A.24}$$

Since such equations are not independent, we use the first one obtaining:

$$u_x = b, u_y = -(a - e^{-i\vartheta}).$$

We remark that this is not a unique solution since the choice of u_x is arbitrary. The real and imaginary part of the eigenvector **u** are:

$$\mathbf{v} = \begin{pmatrix} b \\ -a + \cos\vartheta \end{pmatrix} \tag{A.25}$$

$$\mathbf{w} = \begin{pmatrix} 0 \\ -\sin\vartheta \end{pmatrix} \tag{A.26}$$

and the determinant $D = -b\sin\vartheta$. We can now write the equation of the invariant ellipse in the initial system of coordinates (x, y):

$$\begin{aligned} \xi^2+\eta^2 &= \tfrac{1}{D^2}[(w_y^2+v_y^2)x^2 - 2(w_x w_y + v_x v_y)xy + (w_x^2+v_x^2)x^2] \\ &= \tfrac{1}{D^2}[(\sin^2\vartheta + (\cos\vartheta - a)^2)x^2 + 2b(a-\cos\vartheta)xy + b^2y^2] \\ &= \tfrac{1}{D^2}[(1-ad)x^2 + b(a-d)xy + b^2y^2] \\ &= \tfrac{1}{D^2}[-bcx^2 + b(a-d)xy + b^2y^2] \end{aligned} \tag{A.27}$$

The coefficients of the quadratic form agree with those founded in equation (A.6).

2 **The hyperbolic case**
In the hypothesis: $|TrJ| > 2$, then the eigenvalue equation $J\mathbf{u} = \lambda\mathbf{u}$ gives:

$$\begin{cases} \lambda_1 = (a+d) + \sqrt{(a+d)^2-4} \\ \lambda_2 = (a+d) - \sqrt{(a+d)^2-4} \end{cases} \tag{A.28}$$

We consider the eigenvectors $\mathbf{u_1}$ and $\mathbf{u_2}$ asociated with the eigenvalues $\lambda_1 = 1/\lambda_2$ and the transition matrix P with columns $\mathbf{u_1}$ and $\mathbf{u_2}$. We have as previously the change from coordinates (ξ, η), in the basis $(\mathbf{u_1}, \mathbf{u_2})$ to coordinates (x, y) in the basis $(\mathbf{i}, \mathbf{j})$:

$$\begin{pmatrix} x \\ y \end{pmatrix} = P \begin{pmatrix} \xi \\ \eta \end{pmatrix} \tag{A.29}$$

and recalling that (A.3):

$$\begin{pmatrix} x_{i+1} \\ y_{i+1} \end{pmatrix} = J \begin{pmatrix} x_i \\ y_i \end{pmatrix} \tag{A.30}$$

we obtain:

$$\begin{pmatrix} \xi_{i+1} \\ \eta_{i+1} \end{pmatrix} = P^{-1}JP \begin{pmatrix} \xi_i \\ \eta_i \end{pmatrix} \tag{A.31}$$

and since:

$$P^{-1}JP = \begin{pmatrix} \lambda_1 & 0 \\ 0 & 1/\lambda_1 \end{pmatrix} \tag{A.32}$$

we obtain immediately the relation: $\xi\eta = const$. Now using the inverse formula:

$$\begin{pmatrix} \xi \\ \eta \end{pmatrix} = P^{-1} \begin{pmatrix} x \\ y \end{pmatrix} \tag{A.33}$$

we will get the equation of the hyperbola in the x, y basis.

We remark that the second method is strightforward in the case of an hyperbolic rotation, i.e. $|TrJ| > 2$, since the eigenvectors of the matrix J give directly the asymptotic directions. The first method appear more suited for the case of elliptic rotation.

Acknowledgments We thank Prof. A. Giorgilli for useful discussions and suggestions.

References

Arnold, V. I. (1964). Instability of dynamical systems with several degrees of freedom. *Sov. Math. Dokl.*, **6**:581–585.

Benettin, G., Galgani, L., Giorgilli, A. and Strelcyn, J. M. (1980). Lyapunov characteristic exponents for smooth dynamical systems; a method for computing all of them. *Meccanica*, **15**:Part I: theory, 9–20 – Part 2: Numerical applications, 21–30.

Bessi, U., Chierchia, L. and Valdinoci, E. (2001). Upper bounds on Arnold diffusion time via Mather theory. *J. Math. Pures Apll.*, **80**:105–129.

Celletti, A. and Chierchia, L. (1995). *A constructive theory of Lagrangian tori and computer-assisted applications.* Dynamics Reported, New Series, Springer Verlag 4.

Celletti, A. and Chierchia, L. (1997). On the stability of realistic three-body problems. *Comm. Math. Phys.*, **186**, no. 2:413–449.

Celletti, A., Giorgilli, A. and Locatelli, U. (2000). Improved estimates on the existence of invariant tori for Hamiltonian systems. *Nonlinearity*, **13**, no. 2:397–412.

Chirikov, B. V. (1960). *Plasma Phys.*, **1**:253.

Chirikov, B. V. (1979). A universal instability of many dimensional oscillator system. *Phys. Rep.*, **52**:263–379.

Cincotta, P. M. and Simó, C. (2000). Simple tools to study global dynamics in non-axisymmetric galactic potentials - I. *Astron. and Astroph. suppl.*, **147**:205–228.

Froeschlé, C. (1970,a). Numerical study of dynamical system with three degrees of freedom. I graphical displays of four dimensional sections. *Astron. Astrophys.*, **4**:115.

Froeschlé, C. (1970,b). A numerical study of the stocasticity of dynamical systems with two degrees of freedom. *Astron. Astrophys.*, **9**:15–23.

Froeschlé, C. (1972). Numerical study of a four-dimensional mapping. *Astron. Astrophys.*, **16**:172.

Froeschlé, C. (1984). The Lyapunov characteristic exponents and applications. *Jornal de Méc. théor. et apll.*, Numero spécial:101–132.

Froeschlé, C., Gonczi, R. and Lega, E. (1997). The fast Lyapunov indicator: a simple tool to detect weak chaos. Application to the structure of the main asteroidal belt. *Planetary and space science*, **45**:881–886.

Froeschlé, C., Guzzo, M. and Lega, E. (2000). Graphical evolution of the Arnold's web: from order to chaos. *Science*, **289** N.5487:2108–2110.

Froeschlé, C., Guzzo, M. and Lega, E. (2001). *The Fast Lyapunov Indicator, in The Restless Universe.* B. A. Steves and A. J. Maciejewsky (eds.).

Froeschlé, C. and Lega, E. (1999). *Weak chaos and diffusion in Hamiltonian systems. From Nekhoroshev to Kirkwood.* in the NATO/ASI series Vol. 522: The Dynamics of small bodies in the Solar system: a major key to Solar system studies, A.E.Roy (eds.).

Froeschlé, C. and Lega, E. (1996). On the measure of the structure around the last KAM torus before and after its break-up. *Celest. Mech. and Dynamical Astron.*, **64**:21–31.

Froeschlé, C. and Lega, E. (1998). *Modeling mapping: an aim and a tool for the study of dynamical systems.* in Analysis and Modelling of discrete dynamical systems. Benest D. and Foreschlé C. (eds.), Gordon and Breach.

Froeschlé, C. and Lega, E. (2000). On the structure of symplectic mappings. The fast Lyapunov indicator: a very sensitive tool. *Celest. Mech. and Dynamical Astronomy*, **78**:167–195.

Greene, J. M. (1979). A method for determining a stochastic transition. *J. Math. Phys.*, **20**:1183–1201.

Guzzo, M., Lega, E. and Froeschlé, C. (2002). On the numerical detection of the stability of chaotic motions in quasi–integrable systems. *Physica D*, **163**:1–25.

Hénon, M. and Heiles, C. (1964). The applicability of the third integral of motion: Some numerical experiments. *A. J.*, **1**:73–79.

Laskar, J. (1990). The chaotic motion of the Solar System. A numerical estimate of the size of the chaotic zones. *Icarus*, **88**:266–291.

Laskar, J. (1993). Frequency analysis for multi-dimensional systems. Global dynamics and diffusion. *Physica D*, **67**:257–281.

Laskar, J. (1994). Large scale chaos in the Solar System. *Astron. Astrophys.*, **287**:L9–L12.

Laskar, J. (1994). Frequency map analysis of an Hamiltonian system. *Workshop on Non-Linear Dynamics in Particle Accelerators*, Arcidosso, September 1994.

Laskar, J., Froeschlé, C. and Celletti, A. (1992). The measure of chaos by the numerical analysis of the fundamental frequencies. application to the standard mapping. *Physica D*, **56**:253.

Lega, E. and Froeschlé, C. (1997). *Fast Lyapunov Indicators. Comparison with other chaos indicators. Application to two and four dimensional maps.* in The Dynamical Behaviour of our planetary system., Kluwer academ. publ., J. Henrard and R.Dvorak eds.

Lega, E. and Froeschlé, C. (1996). Numerical investigations of the structure around an invariant KAM torus using the frequency map analysis. *Physica D*, **95**:97–106.

Lega, E. and Froeschlé, C. (2001). On the relationship between fast Lyapunov indicator and periodic orbits for symplectic mappings. *Celest. Mech. and Dynamical Astronomy*, **15**:1–19.

Lega, E., Guzzo, M. and Froeschlé, C. (2003). Detection of Arnold diffusion in Hamiltonian systems. *Physica D*, **182**:179–187.

LeVeque, W. J. (1977). *Fundamentals of Number Theory.* Addison-Wesley Publishing Company.

Lichtenberg, A. J. and Lieberman, M. A. (1983). *Regular and Stochastic motion.* Springer, Berlin, Heidelberg, New York.

Locatelli, U., Froeschlé, C., Lega, E. and Morbidelli, A. (2000). On the relationship between the Bruno function and the breakdown of invariant tori. *Physica D*, **139**: 48–71.

Locatelli, U. and Giorgilli, A. (2000). Invariant tori in the secular motions of the three–body palnetary system. *Cel. Mech. and Dynam. Astr.*, **78**, 47-74.

MacKay, R. S. (1993). *Renormalisation in Area Preserving Maps.* World Scientific.

Mahler, K. (1957). *Lectures on Diophantine Approximations.*

Morbidelli, A. and Giorgilli, A. (1995,a). Superexponential stability of KAM tori. *J. Stat. Phys.*, **78**:1607.

Morbidelli, A. and Giorgilli, A. (1995,b). On a connection between KAM and Nekhoroshev's theorems. *Physica D*, **86**:514–516.

Nekhoroshev, N. N. (1977). Exponential estimates of the stability time of near-integrable Hamiltonian systems. *Russ. Math. Surveys*, **32**:1–65.

Nesvorny, D. and Ferraz-Mello, S. (1997). On the asteroidal population of the first-order Jovian resonances. *Icarus*, **130**:247–258.

Olivera, A. and Simó, C. (1987). An obstruction method for the destruction of invariant curves. *Physica D*, **26**:181.

Sándor, Z., Érdi, B. and Efthymiopoulos, C. (2000). The Phase Space Structure Around L4 in the Restricted Three-Body Problem. *Celestial Mechanics and Dynamical Astronomy*, **78**:113–123.

Zaslavski, G. M. and Edelman, M. (2000). Hierarchical structures in the phase space and fractional kinetics: I classical systems. *Chaos*, **10**:135–145.

Zaslavski, G. M. and Edelman, M. (2001). Weak mixing and anomalous kinetics along filamented surfaces. *chaos*, **11**:295–305.

FOURIER ANALYSIS OF CHAOTIC MOTIONS AND APPLICATIONS TO CELESTIAL MECHANICS

Massimiliano Guzzo
Università degli Studi di Padova
Dipartimento di Matematica Pura ed Applicata
Via Belzoni 7, 35131 Padova, Italy
guzzo@math.unipd.it

Abstract We review results about the Fourier Analysis of chaotic solutions of quasi–integrable systems based on the Nekhoroshev theorem. We describe also an application to Asteroids stability.

Keywords: Fourier Analysis, Nekhoroshev theorem, Asteroids.

1. Introduction

In this article we discuss the problem of understanding the long–term stability properties of a solution of a quasi–integrable Hamiltonian system by means of a Fourier analysis on a short observation time. Precisely, even for resonant chaotic motions, we will show how the combined use of Fourier analysis and Nekhoroshev theorem allows to understand the stability properties on a time: $T' \sim \exp(T)$, where T is a suitable observation time, of the order of the resonant period. To be definite, we will refer to quasi–integrable Hamiltonian systems with Hamiltonian of the form:

$$H(I,\varphi) = h(I) + \varepsilon f(I,\varphi) \ , \qquad (I,\varphi) \in B \times \mathbb{T}^n \ , \tag{1}$$

where $B \subset \mathbb{R}^n$ and $\mathbb{T}^n$ is the n–dimensional torus; h and f are analytic and ε is a small parameter. The Fourier analysis will be performed on the generic observables of the system; i.e. we analyze the function of time:

$$g(t) = \mathcal{G}(I(t), \varphi(t)) \tag{2}$$

B.A. Steves et al. (eds.), Chaotic Worlds: from Order to Disorder in Gravitational N-Body Dynamical Systems, 167–200.

where $\mathcal{G}(I, \varphi)$ is the function representing the observable, and $(I(t), \varphi(t))$ is a generic solution of the Hamilton equations of (1). It is well known that for systems like (1) we can find typically both quasi–periodic and chaotic solutions. In the quasi–periodic case, typical of KAM tori, the function (2) can be represented on the whole time interval by means of a discrete Fourier representation, which can be reconstructed numerically using the well known frequency analysis algorithm (Laskar 1990, Laskar et al. 1992). Because the quasi–periodic case corresponds to perpetual stability, the fact that the spectrum of an observable is a line spectrum is considered a numerical evidence that the solution is perpetually stable (possibly on a KAM torus).

In the chaotic case, discrete representations of (2) are not allowed and it is mandatory to use the continuous Fourier representation. Recent papers have shown that the structure of this continuous spectrum allows to understand the long–term stability properties of the chaotic solution. These results are based on Hamiltonian perturbation theory, and specifically on the Nekhoroshev theorem which proves (with suitable hypotheses) that even chaotic solutions of Hamiltonian system (1) are extremely stable: they can macroscopically diffuse in the action space only on times which grow exponentially with a positive power of $1/\varepsilon$ (Nekhoroshev 1977, 1979, Benettin et al. 1985, Benettin and Gallavotti 1986, Lochak 1992, Pöschel 1993 for non–degenerate systems and Littlewood 1977, Fassò 1995, Benettin and Fassò 1996, Niedermann 1996, Benettin et al. 1997, Guzzo and Morbidelli 1997, Morbidelli and Guzzo 1997, Guzzo 1998, Guzzo 1999 for the degenerate ones). Instead, when ε is increased, many numerical studies (Laskar 1990, Laskar, Froeschlé and Celletti 1992, Lega and Froeschlé 1997, Contopoulos and Voglis 1997, Froeschlé and Lega 1998, Froeschlé, Guzzo and Lega 2000, Guzzo, Lega and Froeschlé 2002) have shown that the system enters a regime of overlapping of resonances, known as Chirikov regime (Chirikov 1979), in which actions can diffuse in times of the order of $1/\varepsilon$. Recent papers (Guzzo and Benettin 2001, Guzzo 2002, Guzzo Lega and Froeschlé 2002) have shown that the Fourier spectrum of the observables of the system keeps trace of these two different situations. The following sections of this article are devoted to illustrate these results, and an application to asteroids described in Guzzo, Knezevic and Milani (2002).

2. The 'peak' structure of Fourier spectra

In this section we illustrate the characterizations which can be given to the Fourier spectra of chaotic solutions of quasi–integrable systems. We first consider the easy integrable case:

$$H(I,\varphi) = h(I)\ , \qquad (I,\varphi) \in B \times \mathbb{T}^n\ . \tag{3}$$

The solutions of the Hamilton equations $I(t) = I(0)$, $\varphi(t) = \varphi(0) + \omega t$, with $\omega = \frac{\partial h}{\partial I}(I(0))$, are quasi–periodic with frequency vector $\omega \in \mathbb{R}^n$. The quasi–periodicity of a solution $(I(t), \varphi(t))$ can be recognized numerically by computing a suitable observable of the system,

$$\mathcal{G}(I,\varphi) = \sum_{k \in \mathbb{Z}^n} \mathcal{G}_k(I) e^{ik\cdot\varphi} \tag{4}$$

on $(I(t), \varphi(t))$. In fact, the function of time $g(t) = \mathcal{G}(I(t), \varphi(t))$ has the discrete Fourier representation:

$$g(t) = \sum_{k \in \mathbb{Z}^n} g_k e^{i\nu_k t} \tag{5}$$

with amplitudes g_k and frequencies $\nu_k = k \cdot \omega$ given by:

$$g_k = \mathcal{G}_k(I(0)) e^{ik\cdot\varphi(0)}, \quad \nu_k = k \cdot \omega. \tag{6}$$

Now, let us suppose that the system is in a black–box, and that all we can know of it is the observable $\mathcal{G}$, sampled up to a finite time interval. This is the typical situation occurring in Physics, where one obtains information on some system on the basis of the output of an experiment. The amplitudes and frequencies (6) can be numerically computed from $g(t)$, for example, by means of the frequency analysis method (Laskar 1990, Laskar et al. 1992). However, if we are interested mainly in recognizing the quasi–periodic nature of the solution, it is not necessary to use a refined frequency analysis, but it is sufficient to compute the Fast Fourier transform of $\Phi(t)g(t)$ on a finite time interval $[-T, T]$, where $\Phi(t)$ is a suitable analytic window on $[-T, T]$ (see Section 4 for details). Figure 1 shows an example of such an analysis. Within the precision of our computation (a line is identified with an error of about 10^{-5} in frequency) we can easily recognize that the spectrum of $g(t)$ is a line spectrum.

Now, we consider the more interesting quasi–integrable Hamiltonian:

$$H(I,\varphi) = h(I) + \varepsilon f(I,\varphi), \qquad (I,\varphi) \in B \times \mathbb{T}^n, \tag{7}$$

where h and f are analytic functions and ε is a small parameter. The KAM theorem applies to (7) if h is non–degenerate and ε is suitably small: in such a case, the domain is filled with a large volume of invariant tori supporting quasi–periodic motions with frequencies $\tilde{\omega}$ which are ε–close to the unperturbed frequency vector ω. Therefore, for solutions $(I(t), \varphi(t))$ on KAM tori, the observable admits a discrete Fourier representation as in the integrable case:

$$g(t) = \sum_{k \in \mathbb{Z}^n} \tilde{g}_k e^{i\tilde{\nu}_k t} \tag{8}$$

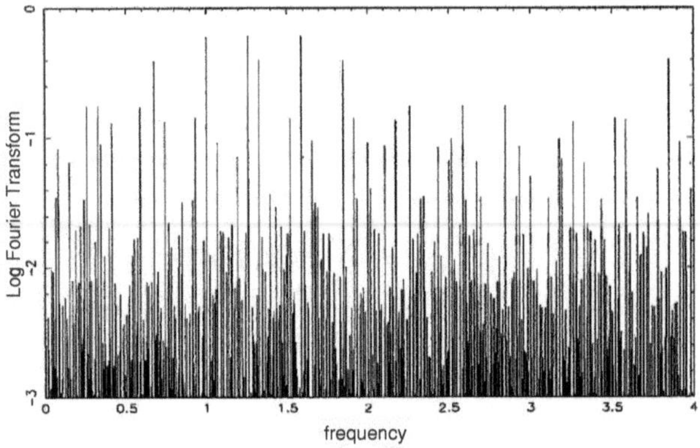

Figure 1. Fast Fourier transform of the function $g(t) = \Phi(t)\mathcal{G}(I(t), \varphi(t))$, where $\mathcal{G} = ((\cos\varphi_1 + \cos\varphi_2 + \cos\varphi_3 + 1)^8 + 1)^{-1}$, $\Phi(t)$ is described in Section 4, $(I(t), \varphi(t))$ is a solution of the Hamilton equations of the integrable Hamiltonian $H = \frac{I_1^2}{2} + \frac{I_2^2}{2} + \frac{I_3^2}{2}$ with $I_1(0) = 1$, $I_2(0) = 2^{\frac{1}{3}}$, $I_1(0) = 2^{\frac{2}{3}}$. The spectrum has been obtained by analyzing the solution up to a time $T \sim 10^5$.

with frequencies $\tilde{\nu}_k = k \cdot \tilde{\omega}$. The analytic computation of amplitudes and frequencies in the case of KAM tori is not straightforward. However, they can be numerically estimated (see Figure 2). In the example of Figure 2, within the precision of our computation (a line is identified with an error of about 10^{-5} in frequency) we can easily recognize that the spectrum of $g(t)$ is a line spectrum.

In quasi–integrable systems we do not find only KAM tori, but also resonant motions, and among resonant motions we find the chaotic ones. If ε is small and h satisfies a suitable geometric condition (convexity of h is sufficient) the Nekhoroshev theorem proves the exponential stability of the actions for all initial conditions, including the resonant ones. More precisely, there exist positive constants $\varepsilon_0, a, b, I_0, t_0$ such that if $\varepsilon \leq \varepsilon_0$, for any $(I(0), \varphi(0)) \in B \times \mathbb{T}^n$ it is $|I(t) - I(0)| \leq I_0 \varepsilon^a$ for any time t satisfying the exponential estimate:

$$|t| \leq t_0 \exp{(\varepsilon_0/\varepsilon)^b}.$$

Moreover, for chaotic solutions it is also proved that in suitable coordinates some of the degrees of freedom behave essentially quasi–periodically while the others behave chaotically. Therefore, the Fourier representation of the observables $\mathcal{G}(I, \varphi)$ for these chaotic solutions should be intermediate between a discrete representation, and a continuous one without structures. In Guzzo and Benettin (2001) we introduced a Fourier

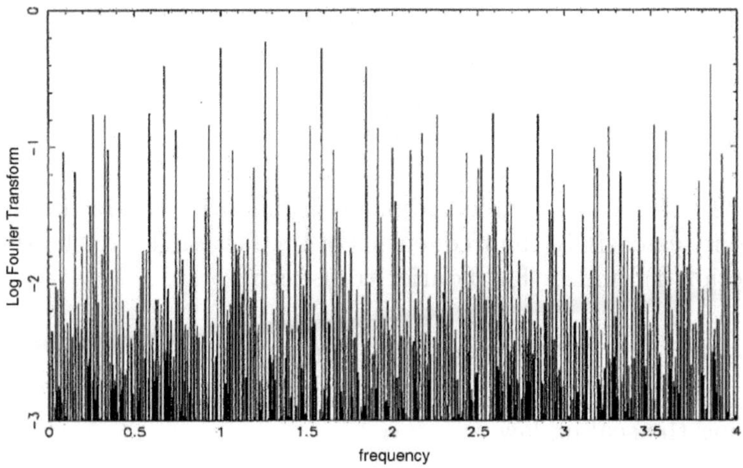

Figure 2. Fast Fourier transform of the function $g(t) = \Phi(t)\mathcal{G}(I(t), \varphi(t))$, where $\mathcal{G}$ and $\Phi(t)$ are as in Figure 1, and $(I(t), \varphi(t))$ is a solution of the Hamilton equations of the quasi–integrable Hamiltonian $H = \frac{I_1^2}{2} + \frac{I_2^2}{2} + \frac{I_3^2}{2} + \varepsilon(\cos\varphi_1 + \sin\varphi_2 + \cos(\varphi_1 + \varphi_2) + \cos(\varphi_2 + \varphi_3) + \sin(\varphi_1 - \varphi_3))$, with $\varepsilon = 10^{-4}$ and initial condition in strong irrational ratio $I_1(0) = 1$, $I_2(0) = 2^{\frac{1}{3}}$, $I_1(0) = 2^{\frac{2}{3}}$. The spectrum has been obtained by analyzing the solution up to a time $T \sim 10^5$.

representation for these intermediate cases. Very shortly, let us consider the Fourier spectrum of

$$G(t) = \Phi(t)\,\mathcal{G}(I(t), \varphi(t)), \tag{9}$$

where $\Phi(t)$ is an analytic "filter function", which is nearly equal to one for a wide time interval $[-T, T]$, and vanishes at infinity (as is necessary, in order for the Fourier transform to be defined). Our precise choice of the filter is (see Section 4 for details and motivations)

$$\Phi(t) = \frac{1}{2}\Big[\tanh\Big(\frac{t+2T}{\sigma}\Big) - \tanh\Big(\frac{t-2T}{\sigma}\Big)\Big], \tag{10}$$

with large T and some $\sigma > 0$. A rough formulation of our theoretical result (for a precise statement see Guzzo and Benettin 2001) is as follows: if ε is suitably small, for any chosen time scale T satisfying

$$\sqrt{(\varepsilon_0/\varepsilon)} \le T \le e^{(\varepsilon_0/\varepsilon)^{1/2n}}, \tag{11}$$

ε_0 being a suitable constant, the function G defined in (9) admits a decomposition

$$G(t) = G^0(t) + G^1(t), \tag{12}$$

depending on T, such that $G^1(t)$ is negligible for $|t| \le T$, while G^0 is a superposition of modulated quasi–periodic terms:

$$G^0(t) = \sum_{k\in\mathbb{Z}^n} g_k(t)\, e^{ik\cdot\Omega t}, \tag{13}$$

where $\Omega \in \mathbb{R}^n$ is a suitable vector determined by the initial datum (it represents essentially the frequencies of the fast angles of the resonance, see Section 4), and g_k has Fourier transform:

$$\tilde{g}_k(\alpha) = \frac{1}{2\pi}\int_{-\infty}^{\infty} dt\; g_k(t)\; e^{-i\alpha t} \tag{14}$$

satisfying the scaling law

$$|\tilde{g}_k(\alpha)| \le a_0\, T\, e^{-a_1|k|} \exp\left(-a_2\frac{|\alpha|}{\varepsilon^{a_3}}\right) \tag{15}$$

for any frequency $\alpha \in \mathbb{R}$; $a_0, \ldots, a_3$ denote here suitable positive constants, and $|k| = |k_1| + \cdots + |k_n|$. The most important constant is a_3, which gives the rate of the exponential decay of the spectrum with respect to the frequency α. We have proved (for convex systems) the uniform bound $a_3 \ge 1/(2n)$, but for a wide class of resonant initial data we got $a_3 = \frac{1}{2}$, and this is indeed the value that we found in numerical experiments. In this case $\tilde{g}_k(\alpha)$ satisfies the stronger estimate:

$$|\tilde{g}_k(\alpha)| \le a_0\, T\, e^{-a_1|k|} \exp\left(-a_2\frac{|\alpha|}{\sqrt{\varepsilon}}\right). \tag{16}$$

As a corollary of the previous result, the observable $g(t)$ admits the representation:

$$g(t) = \sum_{k\in\mathbb{Z}^n} g_k(t) \exp\left(ik\cdot\Omega t\right) + g^1(t) \tag{17}$$

where $g^1(t) = G^1(t) + (1 - \Phi(t))g(t)$ is exponentially small (within the observation time T, i.e. for $|t| \le T$), and $g_k(t)$ and Ω are as above.

The main novelty of representation (17) with respect to the quasi–periodic one consists in having replaced the constant amplitudes g_k with amplitude functions $g_k(t)$. The key point of the result of Guzzo and Benettin (2001) are estimates (15), (16), which state that *in the Nekhoroshev regime the Fourier transform of such amplitude functions decays very fast with respect to the frequency.*

As a consequence, if we represent $|\tilde{g}_k(\alpha)|$ in a semi–log scale we obtain that it fits the shape of a peak, with slope which decreases as $1/\sqrt{\varepsilon}$

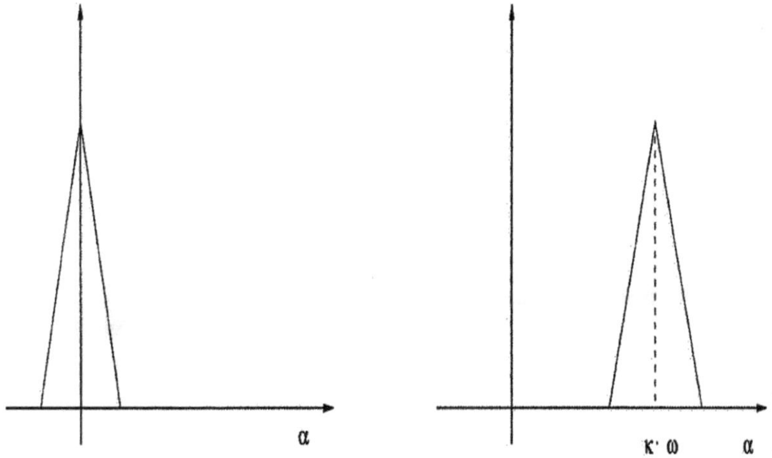

Figure 3. The representation of $|\tilde{g}_k(\alpha)|$ in a semi–log scale fits the shape of a peak (on the left). Instead, the representation of the Fourier transform of $g_k(t) \exp(ik \cdot \Omega t)$ fits the shape of the previous peak, centered around the frequency value $k \cdot \Omega$ (on the right).

(in the case of (16)). If instead we consider the Fourier transform of $g_k(t) \exp(ik \cdot \Omega t)$, then the center of the peak trivially moves around the frequency $k \cdot \Omega$ (see Figure 3). Therefore, if we consider the spectrum of $\sum_{k \in Z^n} g_k(t) \exp(ik \cdot \Omega t)$, which we can easily compute numerically as the Fast Fourier transform of $g(t)$ (g^1 does not really contributes to the FFT), it appears as the superposition of many peaks centered around the frequencies $k \cdot \Omega$: the spectrum assumes a very peculiar peak structure, such as the one represented in Figure 4. In this case the peaks are well separated one from the other one because of the specific resonance considered, which is characterized by only one fast angle. The situation is different when the resonance is characterized by at least two fast angles: the location of all possible peaks $k \cdot \Omega$ can be dense on the frequency space, but on the spectrum we effectively see a finite number of them because the height of each peak decreases exponentially with its order $|k|$, as predicted by (15), (16). This situation is represented in Figure 5.

We now consider what happens to the spectrum of generic observables for suitable high values of ε. It is quite easy to understand that for high values of ε one does not observe the peak structure (see Figure 6). But in Guzzo and Benettin (2001) it is proved that systems in the Littlewood regime are associated to observables with the peak structure of the spectrum. Therefore, we can state the following *instability* criterion:

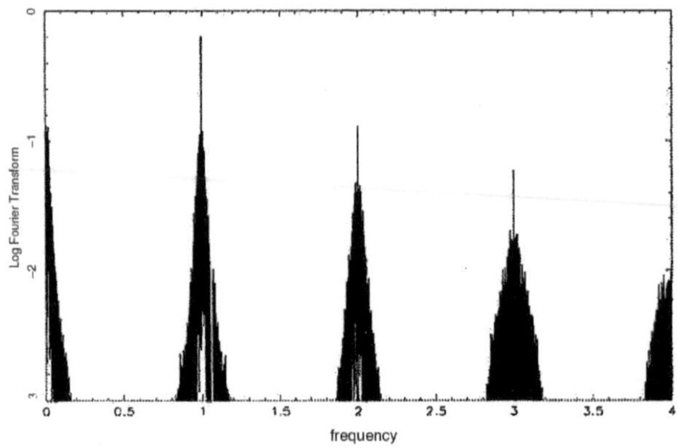

Figure 4. Fast Fourier transform of $g(t) = \Phi(t)\mathcal{G}(I(t), \varphi(t))$, where $(I(t), \varphi(t))$ is a solution of the Hamiltonian system of Figure 2 with $\varepsilon = 10^{-4}$, and the resonant initial condition $I_1(0) = 10^{-2}\sqrt{\varepsilon}$, $I_2(0) = 10^{-2}\sqrt{\varepsilon}$, $I_3(0) = 1$. The fast frequency vector of the resonance is $\Omega = (0, 0, 1)$, so $k \cdot \Omega = k_3$: we expect to find a peak centered on any integer value of the frequency. The observation time is $T \sim 10^5$. On the same time the largest Lyapunov exponent seems to converge to a value $\lambda \sim 0.001$.

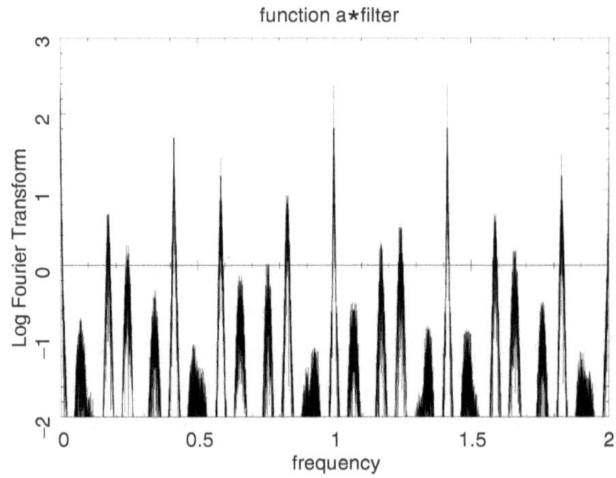

Figure 5. Fast Fourier transform of $g(t) = \Phi(t)\mathcal{G}(I(t), \varphi(t))$ where $\mathcal{G} = 1/((\cos(\varphi_1) + \cos(\varphi_2) + \cos(\varphi_3) + \cos(\varphi_4) + 1)^8 + 100)$, the Hamiltonian is $H = (I_1^2 + I_2^2 + I_3^2)/2 + (\sqrt{5} - 1)I_4/2 + \varepsilon(\cos(\varphi_1) + \sin(\varphi_2) + 0.5\cos(\varphi_1 + \varphi_2) + \sin(\varphi_1 - \varphi_3) + \cos(\varphi_1 + \varphi_2 + \varphi_3 + \varphi_4) + \cos(\varphi_2 + \varphi_3))$ with $\varepsilon = 10^{-4}$, the initial condition has $I_1(0) = 10^{-1}\sqrt{\varepsilon}$, $I_2(0) = 10^{-2}\sqrt{\varepsilon}$, $I_3(0) = 1$, $I_4(0) = \sqrt{2}$. The fast frequency vector of the resonance is $\Omega = (0, 0, 1, (\sqrt{5} - 1)/2)$, so that $k \cdot \Omega = k_3 + (\sqrt{5} - 1)/2k_4$. The observation time is $T \sim 2 \cdot 10^5$.

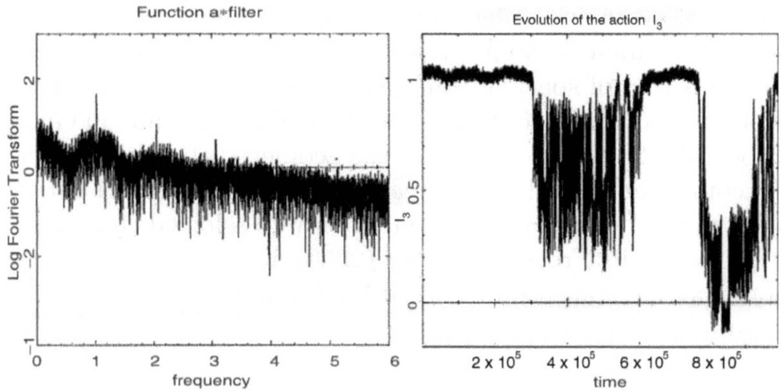

Figure 6. Fast Fourier transform of $g(t) = \Phi(t)\mathcal{G}(I(t), \varphi(t))$, where $\mathcal{G}$ and $\Phi(t)$ are as in Figure 1, and $(I(t), \varphi(t))$ is a solution of the Hamilton equations of $H = \frac{I_1^2}{2} + \frac{I_2^2}{2} + \frac{I_3^2}{2} + \varepsilon(\cos\varphi_1 + \sin\varphi_2 + \cos(\varphi_1+\varphi_2) + \cos(\varphi_2+\varphi_3) + \sin(\varphi_1-\varphi_3))$, with $\varepsilon = 0.015$ and resonant initial condition $I_1(0) = 1$, $I_2(0) = 10^{-2}\sqrt{\varepsilon}$, $I_1(0) = 10^{-2}\sqrt{\varepsilon}$. The observation time is $T \sim 10^4$, in which the action I_3 has not a fast evolution. Nevertheless, we can see that the spectrum does not have the peak structure, and so the system is not in the Nekhoroshev regime . Indeed, on the right, we see that I_3 has big variations in short time at $t \sim 300000$, which is not compatible with the Nekhoroshev regime.

"The fact that the spectrum of an observable of a system computed on a chaotic solution has not the peak structure can be considered a numerical evidence that the system does not satisfy the hypotheses of Nekhoroshev theorem in the neighborhood of that solution, i.e. the Nekhoroshev theorem does not prevent fast evolution of the actions".

It is relevant to remark that the spectra of Figures 4,5 and 6 are computed on a relatively short time scale, in which the actions do not significantly diffuse. Actually, with an observation time T of order $1/\varepsilon$ we get information on the much longer time $t_0 \exp((\varepsilon_0/\varepsilon)^b)$. This means that in the case of Figure 6, we cannot exclude that sooner or later the actions can 'rapidly' diffuse. Indeed, at a much longer time, one of the actions changes of order 1 (Figure 6 on the right). We would like to use also a converse stability criterion, which would give indication of exponential stability in presence of the peak structure of the spectrum. This result would be true if the peak structure of the spectrum suddenly disappears passing from the Littlewood stability regime to the Chirikov diffusive one, which is conjectured in Guzzo and Benettin (2001). Indeed, in the regime of the exponential estimates all resonances match together in such a way that locally it is always possible to average at least one angle. But also the peak structure of the spectra is a consequence of the possibility of averaging at least one angle. Therefore, peak structure and exponential estimates should disappear at the same value of ε.

The transition of spectra from structured to unstructured ones is not described by a theorem, but has been studied numerically in Guzzo, Lega and Froeschlé (2002) by comparing the geometry of resonances of a given system computed with the Fast Lyapunov Indicator with the structure of the spectra of an observable computed on well selected chaotic solutions. More precisely, in Froeschlé et al. (2000) we estimated with the FLI method that the transition between Littlewood and Chirikov regime for the Hamiltonian system:

$$H_\varepsilon = \frac{I_1^2}{2} + \frac{I_2^2}{2} + I_3 + \varepsilon \left(\frac{1}{\cos(\varphi_1) + \cos(\varphi_2) + \cos(\varphi_3) + 4} \right) \tag{18}$$

near $(I_1, I_2) = (1/3, 1/6)$ occurred between $\varepsilon = 0.01$ and $\varepsilon = 0.04$. In Guzzo et al. (2002) we computed the Fourier spectrum of an observable corresponding to the initial actions $I_1(0) = 1/3$ and $I_2(0) = 1/6 - 0.01\sqrt{(\varepsilon/0.06)}$ and different values of ε. For $\varepsilon = 0.001$ and $\varepsilon = 0.00025$ indeed the spectrum of $g(t)$ shows the peak structure, and the slope of the peaks rescales as the expected $1/\sqrt{\varepsilon}$ power law (Figure 7). We then tried to refine the transition interval: it is clear that for $\varepsilon = 0.032$ (Figure 8 right), there is not the peak structure; with a little decreasing of ε, $\varepsilon = 0.031$ (Figure 8 center), some structures appear in the spectrum; with a further little decreasing, $\varepsilon = 0.030$ (Figure 8 left), we recognize in these structure some peaks. These peaks are not so evident as in the case of Figure 7 because we are very near the transition value. Nevertheless, the qualitative difference between the spectra of Figure 8–left and Figure 8–right is sufficient to give as transition interval $[0.030, 0.032]$. This analysis shows that the transition from peak structured spectra to unstructured ones is quite sharp, and agrees with the transition of the geometry of resonances which is expected when passing from the Nekhoroshev to the Chirikov regime. Therefore, it is reasonable to use the following *stability* criterion: *"The fact that the spectrum of an observable of a system computed on a chaotic solution has the peak structure can be considered a numerical evidence that the system satisfies the hypotheses of Nekhoroshev theorem in the neighborhood of that solution, i.e. the Nekhoroshev theorem prevents fast evolution of the actions".*

Of course, even in presence of Nekhoroshev stability, evolution of the actions is possible on the very long exponential times typical of Arnold'd diffusion (Arnold 1964), but this kind of instability is outside the subject of this article (see Lega, Guzzo and Froeschlé 2003 for a numerical study).

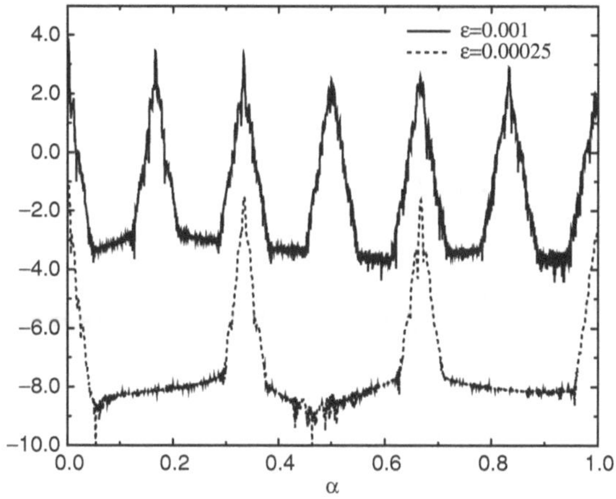

Figure 7. Plot of $\log|\tilde{g}(\alpha)|$ versus α for an orbit of Hamiltonian (18) with initial conditions $I_1 = 1/3$, $I_2 = 1/6 - 0.01\sqrt{(\varepsilon/0.06)}$ ($I_3 = 1$, $\varphi_1 = \varphi_2 = \varphi_3 = 0$) and perturbation parameter $\varepsilon = 0.001$. The spectrum has the peak structure expected in Nekhoroshev's regime. In order to verify the dependency of the slope as $1/\sqrt{\varepsilon}$ it is also plotted the spectrum of $\tilde{g}(\alpha)$ for $\varepsilon = 0.00025$. In this case we have introduced a stretching by a factor of 2 along the α-axis to easily verify that the slope of the peaks for $\varepsilon = 0.001$ are the same as those for $\varepsilon = 0.00025$ in the plot. The spectrum $\tilde{g}(\alpha)$ for the case $\varepsilon = 0.00025$ has also been translated of a factor of -3 in order to more clearly compare it to the spectrum for the case $\varepsilon = 0.001$.

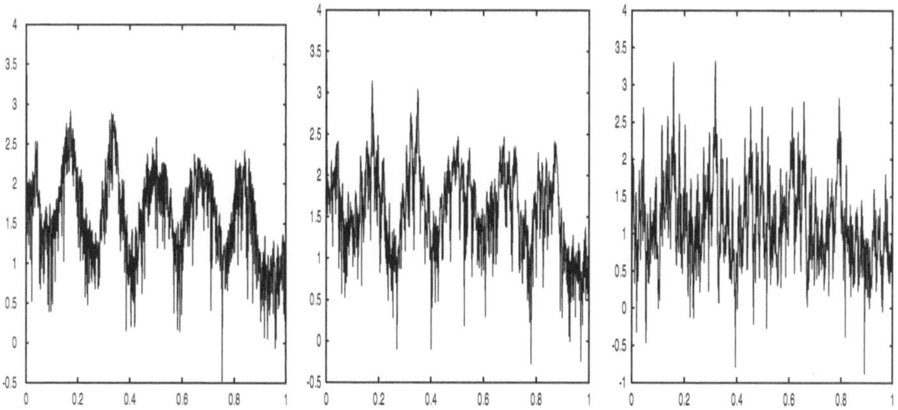

Figure 8. Spectrum of the function $g(t)$ defined in the text for an orbit of Hamiltonian (18) with initial conditions $I_1 = 1/3$, $I_2 = 1/6 - 0.01\sqrt{(\varepsilon/0.06)}$. Left column: $\varepsilon = 0.03$, central column $\varepsilon = 0.031$, right column $\varepsilon = 0.032$ (for all details see Guzzo, Lega and Froeschlé 2002).

All these examples concerned the non–degenerate case. In the next section we describe the spectral formulation of Nekhoroshev theorem also for degenerate cases, providing a method which can be directly applied to the real systems of Celestial Mechanics.

3. The degenerate case

The exponential stability of degenerate systems has been considered since the paper of Nekhoroshev (1977). Many works have shown (Neishtadt 1987, Benettin and Fassò 1996, Bessi et al. 2000) that degenerate systems can typically exhibit short–term diffusive phenomena even in the case of small perturbing parameters. However, Guzzo and Morbidelli (1997), Guzzo (1998) and Guzzo (1999) proved that for a certain class of degenerate systems, which includes many systems of interest for Celestial Mechanics (including the case of Asteroids), such kind of short–term instability can occur only in few very low order resonances, while the rest of the phase space is exponentially stable, provided that the perturbing parameters of the specific system considered are suitably small. More precisely, in Guzzo (1999) (see also Morbidelli and Guzzo 1997, Guzzo and Morbidelli 1997, Guzzo 1998) we considered a degenerate system with Hamiltonian in the form:

$$H = h(I) + \varepsilon \left(K_0(I,J) + f(I,J,\varphi,\psi) + \eta K_1(I,J,\psi) \right) \tag{19}$$

where $(I,\varphi) \in \mathbb{R}^n \times \mathbb{T}^n$ and $(J,\psi) \in \mathbb{R}^m \times \mathbb{T}^m$ are sets of canonically conjugate action–angle coordinates; ε and η are small parameters; the Fourier average of f with respect to the angles φ is zero. The above system is clearly degenerate because for $\varepsilon = 0$ the Hamiltonian does not depend on the actions J. The peculiar structure of the perturbation essentially means that the so–called secular Hamiltonian of H, i.e.

$$h(I) + \varepsilon \left(K_0(I,J) + \eta K_1(I,J,\psi) \right), \tag{20}$$

which is the Fourier average of H on the fast angles φ, is quasi–integrable. Many problems of Celestial Mechanics are in this form, for example the Asteroids (ε represents the mass of the planets; η represents the eccentricity and inclination of the planets, see Morbidelli and Guzzo (1997)). The action–angles variables (I,φ) are called non–degenerate variables, while the (J,ψ) are called degenerate variables. In Guzzo 1999 we proved the following result (see Guzzo and Morbidelli 1997 for the case of Asteroids):

THEOREM 1 *Let Hamiltonian (19) be defined for* $(I, \varphi) \in D \times \mathbb{T}^n$ *and* $(J, \psi) \in S \times \mathbb{T}^m$, *with* $D \subseteq \mathbb{R}^n$ *and* $S \subseteq \mathbb{R}^m$ *open sets, and analytic in some complex neighborhood of* $D \times \mathbb{T}^n \times S \times \mathbb{T}^m$. *Let* h *and* $h + \varepsilon K_0$ *be convex with respect to the* I *and the* (I, J) *respectively. There exist positive constants* $c_1, \ldots, c_4, \alpha, \beta, \varepsilon_0, \eta_0, \zeta_0$ *such that if* $\varepsilon \leq \varepsilon_0$ *and* $\eta \leq \eta_0$, *then for any motion* $I(t), J(t), \varphi(t), \psi(t)$ *with* $I(0)$ *in the domain:*

$$D_\varepsilon = \{I \in D \text{ such that } |k \cdot \omega_0(I)| \geq \varepsilon^{c_1} \text{ for any } k \in \mathbb{Z}^n \backslash 0 \text{ with } |k| \leq c_2 \ln \frac{1}{\varepsilon}\} \tag{21}$$

it is:

$$|I(t) - I(0)| \leq c_3 \varepsilon^{\frac{5}{6}} \zeta^\alpha \tag{22}$$

$$|J(t) - J(0)| \leq c_3 \zeta^\alpha \tag{23}$$

for any time t satisfying:

$$|t| \leq T_* = c_4 \varepsilon^{-1} \eta^{-1} \exp\left(\frac{\zeta_0}{\zeta}\right)^\beta \tag{24}$$

where $\zeta = \max\{\varepsilon, \eta\}$.

According to the above result, a chaotic solution of a degenerate system with Hamiltonian as in (19) can be in one of the three different dynamical situations: i) it belongs to some very low order resonances, where the Nekhoroshev theorem does not apply a priori (i.e. $I(0)$ is not in the domain D_ε); ii) it is outside the very low order resonances (i.e. $I(0)$ is in the domain D_ε), and the perturbing parameters are so small that Theorem 1 provides exponential stability; iii) the perturbing parameters are sufficiently large to prevent an exponential stability result. It is evident that ii) is the only case in which Theorem 1 proves long–term stability, while for i) and iii) the possibility for diffusion of the actions in polynomial times is not excluded. A Fourier Analysis of the observables of the system allows to numerically distinguish these different cases. Correspondingly, the spectra can be of three different types: without any structure, with the main peak structure and with a secondary peak structure.

3.1 The main peak structure

A straightforward application of the usual non-degenerate version of the Nekhoroshev theorem to Hamiltonian (19) allows to prove that if h is quasi–convex (with respect to the I) and ε is suitably small then the non–degenerate actions I are exponentially stable, if in the meantime the non-degenerate actions J do not leave the domain of H (Littlewood 1977). Therefore, on the one hand it is predicted stability for the non–degenerate actions only, on the other if the degenerate actions escape the domain of definition of H (a case that can indeed occur in some situations) then also the stability of the actions I can be compromised. Such a result, has been proved with the usual non–degenerate Littlewood construction, based on the averages of the fast–angles φ in the case of small ε. Also the spectral formulation of Guzzo and Benettin (2001) can be adapted in straightforward manner to Hamiltonian (19): if ε is small, any test function $\mathcal{G}(I, J, \varphi, \psi)$, computed on a solution $(I(t), J(t), \varphi(t), \psi(t))$ of (19) admits the representation:

$$g(t) = \sum_{k \in \mathbb{Z}^n} g_k(t) \exp(ik \cdot \Omega t) + g^1(t) \tag{25}$$

where $g^1(t)$ is exponentially small in any observation time $T \in [1/\sqrt{\varepsilon}, T_*]$, $\Omega \in \mathbb{R}^n$ depends on the initial datum and on the specific resonance considered, the Fourier transform of g_k satisfies the scaling law:

$$|\tilde{g}_k(\alpha)| \leq a_0 T e^{-a_1|k|} \exp\left(-a_2 \frac{|\alpha|}{\varepsilon^{a_3}}\right) \tag{26}$$

for any $\alpha \in \mathbb{R}$. Representation (25) with the scaling law (26) is the same introduced for the non–degenerate systems. However, for non–degenerate systems, such a structure does not give indication of exponential stability, but only of the weak stability result for the non–degenerate actions I described above. We will refer to the representation (25) as the 'main peak structure'. The reason for this weak result is that it does not yet exploit the fact that Theorem 1 provides a stronger stability result outside some very low order resonances. In Guzzo (2002) it is proved that if also η is small and $I(0) \in D_\varepsilon$, then $g(t)$ admits the representation:

$$g(t) = \sum_{k=(k_0,k_1) \in \mathbb{Z}^n \times \mathbb{Z}^m} g_k^0(t) \exp(i\varepsilon k_1 \cdot \Omega_1 t) \exp(ik_0 \cdot \Omega_0 t) + g^1(t) \tag{27}$$

where $g^1(t)$ is exponentially small in the observation time T, $\Omega_0 \in \mathbb{R}^n$ and $\Omega_1 \in \mathbb{R}^m$ are suitable vectors which depend on the initial datum and on the specific resonance considered, the Fourier transform of g_k satisfies the scaling law:

$$\left|\tilde{g}_k^0(\alpha)\right| \leq a_0 T e^{-a_1|k|} \exp\left(-a_2 \frac{|\alpha|}{\varepsilon \zeta^{a_3}}\right) \tag{28}$$

for any $\alpha \in \mathbb{R}$, where $\zeta = \max\{\varepsilon, \eta\}$. The observation time T has to be at least of order $T_0 = 1/(\varepsilon\sqrt{\zeta})$, and smaller than T_* as in (24).

3.2 The secondary peak structure

We will refer to the representation (27) with the estimate (28) as the 'secondary peak structure', which is the novelty with respect to the non–degenerate case and its presence gives indication that all the actions of the system, including the degenerate ones, are exponentially stable. The structure can be easily recognized numerically: from (27) we understand that for any $k_0 \in \mathbb{Z}^n$ there is a multiplet of secondary peaks centered on the frequencies: $k_0 \cdot \Omega_0 + \varepsilon k_1 \cdot \Omega_1$, for any $k_1 \in \mathbb{Z}^m$. The slope of these secondary peaks, which is of order $1/(\varepsilon\zeta^{a_3})$ is much bigger than the slope of the main peaks, which is only of order $1/\sqrt{\varepsilon}$. These secondary peaks are separated in the frequency axis by distances of order ε, and their amplitude decreases exponentially as $\exp(-|k_1|)$. It is therefore clear that these peaks group together to constitute one of the so–called main peaks. By increasing the value of η at fixed ε the slope of the secondary peaks becomes smaller, and when we reach the threshold η_0 at which the exponential stability of the degenerate actions is no longer valid, we expect that the main peak appears as a single peak, no longer resolved in a multiplet of secondary peaks. We summarize this description in Figure 9: on the top we represent the spectrum of a system with small ε but with large η: only the main peak structure is visible with many peaks centered at the values $k_0 \cdot \Omega_0$, for $k_0 \in \mathbb{Z}^n$. On the bottom we represent the spectrum of a system with small ε and small η: the spectrum is made by many secondary peaks centered at $k_0 \cdot \Omega_0 + \varepsilon k_1 \cdot \Omega$ for $k_1 \in \mathbb{Z}^m$, and these peaks group together to constitute the main peak structure (represented by a dashed line in the plot).

3.3 A model example

In Guzzo (2002) we computed numerically the peak structure of a test function of the chaotic solutions of a model degenerate system, detecting the main and secondary peak structures for small values of the perturbing parameters.

We first considered the decoupled system:

$$H_0 = \sum_{i=1}^{3} \frac{I_i^2}{2} + \varepsilon \sum_{i=1}^{3} \frac{J_i^2}{2} + \varepsilon\mu f(\varphi_1, \varphi_2, \varphi_3) + \varepsilon\eta f(\psi_1, \psi_2, \psi_3) \qquad (29)$$

where $(I, \varphi) \in \mathbb{R}^3 \times \mathbb{T}^3$ are the non–degenerate variables, $(J, \psi) \in \mathbb{R}^3 \times \mathbb{T}^3$ are the degenerate variables and

$$f(\varphi_1, \varphi_2, \varphi_3) = \cos(\varphi_1) + \sin(\varphi_2) + \cos(\varphi_1 + \varphi_2) + \sin(\varphi_1 - \varphi_3) + \cos(\varphi_2 + \varphi_3).$$

The non–degenerate variables satisfy the same Hamilton equations of the Hamiltonian system considered in Figure 4, with perturbing parameter

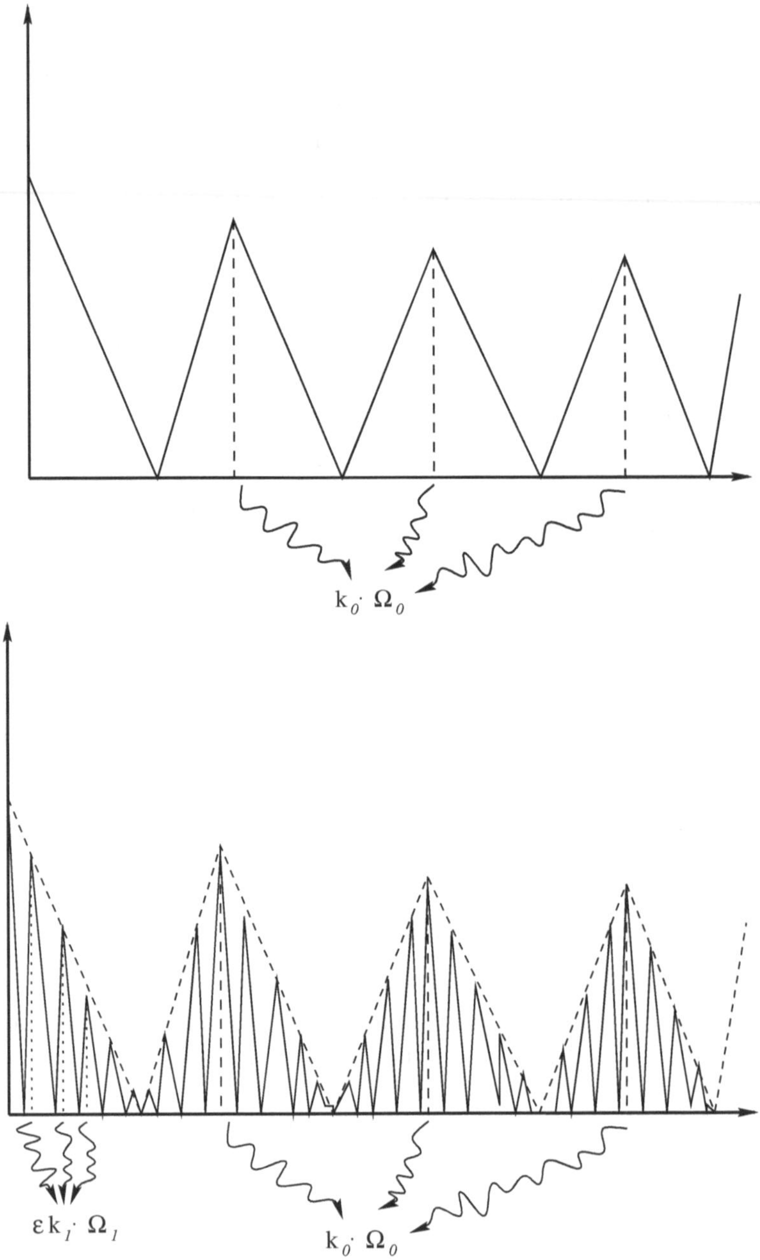

Figure 9. On the top we represent the spectrum of a system with small ε but with large η: only the main peak structure is visible with many peaks centered at the values $k_0 \cdot \Omega_0$, for $k_0 \in \mathbb{Z}^n$. On the bottom we represent the spectrum of a system with small ε and small η: the spectrum is made by many secondary peaks centered at $\varepsilon k_1 \cdot \Omega$ for $k_1 \in \mathbb{Z}^m$, and these peaks group together to constitute the main peak structure (dashed line in the plot).

$\varepsilon\mu$ (instead of ε). Therefore, the initial condition: $I_1(0) = 10^{-2}\sqrt{\varepsilon\mu}$, $I_2(0) = 10^{-2}\sqrt{\varepsilon\mu}$, $I_3(0) = 1$ (the initial phases are $\varphi_1(0) = 1, \varphi_2(0) = 0, \varphi_3(0) = 0$) generates a resonant solution equivalent to that considered in Figure 4, which corresponds to the resonance $\sum_{i=1}^{3} k_0^i I_i = 0$ with $k_0 = (n_1, n_2, 0)$, $n_1, n_2 \in \mathbb{Z}$.

Also the degenerate variables satisfy the same Hamilton equations of the Hamiltonian system considered in Figure 4, with perturbing parameter η (instead of ε), and time scaled by a factor ε. A resonant solution equivalent to that considered in Figure 4 is generated by the initial condition $J_1(0) = 10^{-2}\sqrt{\eta}$, $J_2(0) = 10^{-2}\sqrt{\eta}$, $J_3(0) = 1$ (the initial phases are $\psi_1(0) = 1, \psi_2(0) = 0, \psi_3(0) = 0$). Therefore, we are considering the resonance among the degenerate and non–degenerate variables:

$$\sum_{i=1}^{3} k_0^i I_i + \sum_{i=1}^{3} \varepsilon k_1^i J_i = 0$$

generated by the integer vectors $(k_0, k_1) = (n_1, n_2, 0, n_3, n_4, 0)$, with n_1, n_2, n_3, $n_4 \in \mathbb{Z}$. Correspondingly, with the same notations of Section 3.2, it is:

$$\Omega_0 = (0, 0, 1), \quad \Omega_1 = (0, 0, 1).$$

We now couple the non–degenerate and degenerate variables by considering the Hamiltonian:

$$H = H_0 + a\varepsilon g_1(\varphi, \psi) + \varepsilon\mu g_2(\varphi_1, \varphi_2, \psi_1, \psi_2), \tag{30}$$

where $g_1 = -\sin(\varphi_1 + \varphi_2 + \varphi_3 + \psi_1 + \psi_2 + \psi_3)$ has zero–average with respect to the non–degenerate angles, and $g_2 = -\sin(\varphi_1+\varphi_2+\psi_1+\psi_2)$ is a resonant harmonic. Therefore, the secular Hamiltonian of H is quasi–integrable with perturbation of order $\varepsilon\eta$, and H is in the form (19) and satisfies the hypotheses of Theorem 1. Concerning the initial conditions, we are considering resonances among the non–degenerate angles of order 1, which are not the logarithmically high order resonances considered in Theorem 1, equation (21). However, with minor changes to the proof of Theorem 1, it is possible to show that the exponential estimates for the I, J hold also for this resonance if: $\mu < \varepsilon\eta$ and $a < \sqrt{\eta}$ (see Guzzo 2002).

In the following I report the result of the Fourier Analysis of the solutions corresponding to the above initial conditions, and different values of ε and η. The test function used to mix the different degrees of freedom is:

$$\mathcal{G} = \frac{1}{\left(\sum_{i=1}^{3}(\cos(\varphi_i) + \cos(\psi_i)) + 1\right)^8 + 10}. \tag{31}$$

The data are also filtered before the Fast Fourier transform, with a suitable analytic window $\Phi(t)$, as explained in Section 4.

A first set of numerical integrations concerns a small value of $\varepsilon = 0.02$, three different values of $\eta = 0.0001, 0.0004, 0.01$, and $\mu = \varepsilon\eta/10, a = \sqrt{\eta/10}$. As explained above, if ε and η are suitably small, Theorem 1 proves the exponential stability of the degenerate variables, and correspondingly we expect to observe both the main and the secondary peak structures. Indeed, in the spectra reported in Figure 10 we can clearly distinguish the main peaks located at the frequencies $\alpha = 0, 1, 2, ...$ (pictures on the left), and the secondary peaks at a distance $\varepsilon = 0.02$ one to the other, for the small values of $\eta = 0.0001, 0.0004$ (pictures on the top–right, middle–right). Moreover, the slope of the secondary peaks in the case $\eta = 0.0001$ is bigger by a factor two than the slope of the secondary peaks in the case $\eta = 0.0004$. Instead, for the bigger value $\eta = 0.01$ the secondary peak structure is destroyed (picture on the bottom–right), while the main peak structure is maintained (picture on the bottom–left). These spectra have been obtained by analyzing the solution up to a time $t \sim 10^7$, which is longer than the time used for the non–degenerate system of Figure 4. The reason is that the dynamics of the degenerate variables takes place on times scaled by a factor ε with respect to the non–degenerate variables and therefore, if we want our analysis to include also the effects of the degenerate variables, it is mandatory to compute the solution up to a time multiplied by a factor $1/\varepsilon$. This fact is typical of degenerate systems, and not specific of this numerical analysis. A second set of numerical integrations concerns small values of $\varepsilon = 0.0001, 0.000025$, but $\mu = a = \eta = 1$. As explained above, in these conditions Theorem 1 does not prove the exponential stability of the degenerate variables, and correspondingly we expect to observe only the main peak structure, not the secondary one. In Figure 11 we report the result of this Fourier Analysis. Indeed, the spectrum consists only of the main peaks, which rescale as $1/\sqrt{\varepsilon}$ as expected.

4. Basic Fourier Analysis of chaotic solutions

A. Preliminaries. In this section we sketch the mathematical ideas which are at the basis of the Fourier representation of chaotic solutions. All mathematical details can be found Guzzo and Benettin (2001), Guzzo (2002).

What follows refers to the Fourier Analysis of the functions: $g(t) = \mathcal{G}(I(t), \varphi(t))$ where $\mathcal{G} : B \times \mathbb{T}^n \to \mathbb{R}$ is real analytic, $B \subseteq \mathbb{R}^n$ is open and $I(t), \varphi(t)$ is a solution of system (1).

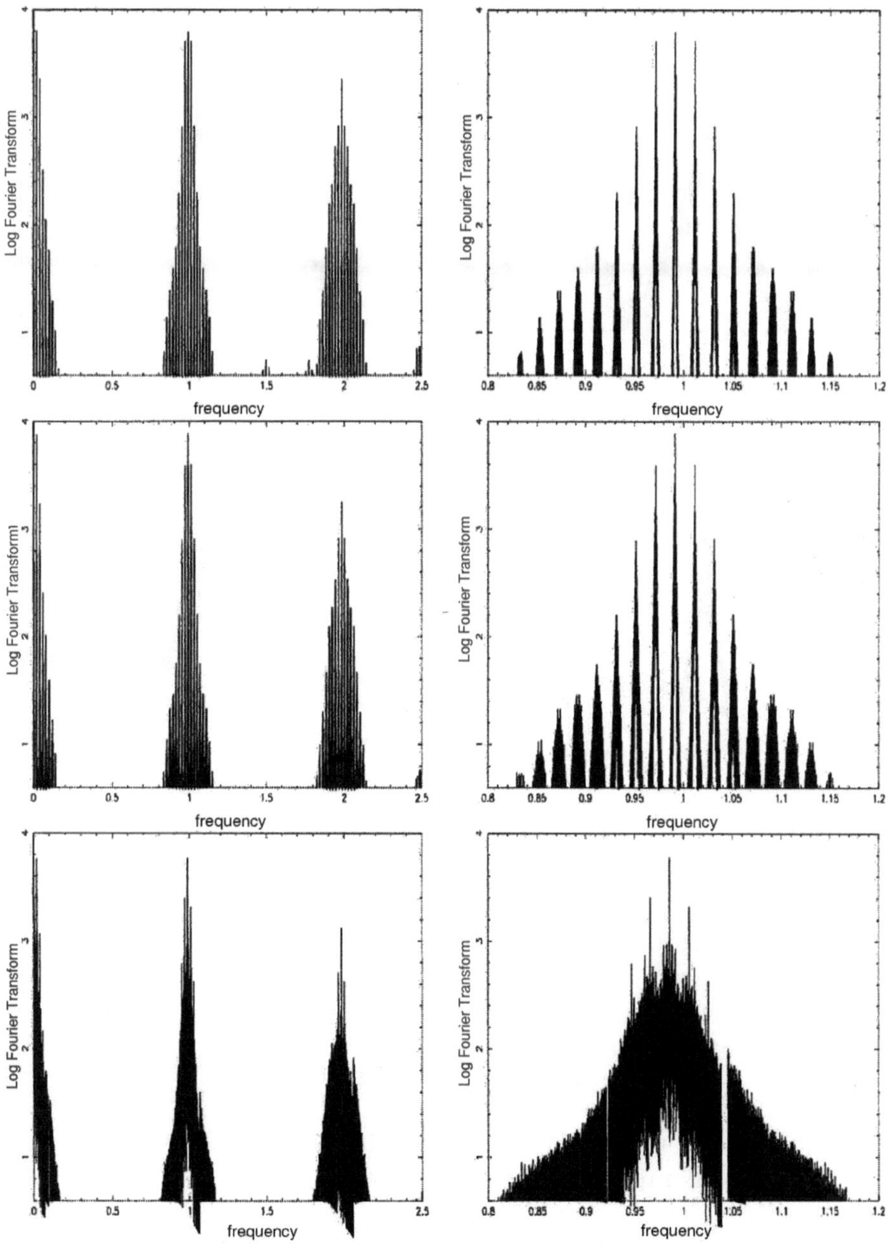

Figure 10. Spectra corresponding to $\varepsilon = 0.02$, and $\eta = 0.0001$ (on the top), $\eta = 0.0004$ (on the middle), $\eta = 0.01$ (on the bottom).

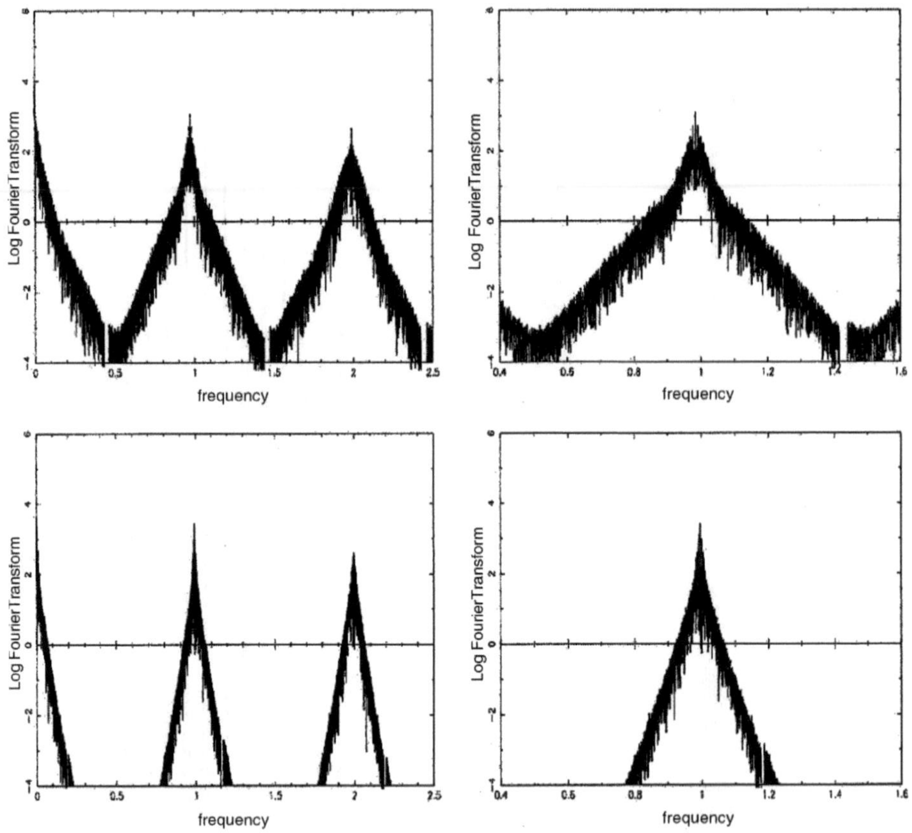

Figure 11. We represent the spectrum of a system with $\eta = \mu = a = 1$ and $\varepsilon = 0.0001$ (on the top), $\varepsilon = 0.000025$ (on the bottom). These spectra have been obtained by analyzing the solution up to a time $t \sim 10^5$.

For any function $u : \mathbb{R} \to \mathbb{R}$, we denote by:

$$u(t) = \int_{-\infty}^{\infty} d\alpha \; \tilde{u}(\alpha) \; e^{i\alpha t} \tag{32}$$

its continuous Fourier representation where $\tilde{u}(\alpha)$ is the usual Fourier transform of u computed on the frequency α:

$$\tilde{u}(\alpha) = \frac{1}{2\pi} \int_{-\infty}^{\infty} dt \; u(t) \; e^{-i\alpha t}. \tag{33}$$

We refer to the spectrum of a function u as to the Cartesian graph of $|\tilde{u}(\alpha)|$ versus the frequency α. The asymptotic behavior of the spectrum of a function u, i.e. its behavior at high frequencies, is determined by the analyticity properties of u, as stated by the following lemma:

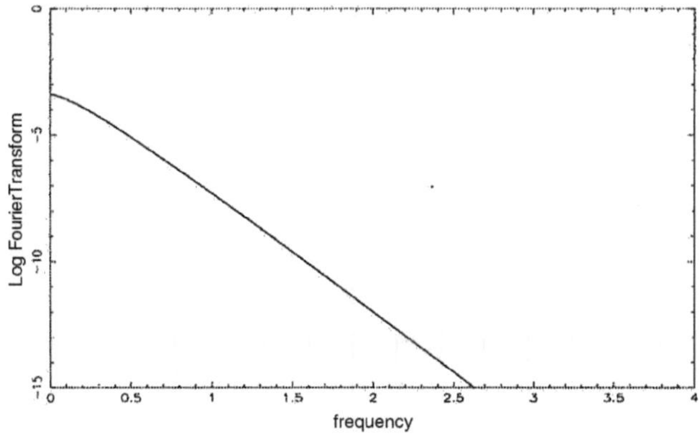

Figure 12. Fast Fourier transform of $u(x) = 1/(25 + x^2)^{\frac{3}{2}}$, represented in semi–logarithmic scale. Indeed, the spectrum is below a straight line of slope -5, according to the analyticity radius which is 5.

LEMMA 1. *Let the complex function $u(t)$ be analytic for $|\mathrm{Im}\, t| \leq \sigma$ and:*

$$\lim_{t\to\infty, t\in\mathbb{R}} \sup_{|\xi|\leq\sigma} |u(t+i\xi)| = 0; \tag{34}$$

let the constant C be such that for any $|\xi| \leq \sigma$ it is: $\int_R |u(t+i\xi)|dt \leq C < \infty$. Then, for any $\alpha \in R$ the Fourier transform of u satisfies:

$$|\tilde{u}(\alpha)| \leq \frac{C}{2\pi} e^{-|\alpha|\sigma}.$$

Therefore the exponential decay of the Fourier transform of a function depends on its analyticity radius σ (see Figure 12).

Being interested in the asymptotic behavior of the spectrum of the observables $g(t)$ of Hamiltonian systems, we consider the analyticity radius of $g(t)$ in the variable t. Standard estimates ensure that the Taylor expansion of $g(t)$ around any $t \in \mathbb{R}$ converges in a finite radius which does not diverge as ε goes to zero, i.e. the analyticity radius does not increase as ε decreases. From lemma 1 it follows that the spectrum of $g(t)$ decays exponentially at a rate which does not increase when ε becomes small. This is confirmed by the three Fourier analysis shown in Figure 13, where we considered the same quasi–integrable system, with the same resonant initial condition, but for three different values of ε: it is clear that increasing ε from 0 (Figure 13 top) to 0.1 (Figure 13 bottom) changes qualitatively the spectrum, but does not change significantly the exponential decay. Therefore, we cannot consider the exponential rate at

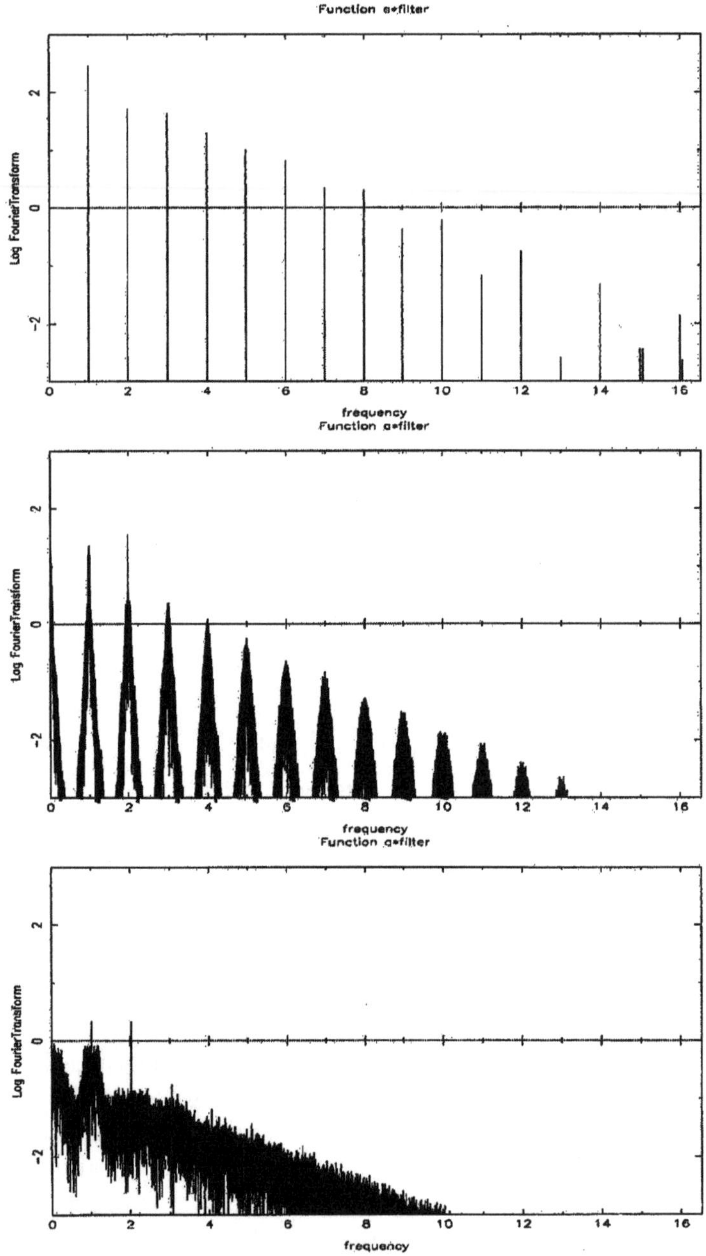

Figure 13. Fast Fourier transform of the function $g(t) = \Phi(t)\mathcal{G}(I(t), \varphi(t))$, where $\mathcal{G}$ and $\Phi(t)$ are as in Figure 2, and $(I(t), \varphi(t))$ is a solution of the Hamilton equations of the quasi–integrable Hamiltonian of Figure 2 with $\varepsilon = 0$ (top), $\varepsilon = 10^{-4}$ (middle), $\varepsilon = 0.01$ (bottom) and the resonant initial condition $I_1(0) = 10^{-2}\sqrt{\varepsilon}$, $I_2(0) = 10^{-2}\sqrt{\varepsilon}$, $I_3(0) = 1$. In all cases, the spectrum decays exponentially with a rate which is approximately 14/5.

which $|\tilde{g}|$ decreases as significant to understand the dynamical regime of the solution. Instead, it is evident from Figure 13 that what qualitatively changes is how the spectrum is concentrated around the multiples of the fast frequency 1. The presence of fast frequencies in resonant motions can be suitably treated using the normal form variables adapted to the resonance properties of the solution.

B. Normal forms variables. In the proof of Nekhoroshev theorem the action space is covered by many domains characterized by all possible resonances generated by integer vectors k of order $|k| = \sum_i |k_i|$ up to a given positive number K, chosen of order $1/\varepsilon^{2n}$ for convex systems (see Lochak 1992, Pöschel 1993). For the definition of the geometry of resonances we follow the definitions given in Pöschel 1993 and we refer to that paper for all details. Here, we just recall that a d-dimensional lattice $\Lambda \subseteq Z^n$ defines a resonance through the relation:

$$k \cdot \frac{\partial h}{\partial I} = 0 \tag{35}$$

for any $k \in \Lambda$, or equivalently: $\Pi_\Lambda \frac{\partial h}{\partial I} = 0$, where Π_Λ denotes the Euclidean projection of a vector onto the linear space spanned by Λ. As already remarked, we only consider resonance lattices $\Lambda \subseteq Z^n$ which are generated by $d \leq n-1$ independent integer vectors $k^{(i)}$, $i \leq d$, with order smaller than a threshold order K, which grows as $1/\varepsilon^{\frac{1}{2n}}$.

According to the definitions given in Pöschel (Pöschel 1993), the resonant domain associated to a lattice Λ is a neighborhood of the resonance defined in the following way: first we require that the action is suitably close to the resonance through the inequality:

$$\|\Pi_\Lambda \frac{\partial h}{\partial I}(I)\| \leq \frac{a_0}{(a_1 K)^{n-d}|\Lambda|} \tag{36}$$

(a_0, a_1 are suitable constants, $|\Lambda|$ is the Euclidean volume of the lattice Λ), second we require that the action I is suitably far from the other resonances; more precisely we require:

$$\|\Pi_{\Lambda'} \frac{\partial h}{\partial I}(I)\| > \frac{a_0}{(a_1 K)^{n-d-1}|\Lambda'|} \tag{37}$$

for any lattice Λ' generated by $d+1$ independent integer vectors of order smaller that K. Then, in the resonant domain of the lattice Λ, if ε is small, one can construct a canonical transformation which averages from the Hamiltonian the non–resonant angles up to an exponentially small remainder. More precisely, there exists a near to the identity canonical transformation:

$$\mathcal{C}_\Lambda : \quad (I', \varphi') \longmapsto (I, \varphi) \tag{38}$$

conjugating (1) to the normal form Hamiltonian:

$$H \circ \mathcal{C}_\Lambda(I', \varphi') = h(I') + \varepsilon f_\Lambda(I', \varphi') + \varepsilon e^{-K} r(I', \varphi'), \tag{39}$$

where the Fourier series of f_Λ contains only harmonics $f_\Lambda^k(I') e^{ik\cdot\varphi'}$ with $k \in \Lambda$. As a first approximation, the motions of (39) can be described following Benettin and Gallavotti (1986), who gave a simplified description neglecting in (39) the exponentially small remainder. In the truncated Hamilton equations:

$$\dot{I}' = -\varepsilon \frac{\partial f_\Lambda}{\partial \varphi}(I', \varphi'), \quad \dot{\varphi}' = \frac{\partial h}{\partial I'} + \varepsilon \frac{\partial f_\Lambda}{\partial I'}(I', \varphi') \tag{40}$$

$\dot{I}'$ is always parallel to a vector in the linear space spanned by Λ, so that $I'(t)$ is contained in the plane spanned by Λ through the point $I'(0)$, which is usually called "fast-drift plane". Because h is convex there is only one point in the fast-drift plane such that $\Pi_\Lambda \partial_I h(I^\Lambda) = 0$ which we denote with I^Λ, and the restriction of the unperturbed energy $h(I)$ to the fast drift plane has a minimum (or a maximum) in I^Λ. Moreover, for any motion $I'(t)$ starting in the resonance, i.e. near I^Λ, it is: $|h(I'(t)) - h(I'(0))| \leq 2\varepsilon \, |f_\Lambda|^\infty$, and therefore it cannot move far from the resonance. For the true Hamilton equations (i.e. the non–truncated ones), these conclusion is valid not for an infinite time, but up to the exponentially long one of Nekhoroshev theorem.

In the following we work out a simplified description of the Fourier spectrum of g neglecting in the normal form the exponentially small remainder.

C. The truncated normal form approximation. Consider a resonant motion with initial condition in the resonant domain of a lattice Λ and replace the true solution with

$$(I(t), \varphi(t)) = \mathcal{C}_\Lambda(I'(t), \varphi'(t)), \tag{41}$$

where $(I'(t), \varphi'(t))$ is the solution of the Hamilton equations (40) of the truncated normal form. It is convenient to introduce the pull–back of the observable $\mathcal{G}$ with respect to the normal form transformation $\mathcal{C}_\Lambda$:

$$\mathcal{F}(I', \varphi') = \mathcal{G} \circ \mathcal{C}_\Lambda(I', \varphi') \tag{42}$$

which has the Fourier expansion: $\mathcal{F} = \sum_{k\in Z^n} \mathcal{F}_k(I')e^{ik\cdot\varphi'}$. Therefore, the function $g(t) = \mathcal{G}(I(t), \varphi(t)) = \mathcal{F}(I'(t), \varphi'(t))$ can be represented in the following way:

$$g(t) = \sum_{k\in Z^n} \mathcal{F}_k(I'(t))e^{ik\cdot\varphi'(t)}. \tag{43}$$

Again, it would be not difficult to show that each term of this expansion:

$$G_k(t) = \mathcal{F}_k(I'(t))e^{ik\cdot\varphi'(t)}, \tag{44}$$

has a convergence radius of order 1 in ε. To be more precise, if we look for constants C_k and σ such that:

$$\left|\tilde{G}_k(\alpha)\right| \leq C_k e^{-|\alpha|\sigma}, \tag{45}$$

for any $\alpha \in \mathbb{R}$, with the constraint that C_k is not divergent in ε, then one obtains only σ of order 1 in ε. The reason is that the presence of quasi–periodic terms in the function implies that its norm grows exponentially with the imaginary part of the complex time t. Consider the toy function: $G(t) = \frac{1}{1+\varepsilon t^2}e^{it}$. The 'amplitude' $\frac{1}{1+\varepsilon t^2}$ has convergence radius $1/\sqrt{\varepsilon}$, and therefore its Fourier spectrum appears in a semi–logarithmic plot as a peak centered around the frequency 1, and slopes $1/\sqrt{\varepsilon}$. If we want an estimate like (45) with $\sigma = 1/\sqrt{\varepsilon}$, we are forced to choose C of order $e^{1/\sqrt{\varepsilon}}$. A more convenient way to represent $G(t)$ is: $G(t) = g(t)e^{it}$, and indeed the Fourier Transform of g trivially satisfies $|\tilde{g}(\alpha)| \leq Ce^{-|\alpha|\sigma}$ with $\sigma = \mathcal{O}(1/\sqrt{\varepsilon})$ and $C = \mathcal{O}(1)$. Therefore, returning to the observable $g(t) = \mathcal{F}(I'(t), \varphi'(t)) = \sum_{k\in\mathbb{Z}^n} \mathcal{F}_k(I'(t))e^{ik\cdot\varphi'(t)}$, we have to avoid to over-estimate the norm of quasi–periodic terms which possibly are produced by the dynamics. This can be done as follows: each Fourier component $\mathcal{F}_k(I'(t))e^{ik\cdot\varphi'(t)}$ can contain of course fast periodic terms. Therefore, *we look for $\Omega \in R^n$ such that the all functions:*

$$g_k(t) = \mathcal{F}_k(I'(t))e^{ik\cdot\varphi'(t)}e^{-ik\cdot\Omega t} \tag{46}$$

are analytic for $|\mathrm{Im}\, t| \leq \frac{1}{\varepsilon^\beta}$, with suitable $\beta > 0$. Such a vector Ω indeed exists and can be chosen as

$$\Omega = (I - P_\Lambda)\omega(I(0)).$$

Indeed, one can prove that the observable $\mathcal{G}$ computed on the Hamiltonian flow of the truncated normal form has the representation:

$$g(t) = \sum_k g_k(t)e^{ik\cdot\Omega t}$$

where the $g_k(t)$, defined as in (46), are analytic in a complex strip of radius: $\sigma = \mathcal{O}(\varepsilon^{\frac{-(n-d)}{2n}})$, where d is the dimension of the resonance lattice. At a distance of order $\sqrt{\varepsilon}$ from the resonance the above estimate improves to the better one (see Guzzo and Benettin 2001): $\sigma = \mathcal{O}(1/\sqrt{\varepsilon})$.

D. The analytic window. Because the amplitude functions $g_k(t)$ in general do not vanish at infinity, they do not satisfy condition (34) of lemma 1. Therefore, we can introduce their Fourier Transform only after multiplication with a suitable window function Φ which vanishes at infinity. It is sufficient to require that:

i) Φ is analytic in a complex strip of radius ϱ with $\varrho > \sigma$;

ii) $\int_R |\Phi(t+is)|\, dt \le C < \infty$ for any $|s| \le \varrho$;

iii) $|\Phi(t) - 1| = \mathcal{O}\Big(\exp\Big(-\frac{1}{\varepsilon^\beta}\Big)\Big)$ if $|t| \le T$;

iv) $|\Phi(t)| = \mathcal{O}\Big(\exp\Big(-\frac{1}{\varepsilon^\beta}\Big)\Big)$ if $|t| \ge 2T$;

A possible choice for Φ is:

$$\Phi(t) = \frac{1}{2}\Big[\tanh\Big(\frac{t+T}{\varrho}\Big) - \tanh\Big(\frac{t-T}{\varrho}\Big)\Big], \tag{47}$$

where ϱ has to be chosen bigger than the analyticity radius of g, i.e.

$$\varrho > \sigma = \frac{1}{\sqrt{\varepsilon}},$$

and T is any time interval (this is because we are considering the truncated normal form dynamics; for the true dynamics T has to be chosen not bigger than the exponentially long stability time predicted by Littlewood theory) satisfying conditions *iii)* and *iv)* above. Then, we write trivially:

$$g_k(t) = \Phi(t) g_k(t) \; + (1 - \Phi(t))\; g_k(t),$$

and the difference between $g_k(t)$ and the windowed quantity $\Phi(t)g_k(t)$ is exponentially small for times t smaller than T. The function of time $\Phi(t)g_k(t)$ has Fourier transform which decays at an exponential rate σ, and therefore defining:

$$g_1(t) = \sum_k (1 - \Phi(t))\; g_k(t) e^{ik\dot{\Omega}t}, \tag{48}$$

the difference $g(t) - g_1(t)$ has the peak structure.

The problem is understood for the truncated normal form dynamics, but the true dynamics is the one given by the complete Hamiltonian, with the exponentially small remainder:

$$\tilde{H} = h(I) + \varepsilon \tilde{f}(I, \varphi) + \varepsilon e^{-\frac{1}{\varepsilon^{\alpha}}} r(I, \varphi).$$

For such a complete Hamiltonian we cannot repeat the same argument as above, because the remainder, though exponentially small, forces the analyticity radius of the $g_k(t)$ to $\mathcal{O}(1)$ in ε! However, in Guzzo and Benettin (2001) it is proved that the part of $g(t)$ which is analytic only for $|\mathrm{Im}\, t| = \mathcal{O}(1)$ is exponentially small for $|t| \leq T$, and therefore, it can be relegated in the exponentially small term g^1.

5. Application to Asteroids

The application of the Fourier Analysis of chaotic solutions to a physically interesting system, namely the dynamics of some numbered Asteroids, has been done in Guzzo, Knezevic and Milani (2002). In this section we briefly describe some of the results published in that paper.

The Hamiltonian of an asteroid perturbed by the four giant planets has the form:

$$H = -\frac{1}{2L^2} + \sum_{j=1,4} m_j \mathcal{P}(L, G, \Theta, l, g, \vartheta, q_j(t)) \tag{49}$$

where $L = \sqrt{a}, G = L\sqrt{1-e^2}, \Theta = G\cos i, l, g = \omega, \vartheta = \Omega$ are the canonical action–angle Delaunay variables where l is the mean anomaly, ω is the argument of perihelion, Ω is the longitude of node, a is the semi major axis, e is the eccentricity and i is the inclination; the gravitational constant and the mass of the Sun are set equal to unity; the $m_j, j \leq 4$ are the masses of the giant planets; the functions $q_j : R \to R^3$ denote the position of the j–th planet as a function of time.

The Hamiltonian (49) is quasi–integrable with integrable part $\mathcal{H}_0 = -1/(2L^2)$ and perturbation depending on all other variables and on time through the osculating elements of the planets. Because $\mathcal{H}_0$ depends only on the action L the problem is degenerate. In Guzzo and Morbidelli (1997) exponential estimates have been proved, except for low order mean motion and secular resonances (among these resonances there are those associated with the Kirkwood gaps, and possibly many others, see Morbidelli and Nesvorny 1999), provided that the perturbation parameters of the problem (the masses of the planets with respect to the Sun, the eccentricity and the inclination of the planets) are suitably small.

THEOREM 2 (Guzzo and Morbidelli 1997). *Let n_j, e_j, i_j, $j = 1, \dots, 4$, denote the frequencies of the mean motion, the eccentricity and the inclination of the planets from Jupiter to Neptune. Let ε denote the mass of Jupiter. There exist positive constants $c_1, c_2, \varepsilon_0, e_0, i_0$ such that if $\varepsilon \leq \varepsilon_0$, $e_j \leq e_0$ and $i_j \leq i_0$ for any $j \leq 4$, then for any motion $L(t)$, $G(t)$, $\Theta(t)$, $g(t)$, $l(t)$, $\vartheta(t)$ with $L(0)$ satisfying:*

$$\left| \nu_0 \frac{1}{L(0)^3} + \sum_{k=1}^{4} \nu_k n_k \right| \geq \varepsilon^{c_1} \tag{50}$$

for any $(\nu_0, \nu_1, \dots, \nu_4) \in Z^5$ with $\sum_{k=0}^{4} |\nu_k| \leq c_2 \ln(1/\varepsilon)$, the semi–major axis, the eccentricity and the inclination of the asteroid are exponentially stable (the stability time grows exponentially as a suitable inverse power of $\varepsilon, e_1, \dots, e_4, i_1, \dots, i_4$).

As explained in detail in Morbidelli and Guzzo (1997), we do not expect that the thresholds ε_0, e_0, i_0 are uniform in the phase–space. Therefore, in principle, we do not know whether outside the low order mean motion resonances the entire phase space is exponentially stable or not, and consequently the search for exponentially stable objects has to be done with reference to specific regions of the phase–space. In Guzzo, Knezevic and Milani (2002) we selected asteroids in different regions of the phase–space where we suspected different long–term behaviors. The motion has been integrated up to 100 Myr with an accurate multi-step integrator of order 12 (Milani and Nobili 1988) with a Runge-Kutta starter (the software package "orbit9" available through the AstDyS service at `http://hamilton.dm.unipi.it/ astdys`). The dynamical model used for the numerical integrations included the direct effects from the four outer major planets, while the indirect effect of the inner planets has been accounted for by applying a barycentric correction to the initial conditions (Milani and Knežević 1992). To avoid excessive output we applied the digital on–line filter described in Knežević and Milani (2000), Carpino et al. (1987). Then, we considered the observable:

$$\begin{aligned} \mathcal{G} &= \Big([\cos(h) + \sin(h) + \cos(k) + \sin(k) + \cos(q) \\ &\quad + \sin(q) + \cos(p) + \sin(p) + \cos(a) + \sin(a)]^{14} + 1 \Big)^{-1}, \end{aligned} \tag{51}$$

where a, l, $h = e \cos \varpi$, $k = e \sin \varpi$, $p = \tan(i/2) \cos \Omega$, $q = \tan(i/2) \sin \Omega$ are the equinoctial elements, and we computed the Fast Fourier transform of:

$$g(t) = \Phi(t) \mathcal{G}(a(t), h(t), k(t), p(t), q(t)), \tag{52}$$

where $\Phi(t)$ is the usual analytic window function. With the choices made (integration time, sampling time, digital filtering: for all details see Guzzo, Knezevic and Milani 2002) we were able to clearly resolve the characteristics of the Fourier spectrum in the domain of the secular frequencies in the frequency range from $10^{-7} rad/yr$ up to $7 \cdot 10^{-4} rad/yr$, and so the FFT of the filtered data is sufficient to detect the presence of the secondary peak structure (while we loose the main peak structure, around the spectral lines at the fast frequencies pertaining the mean motions, but this is not relevant for the long–term stability of the system). Corresponding to the choice of Asteroids with different dynamical behavior, we found objects with very different Fourier spectrum. One of the most interesting cases is asteroid 305 Gordonia, with a moderately large Lyapunov exponent ($\simeq 3 \times 10^{-5}$); it has a spectrum which is continuous and has the peak structure which can be observed from spectra computed using numerical integrations up to 100Myr and 10Myr (but with more frequent sampling), as it can be seen from Figure 15. This is an indication that this object is in the Nekhoroshev regime. As it can be seen from Figure 14, its semi–major axis, eccentricity and inclination, despite the presence of local chaos, and proximity to the secular resonance $g + s - g_6 - s_6$ (period of oscillation $\simeq$ 5 Myr), do not diffuse in the phase–space in 100Myr. Being in the Nekhoroshev regime, we expect that this object does not drift in the phase–space over even longer times. Other interesting cases we found pertain to objects with moderate-to-large Lyapunov exponent ($3 - 8 \times 10^{-5}$), and with continuous spectra without the peak structure. One of these objects is asteroid 165 Loreley, whose spectrum reported in Figure 16 is quite confused so that we could not recognize a peak structure. Therefore, the spectrum gives an indication that the object is not in the Nekhoroshev regime. Nevertheless, its semi–major axis, eccentricity and inclination are quite stable (except for possible local chaos) up to 100Myr. The contradiction is only apparent. The 100Myr integration time on the one hand is sufficient to recognize with Fourier analysis that some weak resonances possibly overlap, on the other it is too small to observe a drift in the phase–space due to these overlapping resonances. All these objects are close to the Veritas family, but their motion is affected also by other resonances (Milani 1995; Knežević and Jovanović 1997).

Acknowledgments In this article I have described works done in collaboration with G. Benettin, C. Froeschlé, Z. Knezevic, E. Lega and A. Milani. I wish to thank all of them.

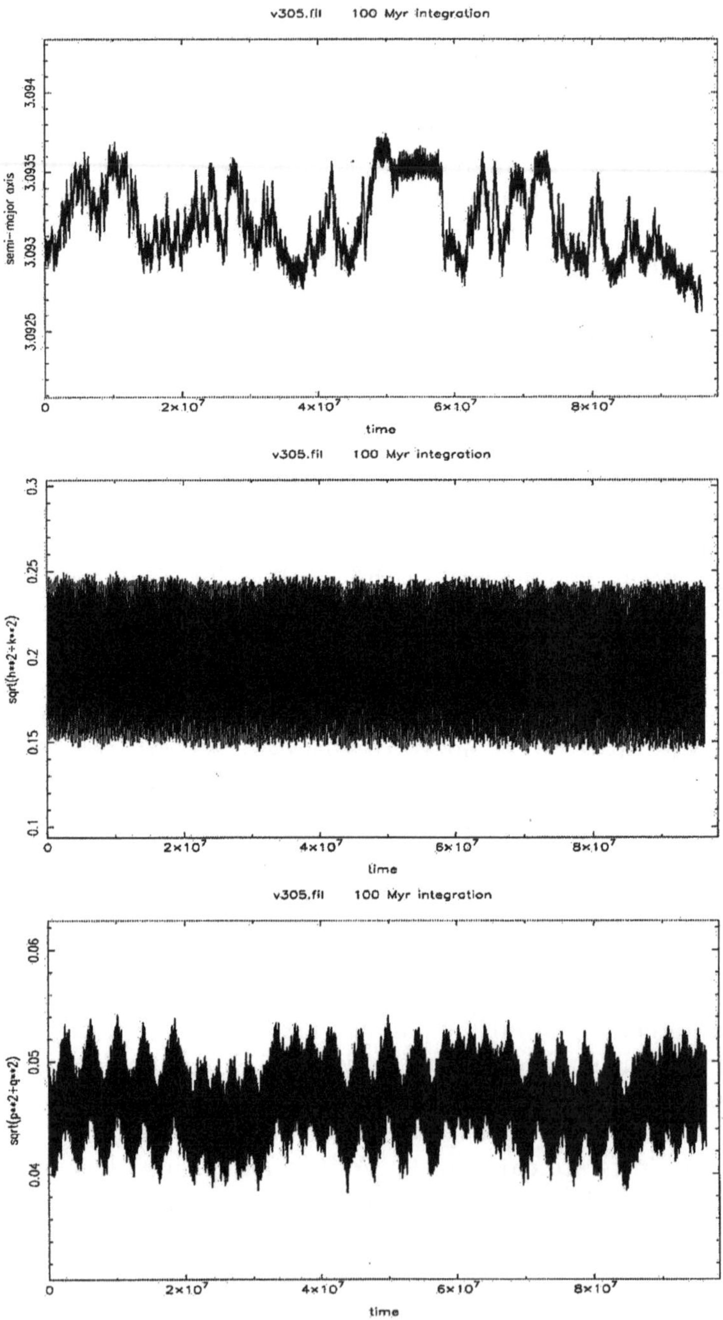

Figure 14. Evolution of semi-major axis (on the top), eccentricity (in the middle) and inclination (at the bottom) for asteroid 305 Gordonia.

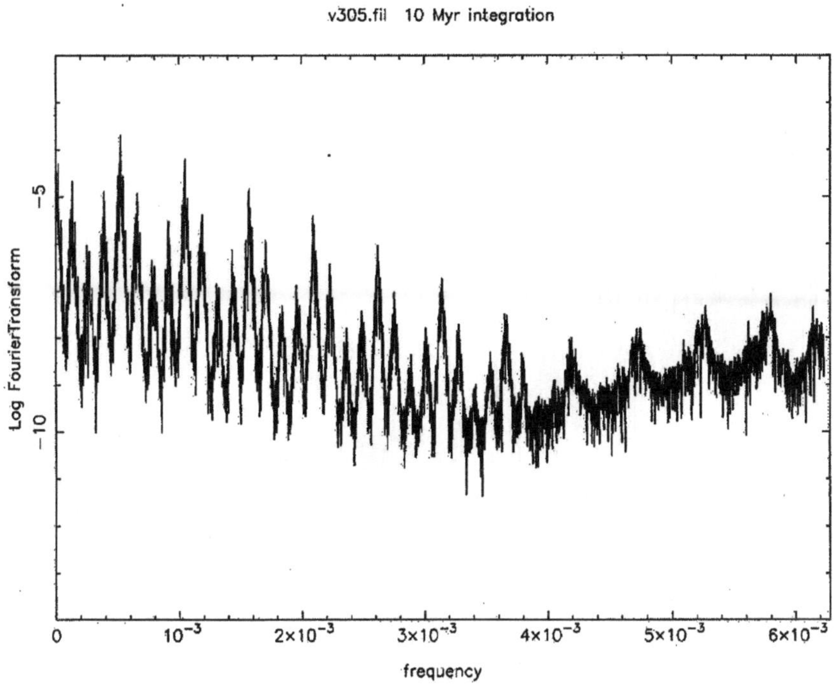

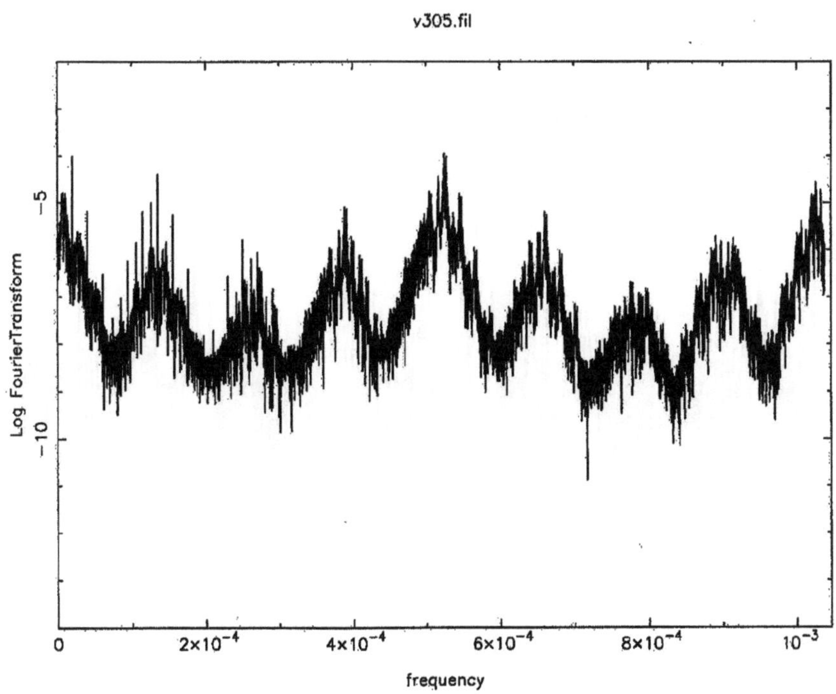

Figure 15. Fourier spectra (52) computed for asteroid 305 Gordonia (lower one computed using a 100 Myr integration, upper one using a 10 Myr integration with more frequent sampling).

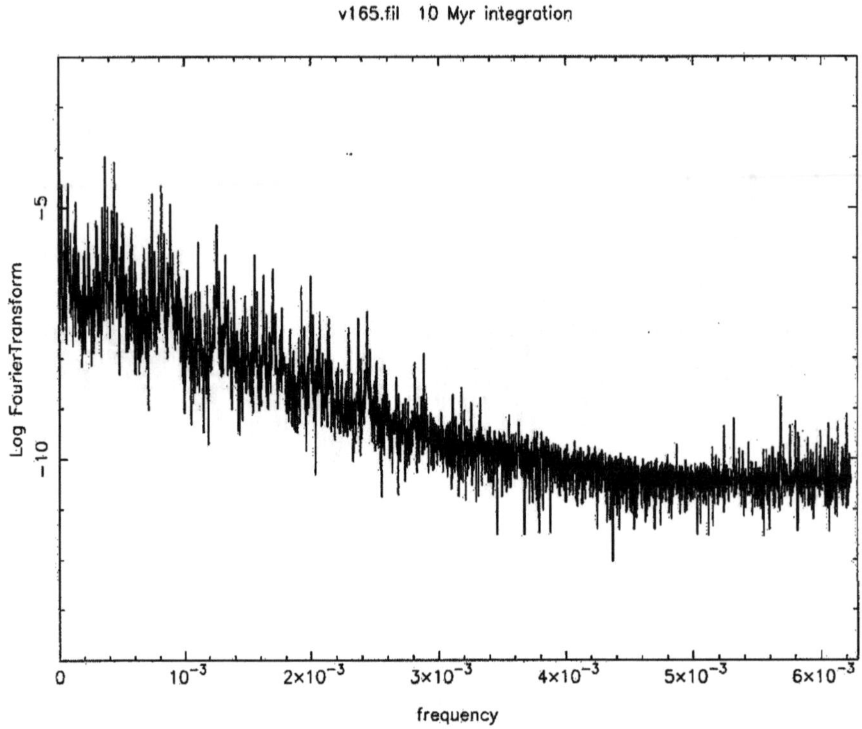

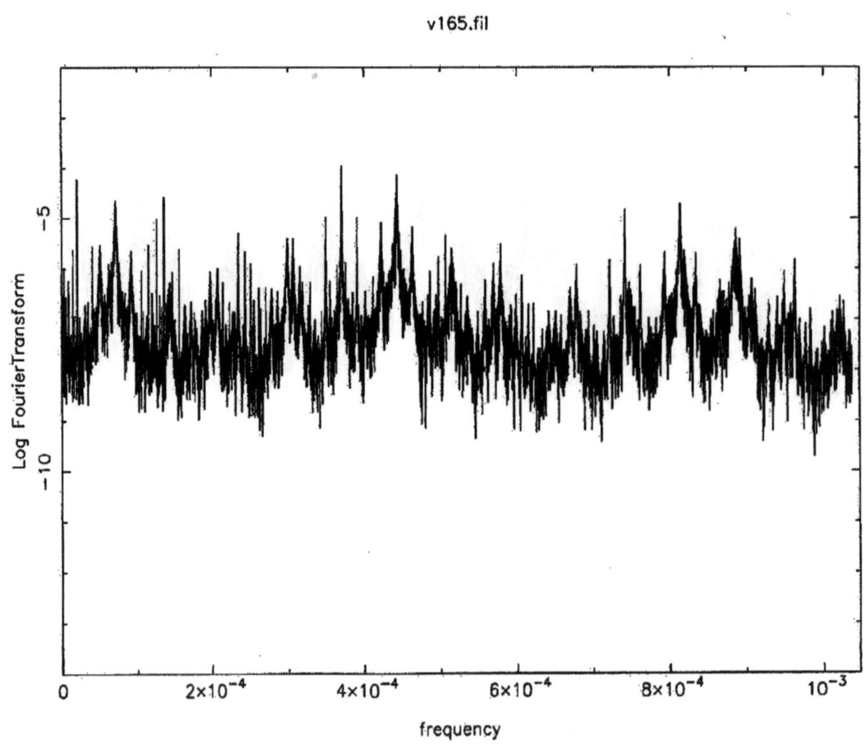

Figure 16. Fourier spectrum of (52) computed for asteroid 165 Loreley (lower one computed using a 100 Myr integration, upper one using a 10 Myr integration with more frequent sampling).

References

Arnold, V.I. (1964): Instability of dynamical systems with several degrees of freedom. *Sov. Math. Dokl.*, Vol. 6, 581–585.

Benettin, G. and Fassò, F. (1996): Fast rotations of the rigid body: a study by Hamiltonian perturbation theory. Part I *Nonlinearity*, Vol. 9, 137–186.

Benettin, G., Fassò, F. and Guzzo, M. (1997): Fast rotations of the rigid body: a study by Hamiltonian perturbation theory. Part II: Gyroscopic Rotations *Nonlinearity*, vol. 10, 1695-1717.

Benettin, G., Galgani, L. and Giorgilli,A. (1985): A proof of Nekhoroshev's theorem for the stability times in nearly integrable Hamiltonian systems. *Cel. Mech.*, vol. 37, pag. 1.

Benettin, G. and Gallavotti,G. (1986): Stability of motions near resonances in quasi–integrable Hamiltonian systems *J. Stat. Phys.*, vol. 44, 293–338.

Bessi, U., Chierchia, L. and Valdinoci, E. (2000): Upper Bounds on Arnold Diffusion Time via Mather Theory. *Preprint.*

Carpino, M., Milani A. and Nobili, A.M. (1987): Long term numerical integrations and synthetic theories for the motion of the outer planets, *Astron. Astrophys.*, vol. 181, 182–194.

Chirikov, B.V. An universal instability of many dimensional oscillator system. *Phys. Rep.*, **52**:263–379, (1979).

Contopoulos, G. and Voglis, N. A fast method for distinguishing between order and chaotic orbits. *Astron. Astrophys.*, **317**:73–82, (1997).

Fassò, F. (1995): Hamiltonian perturbation theory on a manifold. *Cel. Mech. Dyn. Astron.*, vol. 62, pag 43.

Froeschle' C., Guzzo, M. and Lega, E. (2000): Graphical Evolution of the Arnold Web: From Order to chaos. *Science*, Volume 289, n. 5487.

Froeschle' C. and Lega, E. Twist angles: a fast method for distinguishing islands, tori and weak chaotic orbits. Comparison with other methods of analysis. *AA*, **334**:355–362, (1998).

Guzzo, M. (1998): Nekhoroshev stability of quasi–integrable Hamiltonian systems with singularities and degeneracy. *PhD thesis.*

Guzzo, M. (1999): Nekhoroshev stability of quasi–integrable degenerate Hamiltonian systems. *Regular and chaotic dynamics*, Volume 4, n. 2.

Guzzo, M. and Benettin, G. (2001): A spectral formulation of the Nekhoroshev theorem and its relevance for numerical and experimental data analysis. *Discrete and Continuous Dynamical Systems-Series B*, Volume 1, n. 1.

Guzzo, M., Knezevic, Z. and Milani, A. (2002): Probing the Nekhoroshev stability of Asteroids. *Celest. Mech. Dyn. Astr.*, Volume 83, Issues 1-4, 2002.

Guzzo, M., Lega, E. and Froeschlé, C. (2002): On the numerical detection of the effective stability of chaotic motions in quasi-integrable systems. *Physica D*, Volume 163, Issues 1-2, 1-25, 2002.

Guzzo, M. and Morbidelli, A. (1997): Construction of a Nekhoroshev like result for the asteroid belt dynamical system. *Celest. Mech. Dyn. Astron.*, vol. 66, 255–292.

Knežević, Z. and Jovanović, B. (1997): Is the chaotic clock ticking correctly?. *Bull. Astron. Belgrade*, vol. 156, 47–69.

Knežević, Z. and Milani, A. (2000): Synthetic Proper Elements for Outer Main Belt Asteroids. *Celest. Mech. Dyn. Astron.*, vol. 78, 17–46.

Laskar, J. (1990): The chaotic motion of the solar system. A numerical estimate of the size of the chaotic zones. *Icarus*, vol. 88, 266–291.

Laskar, J., Froeschlè, C. and Celletti, A. (1992): The measure of chaos by the numerical analysis of the fundamental frequencies. Application to the standard mapping. *Physica D*, vol. 56, pag. 253.

Lega, E. and Froeschlé, C. (1997): Fast Lyapunov Indicators. Comparison with other chaos indicators. Application to two and for dimensional maps. *The Dynamical Behavior of our planetary system*, Kluwer academ. publ., Henrard, J. and Dvorak, R. eds.

Lega, E., Guzzo, M. and Froeschlé, C. (2003): Detection of Arnold diffusion in Hamiltonian systems. *Preprint.*

Lochak, P. (1992): Canonical perturbation theory via simultaneous approximation. *Russ. Math. Surv.*, vol. 47, 57–133.

Milani, A. (1995): Proper elements and stable chaos. *From Newton to chaos*, Plenum Press, Roy, A.E. and Steves, B.A. eds., pp. 47–78.

Milani, A. and Knežević, Z. (1992): Asteroid proper elements and secular resonances *Icarus*, vol. 98, 211–232.

Milani, A. and Nobili, A.M. (1988) Integration error over very long time spans *Celest. Mech. Dyn. Astron.*, vol. 43, 1–34.

Morbidelli, A. and Guzzo, M. (1997): The Nekhoroshev theorem and the asteroid belt dynamical system. *Celest. Mech. Dyn. Astron.*, vol. 65, 107–136.

Morbidelli, A. and Nesvorny, D. (1999): Numerous weak resonances drive Asteroids toward terrestrial planets orbits. *Icarus*, vol. 139, 295–308.

Nekhoroshev, N.N. (1977): Exponential estimates of the stability time of near–integrable Hamiltonian systems. *Russ. Math. Surveys*, vol. 32, 1–65.

Littlewood, N.N. (1979): Exponential estimates of the stability time of near–integrable Hamiltonian systems, 2 *Trudy Sem. Petrovs.*, vol. 5, 5–50.

Neishtadt A. I. (1987): On the change in the adiabatic invariant on crossing a separatrix in systems with two degrees of freedom. *P.M.M. U.S.S.R.*, vol. 51, 586-592.

Niederman, L. (1996): Stability over exponentially long times in the planetary problem. *Nonlinearity*, vol. 9, 1703–1751.

Pöschel, J. (1993): Littlewood's estimates for quasi–convex Hamiltonian systems. *Math. Z.*, vol. 213, pag. 187.

III

SOURCES OF CHAOS

BASICS OF REGULARIZATION THEORY

Alessandra Celletti
Università di Roma Tor Vergata Via della Ricerca Scientifica 1, I-00133 Roma (Italy)
celletti@mat.uniroma2.it

Abstract We consider the dynamics of three point masses, where we assume that the mass of the third body is so small that it does not affect the motion of the primaries. In the framework of the *restricted three-body problem, we investigate the collisional trajectories, which correspond to a singularity of the equations of motion. We investigate the regularizing techniques known as Levi–Civita, Kustaanheimo–Stiefel and Birkhoff Transformations. The Levi–Civita regularization is adapted to the study of the planar* restricted three-body problem, when considering a single collision with one of the primaries. The Kustaanheimo–Stiefel method concerns the same problem when the bodies are allowed to move in the 3–dimensional space. A simultaneous regularization with both primaries is achieved through the implementation of Birkhoff's transformation.

Keywords: Regularization theory

1. Introduction

The motion of the celestial bodies of the solar system is ruled by Newton's law, which states that the attraction between massive bodies is directly proportional to the product of the masses and inversely proportional to the square of their distance. Indeed, a collision between any two objects is marked by the fact that their distance becomes zero, which corresponds to a singularity of Newton's equations. The aim of regularization theory is to transform the singular differential equations into regular ones, thus providing an efficient mathematical tool to analyze motions leading to collisions. We shall be concerned with the *restricted three-body problem,* dealing with the motion of a small body in the gravitational field of two massive primaries. It is assumed that the primaries move on circular orbits around their common center of

B.A. Steves et al. (eds.), Chaotic Worlds: from Order to Disorder in Gravitational N-Body Dynamical Systems, 203–230.

mass. In this framework, we start our discussion by introducing the basic technique to regularize the dynamics (Stiefel and Scheifele, 1971, Szebehely, 1967), i.e. the so–called *Levi–Civita transformation,* which is convenient when dealing with the planar three-body problem (i.e., when neglecting the mutual inclinations). When the three bodies are allowed to move in the space, a different method must be adopted, i.e. the so–called Kustaanheimo–Stiefel regularization theory (Kustaanheimo and Stiefel, 1965), often denoted as KS–transformation. Both Levi–Civita and KS methods are *local* transformations, in the sense that their application allows to regularize collisions with only one of the two primaries. A suitable extension of such techniques allows to obtain a simultaneous regularization with both primaries, thus obtaining a *global* transformation, known as Birkhoff's method (Stiefel and Scheifele, 1971, Szebehely, 1967). All these techniques rely on a common procedure, which consists in performing a suitable change of variables, a rescaling of time and in using the preservation of energy (see also Celletti, 2002, Stiefeland ossler and Waldvogel and Burdet, 1967, Stiefel and Scheifele, 1971, Stiefel, 1972, Szebehely, 1967, J. Waldvogel, 1972).

Before entering into the intriguing world of the regularizing transformations, we premise some *facts about concrete* collisional events that occurred in the solar system (see also Celletti, 2002). When thinking to impacts of heavy objects with the Earth, one is immediately led back to 65 million years ago: it is widely accepted that the disappearance of dinosaurs was caused by the collision of a large body with the Earth. The astroblame is the so–called Chicxulub's crater of about 180 km of diameter, which was located in the depth of the ocean, close to the Yucatan peninsula. Beside this catastrophic event, many other impacts marked the life of the Earth. Just to give a few examples, other astroblames were found throughout our planet: from Arizona (the Meteor Crater of about 1200 m of diameter), to Australia (the Wolf Creek crater of about 850 m), to Arabia (the Waqar crater of about 100 m of diameter). Live images of an impact in the solar system were provided in 1994 by the spectacular collision of the Shoemaker–Levy 9 comet with Jupiter, which fragmented in several pieces before the impact with the giant planet. This paper is organized as follows. In Section 2 we provide the application of the Levi–Civita regularization theory to a simple example, namely the motion of two bodies on a straight line. In Section 3 we recall the equations of motion governing the two-body problem, while in Section 4 we describe the equations concerning the planar, circular, restricted 3–body problem. TheLevi–Civita and Kustaanheimo–Stiefel regularization techniques are presented, respectively, in Sections 5 and 6. Birkhoff's global transformation is outlined in Section 7.

2. The idea of regularization theory

The standard technique concerning regularization theory is essentially based on three steps:

- a change of coordinates, known as the Levi–Civita transformation;
- the introduction of a *fictitious time,* in order to get rid of the fact that the velocity becomes infinite at the singularity;
- the use of the conservation of energy, in order to transform the singular differential equations into regular ones, through the introduction of the *extended phase space.*

For didactical purposes, it is useful to present an analogy between the three basic steps of regularization theory and some parts of the nice tale by L. Carroll "Alice's adventures in wonderland" (see Figure 1). In order to remind the fundamental steps outlined before, we can imagine that the change of coordinates corresponds to the beginning of the dream, when Alice falls in the well (see Figure 1a). As far as the introduction of the fictitious time is concerned, we recall the main characters of the white rabbit with pink eyes and of the hatter, who was accused to "murder the time" (see Figure 1b). The analogy with the preservation of energy is offered by the continuous transformations that Alice undergoes along her adventures: from normal height, she becomes extremely small after eating a biscuit; then, she reaches again a normal size and afterwards she becomes extremely tall (see Figure 1c). The continuous sequence of stretchings and shortenings ends up with her awakening with normal size, providing a preservation of her height!

We begin the presentation of regularization theory using a very simple example: consider two bodies, $\overline{P}_1$ and $\overline{P}_2$ (with masses, respectively, m_1 and m_2), which interact through Newton's law. We assume that the two bodies move on a straight line; as a consequence, we select a reference frame with the origin located in $\overline{P}_2$ and with the abscissa coinciding with the line of motion. Let us introduce the quantity $K = G(m_1+m_2)$, where G denotes the gravitational constant. The motion of $\overline{P}_1$ with respect to $\overline{P}_2$ is ruled by the equation

$$\ddot{x} + \frac{K}{x^2} = 0 ;$$

the corresponding energy integral is provided by

$$h = \frac{K}{x} - \frac{1}{2}\dot{x}^2 .$$

We remark that the velocity $\dot{x} = \pm\sqrt{2(\frac{K}{x} - h)}$ becomes infinite at the collision, i.e. whenever $x = 0$.

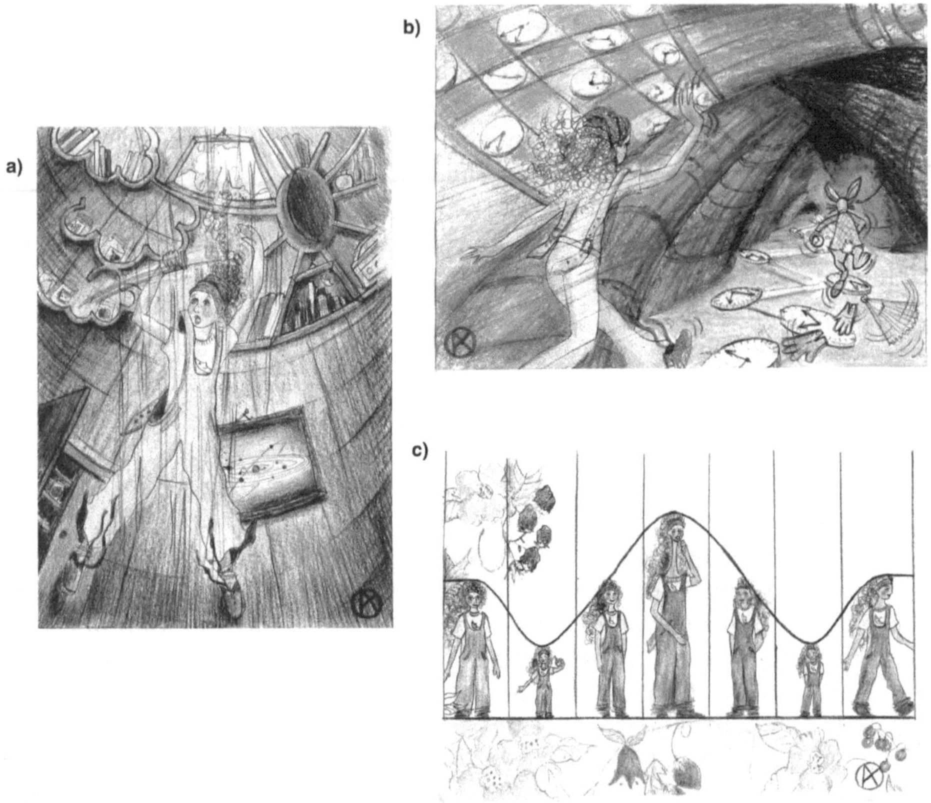

Figure 1. Alice's adventures in wonderland offer a useful analogy to remind the basic steps of regularization theory: a) a change of coordinates, b) the introduction of a fictitious time, c) the preservation of energy.

We start by performing a change of coordinates, the Levi–Civita transformation, which can be written (in the present case) as

$$x = u^2 .$$

The equation of motion in the new coordinate becomes

$$\ddot{u} + \frac{1}{u}\dot{u}^2 + \frac{K}{2u^5} \quad = \ 0,$$

while the energy integral takes the form

$$h = \frac{K}{u^2} - 2u^2\dot{u}^2 .$$

The new velocity is given by

$$\dot{u} = \pm\sqrt{\frac{K}{2u^4} - \frac{h}{2u^2}}\ ;$$

we notice that $\dot{u}$ becomes infinite at collision (i.e., at $u = 0$). Since the equation is still singular, we proceed to apply a change of time by introducing a *fictitious time:* to control the increase of speed at collision, we multiply the velocity by a suitable scaling factor which is zero at the singularity. Therefore, we introduce a fictitious time s defined by

$$dt = x\,ds = u^2\,ds \qquad \text{or} \qquad \frac{dt}{ds} = x = u^2.$$

Denoting by $u' \equiv \frac{du}{ds}$, one has

$$\begin{aligned} \dot{u} &= \frac{du}{dt} = \frac{du}{ds}\frac{ds}{dt} = \frac{1}{u^2}u', \\ \ddot{u} &= \frac{1}{u^2}\frac{d}{ds}\Big(\frac{1}{u^2}u'\Big) = \frac{1}{u^4}u'' - \frac{2}{u^5}u'^2. \end{aligned}$$

The equation of motion and the energy integral become

$$u'' - \frac{1}{u}\,u'^2 + \frac{K}{2u} = 0, \qquad h = \frac{K}{u^2} - \frac{2u'^2}{u^2}.$$

Finally, we make use of the preservation of energy as follows. Let us rewrite the equation of motion as

$$u'' + \frac{1}{2}\Big(\frac{K}{u^2} - \frac{2u'^2}{u^2}\Big)\,u = 0;$$

using the expression of the energy integral, we get the differential equation

$$u'' + \frac{h}{2}\,u = 0,$$

which corresponds to the equation of the harmonic oscillator with frequency $\omega = \sqrt{\frac{h}{2}}$. We conclude by remarking that the above regularizing procedure allowed to obtain a regular differential equation, whose solution is a periodic function of the fictitious time s.

3. The two–body problem

We consider two massive bodies, say $\overline{P}_1$ and $\overline{P}_2$, which attract each other through Newton's law. Since the relative motion takes place on a plane (i.e., the motion is no more constrained on a straight line as in Section 2), we select a reference frame coinciding with the plane of motion and we let (q_1, q_2) be the relative cartesian coordinates. Let us show that, after a suitable normalization of the units of measure, the Hamiltonian function describing the two-body problem is given by

$$H = H(p_1, p_2, q_1, q_2) = \frac{1}{2}(p_1^2 + p_2^2) - \frac{1}{(q_1^2 + q_2^2)^{\frac{1}{2}}}, \tag{1}$$

where we defined the momenta as $p_j \equiv \dot{q}_j$, $j = 1, 2$. We remark that the equations of motion associated to (1) are given by

$$\begin{aligned} \dot{q}_1 &= \frac{\partial H}{\partial p_1} = p_1 \qquad & \dot{p}_1 = -\frac{\partial H}{\partial q_1} = -\frac{q_1}{(q_1^2 + q_2^2)^{\frac{3}{2}}} \\ \dot{q}_2 &= \frac{\partial H}{\partial p_2} = p_2 \qquad & \dot{p}_2 = -\frac{\partial H}{\partial q_2} = -\frac{q_2}{(q_1^2 + q_2^2)^{\frac{3}{2}}}. \end{aligned} \tag{2}$$

In order to derive the Hamiltonian (1), let us denote by R the distance between the center of mass between $\overline{P}_1$ and $\overline{P}_2$ and the origin of the reference frame; let r_1 and r_2 be the distances of the two bodies from the center of mass and define $r \equiv r_2 - r_1$ (see Figure 2). The kinetic energy is the sum of the contributions due to the motion of the center of mass and to the motions of $\overline{P}_1$ and $\overline{P}_2$ relative to the center of mass, i.e.

$$T = \frac{1}{2}(m_1 + m_2)\dot{R}^2 + \frac{1}{2}m_1\dot{r}_1^2 + \frac{1}{2}m_2\dot{r}_2^2.$$

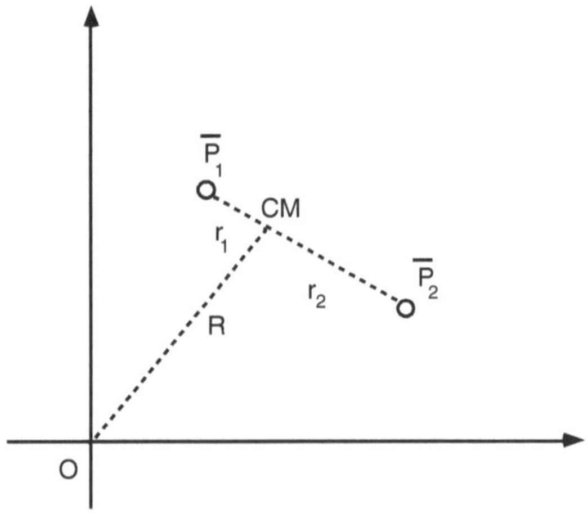

Figure 2. The positions of $\overline{P}_1$ and $\overline{P}_2$ in an inertial reference frame.

Since $r_1 = -\frac{m_2}{m_1+m_2}r$, $r_2 = \frac{m_1}{m_1+m_2}r$, then the Lagrangian function is given by

$$L = \frac{m_1+m_2}{2}\dot{R}^2 + \frac{1}{2}\frac{m_1 m_2}{m_1+m_2}\dot{r}^2 + V(r),$$

where $V(r)$ denotes the Newtonian potential. The first term does not contribute to the equations of motion; the remaining terms can be expressed in cartesian coordinates (q_1, q_2) as

$$L = \frac{1}{2}\mu(\dot{q}_1^2 + \dot{q}_2^2) + \frac{K}{(q_1^2+q_2^2)^{\frac{1}{2}}},$$

where $\mu \equiv \frac{m_1 m_2}{m_1+m_2}$ is the *reduced mass* and K is a suitable constant. Normalizing the units of measure so that $\mu = 1$ and $K = 1$, one obtains the Hamiltonian (1).

Remark: Every elliptic solution of the classical Newtonian equation is *unstable* (Stiefel and Scheifele, 1971). In order to make this statement more precise, let us provide the following

Definition of Lyapunov stability: Consider a reference solution with given initial data at some time t_0; define a second solution, which is obtained by slightly varying the initial data. The reference solution is called *stable*, if for any $t \geq t_0$ the distance between the two solutions can be made smaller than ε by an appropriate choice of the variations of the initial conditions.

In the two–body approximation, Hamilton's equations (1) can be written as

$$\ddot{q} + \omega^2 q = 0,$$

where $q = (q_1, q_2)$, $r = \sqrt{q_1^2 + q_2^2}$ and $\omega = \omega(r) = \frac{1}{r^{3/2}}$. Consider a circular reference solution and a varied solution, which is also circular. The period of revolution varies with r as $\frac{2\pi}{\omega} = 2\pi \cdot r^{3/2}$; therefore, there exists a time when the two particles are opposite to each other with respect to the center of mass, which is incompatible with Lyapunov stability.

4. The planar, circular, restricted 3–body problem

Let S be a body with an infinitesimal mass, subject to the gravitational attraction of $\overline{P}_1$, $\overline{P}_2$ (whose masses are, respectively, μ_1 and μ_2). We assume that the primaries are not affected by S, i.e. we consider the so–called *restricted* three-body problem. *Moreover, we assume that the motion takes place on the same plane (i.e., we neglect the relative inclinations) and that the trajectories of* $\overline{P}_1$ *and* $\overline{P}_2$ *are circular with*

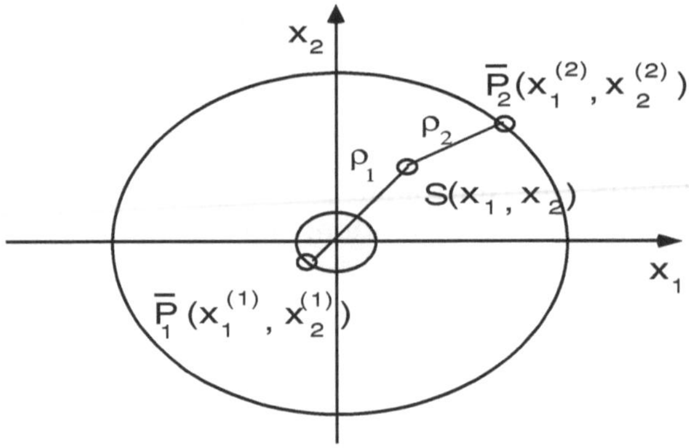

Figure 3. The coordinates of $\overline{P}_1$, $\overline{P}_2$ and S in a fixed reference frame.

origin coinciding with their common center of mass. Let (x_1, x_2) be the coordinates of S in an inertial frame centered at the barycenter of $\overline{P}_1$ and $\overline{P}_2$ and let us normalize the units of measure so that

$$\mu_1 + \mu_2 = 1.$$

We denote by $(x_1^{(1)}, x_2^{(1)})$ and $(x_1^{(2)}, x_2^{(2)})$ the coordinates of $\overline{P}_1$ and $\overline{P}_2$ (see Figure 3).

The motion of S is described by the Lagrangian function

$$L = L(\dot{x}_1, \dot{x}_2, x_1, x_2, t) = \frac{1}{2}(\dot{x}_1^2 + \dot{x}_2^2) + V(x_1, x_2, t),$$

where $V(x_1, x_2, t) \equiv \frac{\mu_1}{\varrho_1} + \frac{\mu_2}{\varrho_2}$ and

$$\begin{aligned} \varrho_1 &\equiv \sqrt{(x_1 - x_1^{(1)})^2 + (x_2 - x_2^{(1)})^2}, \\ \varrho_2 &\equiv \sqrt{(x_1 - x_1^{(2)})^2 + (x_2 - x_2^{(2)})^2}. \end{aligned}$$

We remark that $x_1^{(1)}$, $x_2^{(1)}$, $x_1^{(2)}$, $x_2^{(2)}$ are explicit functions of the time. Denoting by y_i $(i = 1, 2)$ the kinetic moments conjugated to x_i, the Hamiltonian function is given by

$$H(y_1, y_2, x_1, x_2, t) = \frac{1}{2}(y_1^2 + y_2^2) - V(x_1, x_2, t).$$

Remark: In the spatial case, let $(x_1^{(1)}, x_2^{(1)}, x_3^{(1)})$, $(x_1^{(2)}, x_2^{(2)}, x_3^{(2)})$ be the coordinates of $\overline{P}_1$ and $\overline{P}_2$; denoting by y_i $(i = 1, 2, 3)$ the kinetic moments conjugated to x_i, the Hamiltonian function reads as

$$H(y_1, y_2, y_3, x_1, x_2, x_3, t) = \frac{1}{2}(y_1^2 + y_2^2 + y_3^3) - V(x_1, x_2, x_3, t),$$

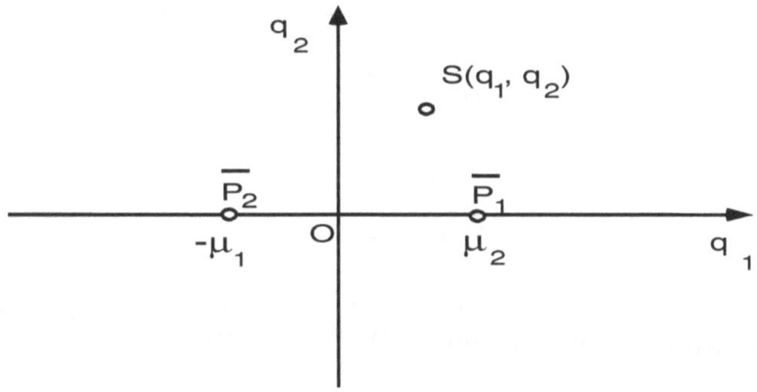

Figure 4. The coordinates of $\overline{P}_1$, $\overline{P}_2$ and S in a synodic reference frame.

where

$$\begin{aligned} V(x_1,x_2,x_3,t) &= \frac{\mu_1}{\sqrt{(x_1-x_1^{(1)})^2+(x_2-x_2^{(1)})^2+(x_3-x_3^{(1)})^2}} \\ &+ \frac{\mu_2}{\sqrt{(x_1-x_1^{(2)})^2+(x_2-x_2^{(2)})^2+(x_3-x_3^{(2)})^2}}. \end{aligned}$$

Let us consider a rotating or *synodic* reference frame centered at the barycenter of $\overline{P}_1$ and $\overline{P}_2$; assume that the units of measure are such that the relative angular velocity of $\overline{P}_1$ and $\overline{P}_2$ is unity. Then, the coordinates of $\overline{P}_1$ and $\overline{P}_2$ are, respectively, $\overline{P}_1(\mu_2,0)$ and $\overline{P}_2(-\mu_1,0)$. Let (q_1,q_2) be the coordinates of S in the synodic frame (see Figure 4). In order to derive the synodic Hamiltonian, we consider the generating function

$$\begin{aligned} W(y_1,y_2,q_1,q_2,t) &= y_1q_1\cos t - y_1q_2\sin t \\ &+ y_2q_1\sin t + y_2q_2\cos t, \end{aligned}$$

which provides the characteristic equations

$$\begin{aligned} x_1 &= \frac{\partial W}{\partial y_1} = q_1\cos t - q_2\sin t \\ p_1 &= \frac{\partial W}{\partial q_1} = y_1\cos t + y_2\sin t \\ x_2 &= \frac{\partial W}{\partial y_2} = q_1\sin t + q_2\cos t \\ p_2 &= \frac{\partial W}{\partial q_2} = -y_1\sin t + y_2\cos t. \end{aligned}$$

Inverting the above equations, one obtains

$$\begin{aligned} q_1 &= x_1\cos t + x_2\sin t \qquad & y_1 = p_1\cos t - p_2\sin t \\ q_2 &= -x_1\sin t + x_2\cos t \qquad & y_2 = p_1\sin t + p_2\cos t. \end{aligned}$$

It is trivial to check that the synodic Hamiltonian takes the form

$$\begin{aligned}\tilde{H}(p_1,p_2,q_1,q_2,t) &= H - \frac{\partial W}{\partial t}\\ &= \frac{1}{2}(p_1^2+p_2^2)+q_2p_1-q_1p_2\\ &- V(q_1\cos t - q_2\sin t, q_1\sin t + q_2\cos t, t).\end{aligned}$$

Recalling that in the fixed frame the bodies $\overline{P}_1$ and $\overline{P}_2$ describe circles of radius μ_2 and μ_1, we can write their coordinates as

$$\begin{aligned}x_1^{(1)} &= \mu_2\cos t \qquad & x_1^{(2)} &= -\mu_1\cos t\\ x_2^{(1)} &= \mu_2\sin t \qquad & x_2^{(2)} &= -\mu_1\sin t\ .\end{aligned}$$

The expression of the perturbing function in the rotating coordinates becomes

$$V(q_1,q_2) = \frac{\mu_1}{\sqrt{(q_1-\mu_2)^2+q_2^2}} + \frac{\mu_2}{\sqrt{(q_1+\mu_1)^2+q_2^2}}.$$

Therefore we are led to the following synodic Hamiltonian:

$$\tilde{H}(p_1,p_2,q_1,q_2) = \frac{1}{2}(p_1^2+p_2^2)+q_2p_1-q_1p_2-V(q_1,q_2), \tag{3}$$

whose associated Hamilton's equations are

$$\begin{aligned}\dot{q}_1 &= p_1+q_2 \qquad & \dot{p}_1 &= p_2+V_{q_1}\\ \dot{q}_2 &= p_2-q_1 \qquad & \dot{p}_2 &= -p_1+V_{q_2}.\end{aligned}$$

Denoting by

$$\overline{\Omega} \equiv \frac{1}{2}(q_1^2+q_2^2)+V, \tag{4}$$

we can write the previous equations in the form

$$\begin{aligned}\ddot{q}_1-2\dot{q}_2 &= \dot{p}_1+\dot{q}_2-2\dot{q}_2 = \dot{p}_1-\dot{q}_2 = q_1+V_{q_1} \equiv \overline{\Omega}_{q_1}\\ \ddot{q}_2+2\dot{q}_1 &= \dot{p}_2-\dot{q}_1+2\dot{q}_1 = \dot{p}_2+\dot{q}_1 = q_2+V_{q_2} \equiv \overline{\Omega}_{q_2}.\end{aligned} \tag{5}$$

For notational convenience, we define $\Omega \equiv \overline{\Omega} + \frac{1}{2}\mu_1\mu_2$. The expression of the so–called *Jacobi integral* is obtained as follows: let us multiply the first equation in (4) by $\dot{q}_1$ and the second by $\dot{q}_2$; over summation one obtains:

$$\dot{q}_1\ddot{q}_1+\dot{q}_2\ddot{q}_2 = \dot{q}_1\overline{\Omega}_{q_1}+\dot{q}_2\overline{\Omega}_{q_2}.$$

Therefore one has $\frac{1}{2}\frac{d}{dt}(\dot{q}_1^2+\dot{q}_2^2) = \frac{d}{dt}(\overline{\Omega})$, from which it follows that

$$\dot{q}_1^2+\dot{q}_2^2 = 2\overline{\Omega} - C' = 2\Omega - C.$$

We define the *Jacobi constant* as

$$C \equiv 2\Omega - (\dot{q}_1^2 + \dot{q}_2^2).$$

Since $\dot{q}_1 = p_1 + q_2$ and $\dot{q}_2 = p_2 - q_1$, then $p_1 = \dot{q}_1 - q_2$, $p_2 = \dot{q}_2 + q_1$. Hence, it is useful to rewrite (3) as

$$\begin{aligned} \tilde{H} &= \frac{1}{2}(\dot{q}_1^2 + \dot{q}_2^2) - \frac{1}{2}(q_1^2 + q_2^2) - V(q_1, q_2) \\ &= \frac{1}{2}(\dot{q}_1^2 + \dot{q}_2^2) - \overline{\Omega}. \end{aligned}$$

Making use of the Jacobi integral one gets

$$\tilde{H} + \Omega = \frac{1}{2}\mu_1\mu_2 + \frac{1}{2}(\dot{q}_1^2 + \dot{q}_2^2) = \frac{1}{2}\mu_1\mu_2 + \Omega - \frac{C}{2},$$

which provides the relation

$$\tilde{H} = \frac{\mu_1\mu_2 - C}{2}.$$

5. The Levi–Civita transformation

5.1 The regularization of the two–body problem

Recall that the Hamiltonian function describing the two-body problem is given (in normalized units of measure) by

$$H(p_1, p_2, q_1, q_2) = \frac{1}{2}(p_1^2 + p_2^2) - \frac{1}{(q_1^2 + q_2^2)^{\frac{1}{2}}}.$$

Let us consider a canonical transformation with generating function of the form

$$W(p_1, p_2, Q_1, Q_2) = p_1 f(Q_1, Q_2) + p_2 g(Q_1, Q_2).$$

Denoting by i the imaginary unit, the *Levi–Civita transformation* is obtained setting

$$f + ig \equiv (Q_1 + iQ_2)^2 = Q_1^2 - Q_2^2 + i \cdot 2Q_1Q_2,$$

namely

$$f(Q_1, Q_2) \equiv Q_1^2 - Q_2^2, \qquad g(Q_1, Q_2) \equiv 2Q_1Q_2.$$

The change of variables associated to the generating function W is given by

$$\begin{aligned} q_1 &= \frac{\partial W}{\partial p_1} = f(Q_1,Q_2) = Q_1^2 - Q_2^2 \\ q_2 &= \frac{\partial W}{\partial p_2} = g(Q_1,Q_2) = 2Q_1Q_2 \\ P_1 &= \frac{\partial W}{\partial Q_1} = p_1\frac{\partial f}{\partial Q_1} + p_2\frac{\partial g}{\partial Q_1} = 2p_1Q_1 + 2p_2Q_2 \\ P_2 &= \frac{\partial W}{\partial Q_2} = p_1\frac{\partial f}{\partial Q_2} + p_2\frac{\partial g}{\partial Q_2} = -2p_1Q_2 + 2p_2Q_1 . \end{aligned} \qquad (6)$$

We refer to (q_1, q_2) as the *physical plane* and to (Q_1, Q_2) as the *parametric plane* with $q_1 + iq_2 = f + ig = (Q_1 + iQ_2)^2$. Using matrix notation, we remark that the Levi–Civita transformation can be written as

$$\begin{pmatrix} q_1 \\ q_2 \end{pmatrix} = \begin{pmatrix} Q_1^2 - Q_2^2 \\ 2Q_1Q_2 \end{pmatrix} = \begin{pmatrix} Q_1 & -Q_2 \\ Q_2 & Q_1 \end{pmatrix}\begin{pmatrix} Q_1 \\ Q_2 \end{pmatrix} = A_0 \begin{pmatrix} Q_1 \\ Q_2 \end{pmatrix},$$

where we defined the matrix

$$A_0 = \begin{pmatrix} Q_1 & -Q_2 \\ Q_2 & Q_1 \end{pmatrix}.$$

It is immediate to check that the Levi–Civita transformation is characterized by the following properties:
(1) the matrix A_0 is orthogonal;
(2) the elements of A_0 are linear and homogeneous functions of Q_1, Q_2;
(3) the first column of the matrix A_0 coincides with the Q–vector.
Another important property of the Levi–Civita transformation is that the angles at the origin are doubled (see Figure 5).
Indeed, when the orbital eccentricity tends to 1, the ellipse degenerates into a straight line. In the physical plane, the position vector makes a sharp bend of angle 2π at the origin. In the parametric plane, the singularity is removed allowing the particle to pass through the origin; in this case, the position vector makes an angle π at the origin. More specifically, denoting by ϑ and ψ the angles formed by the position vector at the origin, one obtains

$$\tan\vartheta = \frac{q_2}{q_1} = \frac{2Q_1Q_2}{Q_1^2 - Q_2^2} = \frac{2Q_2/Q_1}{1-(\frac{Q_2}{Q_1})^2} = \frac{2\tan\psi}{1-\tan^2\psi} = \tan 2\psi .$$

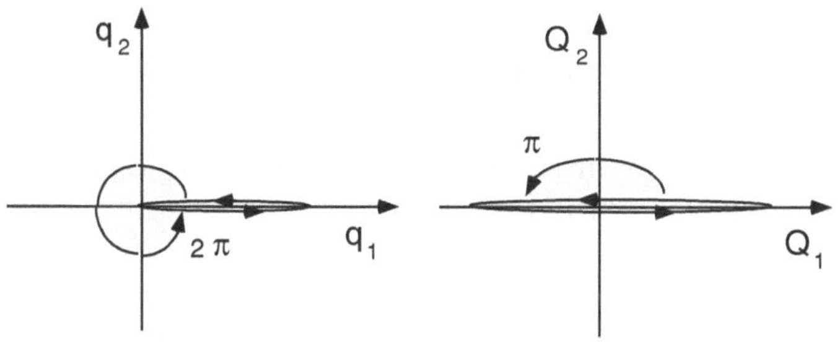

Figure 5. The Levi–Civita transformation doubles the angles at the origin.

Coming back to the Levi–Civita transformation (5), we observe that

$$P = 2A_0^T\, p$$

and that the inverse of the matrix A_0 is provided by $A_0^{-1} = \frac{1}{\det A_0} A_0^T$. Let us define $D \equiv 4 \det A_0 = 4(Q_1^2 + Q_2^2) > 0$; then one obtains

$$P_1^2 + P_2^2 = D(p_1^2 + p_2^2).$$

As a consequence, the new Hamiltonian becomes

$$\tilde{H} = \tilde{H}(P_1, P_2, Q_1, Q_2) = \frac{1}{2D}(P_1^2 + P_2^2) - \frac{1}{(f(Q_1, Q_2)^2 + g(Q_1, Q_2)^2)^{\frac{1}{2}}}.$$

The corresponding Hamilton's equations are

$$\begin{aligned}
\dot{Q}_1 &= \frac{P_1}{D} \\
\dot{Q}_2 &= \frac{P_2}{D} \\
\dot{P}_1 &= \frac{1}{2D}(P_1^2 + P_2^2)\frac{\partial D}{\partial Q_1} - \frac{1}{2}\frac{1}{(f^2 + g^2)^{\frac{3}{2}}}\frac{\partial (f^2 + g^2)}{\partial Q_1} \\
\dot{P}_2 &= \frac{1}{2D}(P_1^2 + P_2^2)\frac{\partial D}{\partial Q_2} - \frac{1}{2}\frac{1}{(f^2 + g^2)^{\frac{3}{2}}}\frac{\partial (f^2 + g^2)}{\partial Q_2}.
\end{aligned}$$

Let us introduce the *extended phase space* (see Appendix A) by defining a new variable T conjugated to time, such that the extended Hamiltonian becomes

$$\Gamma(P_1, P_2, T, Q_1, Q_2, t) = \frac{1}{2D}(P_1^2 + P_2^2) + T - \frac{1}{(f(Q_1, Q_2)^2 + g(Q_1, Q_2)^2)^{\frac{1}{2}}}.$$

Since $\dot{t} = \frac{\partial\Gamma}{\partial T} = 1$ and $\dot{T} = -\frac{\partial\Gamma}{\partial t} = 0$, one obtains that T is constant and in particular, along any solution, one has $T(t) \equiv T = -\tilde{H}$.

Next step consists in introducing a *fictitious (or regularized) time* s defined through the relation

$$dt = D(Q_1, Q_2)\, ds \qquad \text{or} \qquad \frac{d}{dt} = \frac{1}{D}\frac{d}{ds}.$$

Since $\dot{Q} = \frac{\partial\Gamma}{\partial P}$, it follows that

$$\dot{Q} = \frac{dQ}{dt} = \frac{dQ}{ds}\frac{ds}{dt} = \frac{1}{D}\frac{dQ}{ds} = \frac{\partial\Gamma}{\partial P}\ ;$$

the above relation implies that $\frac{dQ}{ds} = \frac{\partial\Gamma^*}{\partial P}$ with $\Gamma^* \equiv D\Gamma$. Similarly, we use $\dot{P} = -\frac{\partial\Gamma}{\partial Q}$ to obtain

$$\dot{P} = \frac{dP}{dt} = \frac{dP}{ds}\frac{ds}{dt} = \frac{1}{D}\frac{dP}{ds} = -\frac{\partial\Gamma}{\partial Q},$$

which provides

$$\frac{dP}{ds} = -\frac{\partial\Gamma^*}{\partial Q},$$

where $\Gamma^* \equiv D\Gamma$; in fact, we observe that

$$\frac{\partial\Gamma^*}{\partial Q} = \frac{\partial D}{\partial Q}\Gamma + D\frac{\partial\Gamma}{\partial Q} = D\frac{\partial\Gamma}{\partial Q},$$

being $\Gamma = 0$ along a solution. Finally, the new Hamiltonian Γ^* is given by

$$\Gamma^* \equiv D\Gamma = DT + \frac{1}{2}(P_1^2 + P_2^2) - \frac{D}{(f^2+g^2)^{\frac{1}{2}}}.$$

The associated Hamilton's equations ($j = 1, 2$) are

$$\begin{aligned}
\frac{dQ_j}{ds} &= P_j \\
\frac{dP_j}{ds} &= -\frac{\partial}{\partial Q_j}\Big[DT - \frac{D}{(f^2+g^2)^{\frac{1}{2}}}\Big] \\
\frac{dt}{ds} &= D \\
\frac{dT}{ds} &= 0.
\end{aligned}$$

Notice that the singularity of the problem is associated to the term $\frac{D}{(f^2+g^2)^{\frac{1}{2}}}$, which is transformed as

$$\begin{aligned} \frac{D}{(f^2+g^2)^{\frac{1}{2}}} &= \frac{D}{r} = \frac{4(Q_1^2+Q_2^2)}{(Q_1^4+Q_2^4-2Q_1^2Q_2^2+4Q_1^2Q_2^2)^{\frac{1}{2}}} \\ &= \frac{4(Q_1^2+Q_2^2)}{(Q_1^2+Q_2^2)} = 4, \end{aligned}$$

where we used $f = Q_1^2 - Q_2^2$ and $g = 2Q_1Q_2$.

Denoting by a prime the derivative with respect to s and recalling that $T = -\tilde{H}$, the equations of motion are $(j = 1, 2)$:

$$\begin{aligned} Q_j' &= P_j \\ P_j' &= -T\frac{\partial D}{\partial Q_j} = -T \cdot 8Q_j = 8\tilde{H}Q_j \end{aligned}$$

Therefore, one gets the second order differential equation

$$Q_j'' = 8\tilde{H}Q_j \qquad (j = 1, 2). \tag{7}$$

Notice that if $\tilde{H} < 0$ (corresponding to an elliptic orbit), one obtains the equation of an harmonic oscillator.

Remark: In relation to the remark of Section 3, we observe that the equation (7) describes pure harmonic oscillations with fixed frequency; therefore, one gets regular differential equations and the solution is **stable.**

We conclude this section by showing the relation between the fictitious time and the eccentric anomaly. To this end, setting $\omega^2 = -8\tilde{H}$, let us write equation (7) as

$$Q_1'' + \omega^2 Q_1 = 0, \qquad Q_2'' + \omega^2 Q_2 = 0.$$

Assuming the initial conditions

$$\begin{aligned} Q_1(0) &= \alpha \qquad & Q_2(0) = 0 \\ Q_1'(0) &= 0 \qquad & Q_2'(0) = \beta\omega, \end{aligned}$$

the solution is given by

$$Q_1(s) = \alpha\cos(\omega s), \qquad Q_2(s) = \beta\sin(\omega s)\ .$$

Defining $E = 2\omega s$, we get $Q_1(E) = \alpha\cos(\frac{E}{2})$, $Q_2(E) = \beta\sin(\frac{E}{2})$. Using the transformation $(Q_1, Q_2) \to (q_1, q_2)$, we get

$$\begin{aligned} q_1 &= Q_1^2 - Q_2^2 = -\frac{\beta^2-\alpha^2}{2} + \frac{\beta^2+\alpha^2}{2}\cos E \\ q_2 &= 2Q_1Q_2 = \alpha\beta\sin E, \end{aligned}$$

which describes an ellipse with center in $(-\frac{\beta^2-\alpha^2}{2}, 0)$. Moreover, the major semiaxis is $a = \frac{\beta^2+\alpha^2}{2}$ and the distance of the focus from the center of the ellipse is given by $ae = \frac{\beta^2-\alpha^2}{2}$. Therefore we obtain

$$\begin{aligned} r &= \sqrt{q_1^2+q_2^2} = Q_1^2 + Q_2^2 \\ &= \frac{\beta^2+\alpha^2}{2} - \frac{\beta^2-\alpha^2}{2}\cos E = a(1 - e\cos E), \end{aligned}$$

which corresponds to the standard Keplerian relation between the radial distance and the eccentric anomaly.

5.2 The regularization of the planar, circular, restricted three-body problem

In a synodic reference frame, the Hamiltonian of the planar, circular, restricted three–body problem is given by

$$H(p_1, p_2, q_1, q_2) = \frac{1}{2}(p_1^2 + p_2^2) + q_2p_1 - q_1p_2 - V(q_1, q_2),$$

where $V(q_1, q_2) = \frac{\mu_1}{r_1} + \frac{\mu_2}{r_2}$ and

$$r_1 = [(q_1 - \mu_2)^2 + q_2^2]^{\frac{1}{2}}, \qquad r_2 = [(q_1 + \mu_1)^2 + q_2^2]^{\frac{1}{2}}.$$

To regularize collisions with $\overline{P}_1$, we consider again a generating function of the form

$$W(p_1, p_2, Q_1, Q_2) = p_1 f(Q_1, Q_2) + p_2 g(Q_1, Q_2),$$

with $f(Q_1, Q_2) = Q_1^2 - Q_2^2 + \mu_2$, $g(Q_1, Q_2) = 2Q_1Q_2$ and characteristic equations

$$\begin{aligned} q_1 &= \frac{\partial W}{\partial p_1} = f(Q_1, Q_2) \\ q_2 &= \frac{\partial W}{\partial p_2} = g(Q_1, Q_2) \\ P_1 &= \frac{\partial W}{\partial Q_1} = p_1\frac{\partial f}{\partial Q_1} + p_2\frac{\partial g}{\partial Q_1} \\ P_2 &= \frac{\partial W}{\partial Q_2} = p_1\frac{\partial f}{\partial Q_2} + p_2\frac{\partial g}{\partial Q_2}. \end{aligned}$$

Remark: To regularize collisions with $\overline{P}_2$, it suffices to substitute f with $f(Q_1, Q_2) = Q_1^2 - Q_2^2 - \mu_1$.

The above change of coordinates transforms $p_1^2 + p_2^2$ into $\frac{1}{D}(P_1^2 + P_2^2)$, while the term $q_2p_1 - p_2q_1$ becomes

$$q_2p_1 - p_2q_1 = \frac{1}{2D}\,[P_1\frac{\partial}{\partial Q_2}(f^2+g^2) - P_2\frac{\partial}{\partial Q_1}(f^2+g^2)].$$

Therefore, the transformed Hamiltonian is given by

$$\begin{aligned}\tilde{H}(P_1,P_2,Q_1,Q_2) &= \frac{1}{2D}[P_1^2 + P_2^2 + P_1\frac{\partial}{\partial Q_2}(f^2+g^2) \\ &\quad -P_2\frac{\partial}{\partial Q_1}(f^2+g^2)] - \tilde{V}(Q_1,Q_2),\end{aligned}$$

where $\tilde{V}$ corresponds to V with q_1 replaced by $f(Q_1, Q_2)$ and q_2 replaced by $g(Q_1, Q_2)$. The Hamiltonian in the extended phase space becomes

$$\Gamma = T + \frac{1}{2D}[P_1^2 + P_2^2 + P_1\frac{\partial}{\partial Q_2}(f^2+g^2) - P_2\frac{\partial}{\partial Q_1}(f^2+g^2)] - \tilde{V}(Q_1,Q_2).$$

Next, we introduce the fictitious time $dt = D\,ds$, obtaining the Hamiltonian

$$\begin{aligned}\Gamma^* \equiv D\Gamma &= DT + \frac{1}{2}[P_1^2 + P_2^2 + P_1\frac{\partial}{\partial Q_2}(f^2+g^2) - P_2\frac{\partial}{\partial Q_1}(f^2+g^2)] \\ &\quad -D\tilde{V}(Q_1,Q_2).\end{aligned}$$

In the following, it will be useful to define the function $\Phi(Q_1, Q_2)$ as

$$\Phi(Q_1,Q_2) \equiv f(Q_1,Q_2) + ig(Q_1,Q_2).$$

From equation (4) one gets $\overline{\Omega} = \frac{1}{2}(q_1^2 + q_2^2) + V = \Omega - \frac{1}{2}\mu_1\mu_2$ and $\frac{1}{2}(f^2 + g^2) + \tilde{V} = \Omega - \frac{1}{2}\mu_1\mu_2$ (with $q_1 = f$ and $q_2 = g$). Observing that $|\Phi|^2 = f^2 + g^2$, one has

$$\frac{1}{2}|\Phi|^2 + \tilde{V} = \Omega - \frac{1}{2}\mu_1\mu_2.$$

From the definition of the Jacobi integral, we obtain

$$\tilde{H} = -T = \frac{\mu_1\mu_2 - C}{2},$$

namely

$$\frac{1}{2}|\Phi|^2 - T + \tilde{V} = \Omega - \frac{C}{2}.$$

Since $D\tilde{V} = D(\Omega - \frac{C}{2}) - \frac{1}{2}D|\Phi|^2 + DT$, we notice that the critical term is just $D(\Omega - \frac{C}{2})$. In order to achieve the desired regularization, let us define the complex physical and parametric coordinates as $z = q_1 + iq_2$ and $w = Q_1 + iQ_2$. The physical coordinates of the primaries are $z_1 = \mu_2$ and $z_2 = -\mu_1$. The regularizing transformation at $\overline{P}_1$ can be written as $z = w^2 + \mu_2$, while the transformation $z = w^2 - \mu_1$ regularizes the singularity at $\overline{P}_2$. By means of the first transformation, the primary $\overline{P}_1$ is moved to the origin of the w–plane, while $\overline{P}_2$ has coordinates $w_{1,2} = \pm i$. Since $r_1 = |w|^2$, $r_2 = |1 + w^2|$ and since

$$\mu_1 r_1^2 + \mu_2 r_2^2 = \mu_1(z - \mu_2)^2 + \mu_2(z + \mu_1)^2 = z^2 + \mu_1\mu_2 ,$$

we obtain that

$$\begin{aligned} U &= \Omega - \frac{C}{2} = \frac{1}{2}\mu_1\mu_2 + \frac{1}{2}(q_1^2 + q_2^2) + V - \frac{C}{2} \\ &= \frac{1}{2}(\mu_1 r_1^2 + \mu_2 r_2^2) + \frac{\mu_1}{r_1} + \frac{\mu_2}{r_2} - \frac{C}{2} \\ &= \frac{1}{2}[\mu_1|w|^4 + \mu_2|1 + w^2|^2] + \frac{\mu_1}{|w|^2} + \frac{\mu_2}{|1 + w^2|} - \frac{C}{2}. \end{aligned}$$

Observing that $D = 4(Q_1^2 + Q_2^2) = 4|w|^2$, we find that the term $DU = D(\Omega - \frac{C}{2})$ does not contain singularities at $\overline{P}_1$; in fact, we get

$$DU = D(\Omega - \frac{C}{2}) = 2|w|^2\ [\mu_1|w|^4 + \mu_2|1 + w^2|^2] + 4\mu_1 + \frac{4\mu_2|w|^2}{|1 + w^2|} - 2C|w|^2,$$

which is regular as far as $w \neq \pm i$, corresponding to the location of the other primary $\overline{P}_2$.

Remark: The expression of the velocity in terms of the fictitious time is obtained as follows. In the physical space the Jacobi integral is $|\dot{z}|^2 = 2U$, while in the parametric space it takes the form

$$|w'|^2 = 8|w|^2 U,$$

where we used $D = 4|w|^2$, $z = w^2 + \mu_2$, $\dot{z} = 2w\dot{w} = \frac{2}{D}ww'$, namely $|w'|^2 = \frac{D^2}{4|w|^2}|\dot{z}|^2$. Therefore, we have:

$$|w'|^2 = 8\mu_1 + |w|^2\Big[\frac{8\mu_2}{|1 + w|^2} + 4\mu_1|w|^4 + 4\mu_2|1 + w^2|^2 - 4C\Big].$$

From the previous relation, we conclude that

i) in $\overline{P}_1$ one has $r_1 = 0$, namely $w = 0$, while $|w'|^2 = 8\mu_1$ and the velocity is *finite;*

ii) in $\overline{P}_2$ one has $r_2 = 0$, namely $w = \pm i$, while $|w'|^2 = \infty$ and the velocity is *infinite.*

6. The Kustaanheimo–Stiefel transformation

In this section we outline the procedure which allows to regularize the singularities in the *spatial* case.

6.1 The equations of motion and the Hamiltonian

In the framework of the circular, restricted three-body problem, let us consider the motion in the 3–dimensional space of the three bodies S, $\overline{P}_1$ and $\overline{P}_2$. The primaries move in the q_1q_2–plane around their common center of mass, while in the synodic frame their coordinates become $\overline{P}_1(\mu_2, 0, 0)$, $\overline{P}_2(-\mu_1, 0, 0)$. Assume that the q_1q_2–plane rotates with unit angular velocity about the vertical axis. Then, the Hamiltonian function is given by

$$H(p_1, p_2, p_3, q_1, q_2, q_3) = \frac{1}{2}(p_1^2 + p_2^2 + p_3^2) + q_2p_1 - q_1p_2 - V(q_1, q_2, q_3),$$

where p_1, p_2, p_3 are the momenta conjugated to the coordinates q_1, q_2, q_3. The equations of motion of S are provided by the differential equations

$$\begin{aligned} \ddot{q}_1 - 2\dot{q}_2 &= \Omega_{q1} \\ \ddot{q}_2 + 2\dot{q}_1 &= \Omega_{q2} \\ \ddot{q}_3 &= \Omega_{q3}, \end{aligned}$$

where

$$\Omega = \frac{1}{2}(q_1^2 + q_2^2) + \frac{\mu_1}{r_1} + \frac{\mu_2}{r_2} + \frac{1}{2}\mu_1\mu_2,$$

with $r_1^2 \equiv (q_1 - \mu_2)^2 + q_2^2 + q_3^2$ and $r_2^2 \equiv (q_1 + \mu_1)^2 + q_2^2 + q_3^2$. More explicitly, the equations of motion are

$$\begin{aligned} \ddot{q}_1 - 2\dot{q}_2 &= q_1 - \frac{\mu_1}{r_1^3}(q_1 - \mu_2) - \frac{\mu_2}{r_2^3}(q_1 + \mu_1) \\ \ddot{q}_2 + 2\dot{q}_1 &= q_2 - \frac{\mu_1}{r_1^3}q_2 - \frac{\mu_2}{r_2^3}q_2 \\ \ddot{q}_3 &= -\frac{\mu_1}{r_1^3}q_3 - \frac{\mu_2}{r_2^3}q_3. \end{aligned}$$

6.2 The KS–transformation

As in the Levi–Civita, we define the fictitious time s as

$$dt = D\ ds,$$

for some factor D to be defined later. The relation between the second derivatives with respect to t and s is given by

$$\begin{aligned}\frac{d^2}{dt^2} &= \frac{d}{dt}(\frac{1}{D}\frac{d}{ds}) = \frac{1}{D}\frac{d}{ds}(\frac{1}{D}\frac{d}{ds}) \\ &= \frac{1}{D^3}(D\frac{d^2}{ds^2} - \frac{dD}{ds}\frac{d}{ds}) = \frac{1}{D^2}\frac{d^2}{ds^2} - \frac{1}{D^3}\frac{dD}{ds}\frac{d}{ds}.\end{aligned}$$

In terms of the fictitious time, the equations of motion are

$$\begin{aligned} Dq_1'' - D'q_1' - 2D^2 q_2' &= D^3\Omega_{q_1} \\ Dq_2'' - D'q_2' + 2D^2 q_1' &= D^3\Omega_{q_2} \\ Dq_3'' &= D^3\Omega_{q_3}, \end{aligned} \tag{8}$$

where the singular terms are contained in the right hand sides of the previous equations. Notice that Ω_{q_1}, Ω_{q_2}, $\Omega_{q_3} \sim O(\frac{1}{r_1^3})$.

Remark: We recall that in the planar case the Levi–Civita transformation is given by

$$\begin{pmatrix} q_1 \\ q_2 \end{pmatrix} = \begin{pmatrix} Q_1 & -Q_2 \\ Q_2 & Q_1 \end{pmatrix} \begin{pmatrix} Q_1 \\ Q_2 \end{pmatrix} = \begin{pmatrix} Q_1^2 - Q_2^2 \\ 2Q_1Q_2 \end{pmatrix},$$

where every element of the matrix $A_0(Q) \equiv$

$$\begin{pmatrix} Q_1 & -Q_2 \\ Q_2 & Q_1 \end{pmatrix}$$

is linear in Q_1, Q_2, with the matrix $A_0(Q)$ being orthogonal.

In order to achieve the regularization in space, we start by investigating the existence of a generalization $A(Q)$ of the matrix $A_0(Q)$ in $\mathbf{R}^n$, with the following properties:

i) the elements of $A(Q)$ must be linear homogeneous functions of the Q_i;

ii) the matrix $A(Q)$ must be orthogonal, namely

a) the scalar product of different rows must vanish;

b) each row must have norm $Q_1^2 + ... + Q_n^2$.

A result by A. Hurwitz (Stiefel and Scheifele, 1971) proves that such matrix exists only within spaces of dimensions $n = 1, 2, 4$ or 8. Therefore, it becomes necessary to map the 3–dimensional physical space into a 4–dimensional parametric space by defining the matrix

$$A(Q) = \begin{pmatrix} Q_1 & -Q_2 & -Q_3 & Q_4 \\ Q_2 & Q_1 & -Q_4 & -Q_3 \\ Q_3 & Q_4 & Q_1 & Q_2 \\ Q_4 & -Q_3 & Q_2 & -Q_1 \end{pmatrix}.$$

Consistently, we will extend the physical space by setting the fourth component equal to zero: $(q_1, q_2, q_3, 0)$.
Considering a collision with the primary $\overline{P}_1$, the Kustaanheimo–Stiefel (KS) regularization is defined as follows. Let

$$\begin{pmatrix} q_1 \\ q_2 \\ q_3 \\ 0 \end{pmatrix} = A(Q) \begin{pmatrix} Q_1 \\ Q_2 \\ Q_3 \\ Q_4 \end{pmatrix} + \begin{pmatrix} \mu_2 \\ 0 \\ 0 \\ 0 \end{pmatrix}$$
$$= \begin{pmatrix} Q_1 & -Q_2 & -Q_3 & Q_4 \\ Q_2 & Q_1 & -Q_4 & -Q_3 \\ Q_3 & Q_4 & Q_1 & Q_2 \\ Q_4 & -Q_3 & Q_2 & -Q_1 \end{pmatrix} \begin{pmatrix} Q_1 \\ Q_2 \\ Q_3 \\ Q_4 \end{pmatrix} + \begin{pmatrix} \mu_2 \\ 0 \\ 0 \\ 0 \end{pmatrix},$$

namely

$$\begin{aligned} q_1 &= Q_1^2 - Q_2^2 - Q_3^2 + Q_4^2 + \mu_2 \\ q_2 &= 2Q_1Q_2 - 2Q_3Q_4 \\ q_3 &= 2Q_1Q_3 + 2Q_2Q_4. \end{aligned}$$

Remarks:
1) Whenever $Q_3 = Q_4 = 0$, the KS–transformation reduces to the planar Levi–Civita.
2) The norms of each row (or column) of the matrix A are equal to $|Q|^2 \equiv Q_1^2 + Q_2^2 + Q_3^2 + Q_4^2$.
3) The regularization with $\overline{P}_2$ is obtained by replacing the constant vector $(\mu_2, 0, 0, 0)$ with $(-\mu_1, 0, 0, 0)$.
4) The matrix A is orthogonal: $A^T(Q)A(Q) = (Q,Q) \cdot Id$. Therefore, denoting by $q \equiv (q_1 - \mu_2, q_2, q_3, 0)$, one obtains

$$\begin{aligned} r_1^2 &= (q,q) = q^T q = Q^T A^T(Q) A(Q) Q \\ &= Q^T Q (Q,Q) = (Q,Q)^2, \end{aligned}$$

namely $r_1 = (Q,Q) = |Q|^2 = Q_1^2 + Q_2^2 + Q_3^2 + Q_4^2$.
5) A trivial computation shows that $A(Q)' = A(Q')$. As a consequence,

$$q' = A(Q')Q + A(Q)Q' = 2A(Q)Q',$$

which yields

$$\begin{pmatrix} q_1' \\ q_2' \\ q_3' \\ 0 \end{pmatrix} = 2A(Q)Q' = 2 \begin{pmatrix} Q_1Q_1' - Q_2Q_2' - Q_3Q_3' + Q_4Q_4' \\ Q_2Q_1' + Q_1Q_2' - Q_4Q_3' - Q_3Q_4' \\ Q_3Q_1' + Q_1Q_3' + Q_4Q_2' + Q_2Q_4' \\ Q_4Q_1' - Q_3Q_2' + Q_2Q_3' - Q_1Q_4' \end{pmatrix}.$$

The last equation is known as the *bilinear relation:*

$$Q_4Q'_1 - Q_3Q'_2 + Q_2Q'_3 - Q_1Q'_4 = 0.$$

In order to prove the canonicity of the transformation induced by the KS–procedure, it is necessary to choose the initial conditions in order that the bilinear equation is satisfied. 6) The second derivative with respect to the fictitious time of the physical coordinates is given by

$$q'' = 2A(Q)Q'' + 2A(Q')Q'.$$

In order to regularize the equations of motion, it is convenient to select the scaling factor D as

$$D \equiv 4r_1 = 4(Q,Q) = 4(Q_1^2 + Q_2^2 + Q_3^2 + Q_4^2).$$

The regularization is finally obtained mimicking the Levi–Civita procedure. More specifically, the scheme is the following: express the coordinates and their first and second derivatives in terms of Q, Q', Q''; recall that the singular part of the equations (7) is given by $D^3\Omega_{q_1}$, $D^3\Omega_{q2}$, $D^3\Omega_{q3}$. In a qualitative way, we proceed to remark that due to the fact that $\Omega_{q_1} \propto \frac{1}{r_1^3}$ and that $D \propto r_1$, one obtains that $D^3\Omega_{q_1} = O(1)$. Therefore we achieved the regularization of the singularity in $\overline{P}_1$. We refer the reader to Stiefel and Scheifele, 1971 for complete details.

7. Birkhoff Transformation

Let us consider two bodies $\overline{P}_1$, $\overline{P}_2$ with masses $1-\mu$, μ, moving on circular orbits around the barycenter O. Let us normalize to unity their distance. In the framework of the circular, restricted three-body problem, let us consider the motion of a third body S, moving in the same plane of the primaries. Let its coordinates be (x_1, x_2) in the synodic reference frame.

The Hamiltonian function governing the motion of S is given by

$$H(y_1, y_2, x_1, x_2) = \frac{1}{2}(y_1^2 + y_2^2) + x_2y_1 - x_1y_2 - \frac{1-\mu}{r_1} - \frac{\mu}{r_2},$$

where

$$r_1 = \sqrt{(x_1+\mu)^2 + x_2^2}, \qquad r_1 = \sqrt{(x_1 - 1 + \mu)^2 + x_2^2}.$$

We shift the origin of the reference frame to the midpoint between $\overline{P}_1$ and $\overline{P}_2$, by means of the complex transformation

$$q_1 + iq_2 = x_1 + ix_2 - \frac{1}{2} + \mu. \tag{9}$$

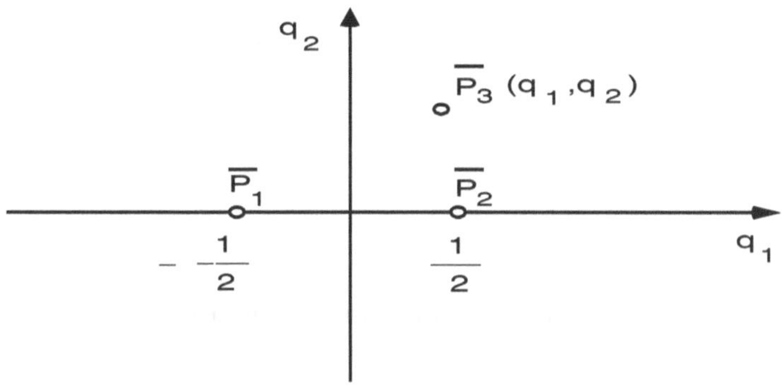

Figure 6. The coordinates of the three bodies after the transformation (9).

Therefore, the primaries will be located at $\overline{P}_{1,2}(\pm\frac{1}{2}, 0)$ (see Figure 6). Let us write the change of coordinates (9) as

$$\begin{aligned} p_1 &= y_1 \qquad q_1 = x_1 - \frac{1}{2} + \mu \\ p_2 &= y_2 \qquad q_2 = x_2; \end{aligned}$$

we obtain the Hamiltonian function

$$H_1(p_1, p_2, q_1, q_2) = \frac{1}{2}({p_1}^2 + {p_2}^2) + q_2 p_1 - (q_1 + \frac{1}{2} - \mu)p_2 - \frac{1-\mu}{r_1} - \frac{\mu}{r_2},$$

where

$$r_1 = \sqrt{(q_1 + \frac{1}{2})^2 + {q_2}^2}, \qquad r_2 = \sqrt{(q_1 - \frac{1}{2})^2 + {q_2}^2}.$$

We remark that the singularities are now located at $\overline{P}_1(-\frac{1}{2}, 0)$, $\overline{P}_2(\frac{1}{2}, 0)$. The aim of the Birkhoff Transformation will be to regularize simultaneously both collisions with $\overline{P}_1$ and $\overline{P}_2$. To this end, let us write the equations of motion as

$$\ddot{q} + 2i\dot{q} = \nabla_q U(q), \tag{10}$$

where $q = q_1 + iq_2$. Denoting by C the Jacobi constant, one has

$$U(q) = \Omega(q) - \frac{C}{2},$$

where

$$\begin{aligned} \Omega(q) &= \frac{1}{2}[(1-\mu){r_1}^2 + \mu {r_2}^2] + \frac{1-\mu}{r_1} + \frac{\mu}{r_2} \\ &= \frac{1}{2}[(1-\mu){r_1}^2 + \mu {r_2}^2] + \Omega_c(q). \end{aligned}$$

Let us define $\Omega_c(q)$ as the *critical part* given by the expression

$$\Omega_c(q) \equiv \frac{1-\mu}{r_1} + \frac{\mu}{r_2}$$

and let us write the Jacobi integral as

$$|\dot{q}|^2 = 2U(q) = 2\Omega(q) - C.$$

In order to determine the regularizing transformation, we perform a change of variables setting the complex parametric coordinates as $w = Q_1 + iQ_2$ and defining

$$q = h(w) = \alpha w + \frac{\beta}{w} ;$$

the unknown expressions for α and β must be determined in order to achieve the desired regularization.
We start by implementing a time transformation from the ordinary time t to a fictitious time s by means of the expression

$$\frac{dt}{ds} = g(w) \equiv |k(w)|^2 = k(w)\overline{k(w)},$$

where the function $g(w)$, or equivalently $k(w)$, must be suitably determined. One easily finds the following relations:

$$\begin{aligned}
\dot{q} &= \frac{dq}{dt} = \frac{dh(w)}{dw}\frac{dw}{ds}\frac{ds}{dt} = h'(w)w'\dot{s} \\
\ddot{q} &= h'(w)w'\ddot{s} + (h''(w)w'^2 + h'(w)w'')\dot{s}^2 \\
\nabla_w U &= \overline{h'}\ \nabla_q U.
\end{aligned}$$

As a consequence, the equations of motion (10) become

$$w'' + 2i\frac{w'}{\dot{s}} + w'\frac{\ddot{s}}{\dot{s}^2} + w'^2\frac{h''}{h'} = \nabla_w U\ \frac{1}{|h'|^2\dot{s}^2}.$$

Using the relations $\dot{s} = \frac{1}{g} = \frac{1}{k\bar{k}}$, $\ddot{s} = -\frac{\dot{g}}{g^2}$, one finds that $\frac{\ddot{s}}{\dot{s}^2} = -\dot{g}$. Moreover, from

$$-\dot{g} = \Big[k\frac{d\bar{k}}{dw}\frac{d\bar{w}}{ds} + \bar{k}\frac{dk}{dw}\frac{dw}{ds}\Big]\dot{s} = -\Big[\frac{\bar{k}'\bar{w}'}{\bar{k}} + \frac{k'w'}{k}\Big],$$

one obtains

$$w'' + 2ik\bar{k}w' - \frac{|w'|^2}{\bar{k}}\bar{k}' + w'^2\Big(\frac{h''}{h'} - \frac{k'}{k}\Big) = \frac{|k|^4}{|h'|^2}\ \nabla_w U.$$

Finally, we make use of the energy integral to obtain

$$|\dot{q}|^2 = 2\Omega(q) - C = 2U(q) = |h'|^2 \; |w'|^2 \frac{1}{|k|^4},$$

which implies that

$$|w'|^2 \;= 2U \; \frac{|k|^4}{|h'|^2}.$$

Since

$$\frac{h''}{h'} - \frac{k'}{k} = \frac{d}{dw} \left(\log \frac{h'}{k}\right),$$

we obtain

$$w'' + 2ik\bar{k}w' + |w'|^2 \; \frac{d}{dw}\left(\log \frac{h'}{k}\right) = \frac{|k|^4}{|h'|^2} \; \left[2U \; \frac{d \log \bar{k}}{dw} + \nabla_w U\right].$$

A suitable choice for the functions k and h is provided by the relation

$$k \;=\; h',$$

from which it follows that (10) becomes

$$w'' + 2i|h'|^2 w' \;=\; \nabla_w \Big(|h'|^2 U\Big).$$

Concerning the choice of α and β, we require that
i) the transformation involving h must eliminate both singularities;
ii) $\overline{P}_1$, $\overline{P}_2$ must stay fixed.
In order to meet the above requirements, we proceed as follows. Concerning statement *i*), we consider the singular term $\Omega_c(w) \; |h'(w)|^2$, where

$$\Omega_c(w) = \frac{1-\mu}{r_1} + \frac{\mu}{r_2} = \frac{1-\mu}{|\alpha w + \frac{\beta}{w} + \frac{1}{2}|} + \frac{\mu}{|\alpha w + \frac{\beta}{w} - \frac{1}{2}|}$$

and

$$|h'(w)|^2 = \frac{|\alpha w^2 - \beta|^2}{|w|^4},$$

namely

$$\Omega_c(w) \; |h'(w)|^2 = \frac{1}{|w|^3} \; \left[\frac{(1-\mu)|\alpha w^2 - \beta|^2}{|\alpha w^2 + \beta + \frac{w}{2}|} + \frac{\mu|\alpha w^2 - \beta|^2}{|\alpha w^2 + \beta - \frac{w}{2}|}\right].$$

We remark that the singularity at $q = \frac{1}{2}$ corresponds to the solutions of $|\alpha w^2 + \beta - \frac{w}{2}| = 0$, which are given by

$$w_{1,2} = \frac{1}{4\alpha} \; \left[1 \pm \sqrt{1 - 16\alpha\beta}\right].$$

Therefore, the roots of the numerator $|\alpha w^2 - \beta| = 0$ must coincide with $w_{1,2}$:

$$\frac{1}{4\alpha}\left[1 \pm \sqrt{1-16\alpha\beta}\right] = \pm\sqrt{\frac{\beta}{\alpha}},$$

i.e.

$$\alpha\beta\,(1-16\alpha\beta) = 0.$$

Since α and β are different from zero, it follows that

$$16\alpha\beta = 1,$$

which implies that

$$w_{1,2} = \frac{1}{4\alpha}.$$

Concerning statement ii), since $\overline{P}_2(\frac{1}{2},0)$ is transformed to $\overline{P}_2(\frac{1}{4\alpha},0)$, one needs to require that $\frac{1}{4\alpha} = \frac{1}{2}$. Therefore, one finds

$$\alpha = \frac{1}{2} \qquad \& \qquad \beta = \frac{1}{8}.$$

In order to regularize the singularity at $\overline{P}_1$, one needs to repeat the above procedure, which leads to exactly the same results, namely $\alpha = \frac{1}{2}$, $\beta = \frac{1}{8}$.

Notice that the equations of motion contain also the singular term $\frac{1}{|w|^3}$; however the singularity $w = 0$ corresponds to $q \to \infty$, which does not have a physical meaning.

In summary, the regularization steps performed to achieve Birkhoff Transformation are the following. Let

$$q = \frac{1}{2}(w + \frac{1}{4w}), \qquad w = Q_1 + iQ_2,$$

namely

$$\begin{aligned} q_1 &= \frac{1}{2}\,(Q_1 + \frac{Q_1}{4(Q_1^2+Q_2^2)}) \equiv f(Q_1,Q_2) \\ q_2 &= \frac{1}{2}\,(Q_2 - \frac{Q_2}{4(Q_1^2+Q_2^2)}) \equiv g(Q_1,Q_2). \end{aligned}$$

We perform a change of coordinates with generating function

$$W(p_1,p_2,Q_1,Q_2) = p_1 f(Q_1,Q_2) + p_2 g(Q_1,Q_2).$$

Let us define the matrix A as

$$A = \begin{pmatrix} f_{Q_1} & g_{Q_1} \\ f_{Q_2} & g_{Q_2} \end{pmatrix} \ker -3pt;$$

whose determinant takes the form

$$\det(A) = \frac{1}{64(Q_1^2+Q_2^2)^2}\left[16(Q_1^2+Q_2^2)^2+1+8(Q_2^2-Q_1^2)\right].$$

We introduce a fictitious time as

$$dt = D\, ds, \qquad D = \det(A).$$

The regularization is finally obtained considering the equations of motion in the extended phase space.

Acknowledgements: The author is deeply greatful to Paola Celletti for hand–drawing the pictures concerning "Alice's adventures in wonderland", whose aim was to help the reader to remind easily the basics of regularization theory.

Appendix A: The extended phase space.

We discuss the introduction of the *extended phase space.* If the Hamiltonian function $\tilde{H} = \tilde{H}(P, Q, t)$ depends explicitly on the time, one can introduce a time-independent Hamiltonian, defined as

$$\Gamma = \Gamma(P, Q, T, t) = \tilde{H}(P, Q, t) + T,$$

where T is conjugated to t. We show that Γ is identically zero along any solution.

In fact, if we select the initial conditions such that $T(0) = -\tilde{H}(P(0), Q(0), 0)$, one obtains that $T(t) = -\tilde{H}(t)$ along a solution for any t. To this end, we first notice that $\frac{d\tilde{H}}{dt} = \frac{\partial\tilde{H}}{\partial t}$, since by Hamilton's equations:

$$\begin{aligned}\frac{d\tilde{H}}{dt} = \frac{d\tilde{H}(P,Q,t)}{dt} &= \frac{\partial\tilde{H}}{\partial t} + \frac{\partial\tilde{H}}{\partial Q}\frac{dQ}{dt} + \frac{\partial\tilde{H}}{\partial P}\frac{dP}{dt} \\ &= \frac{\partial\tilde{H}}{\partial t} + \frac{\partial\tilde{H}}{\partial Q}\frac{\partial\tilde{H}}{\partial P} - \frac{\partial\tilde{H}}{\partial P}\frac{\partial\tilde{H}}{\partial Q} = \frac{\partial\tilde{H}}{\partial t}.\end{aligned}$$

Therefore, by Hamilton's equations one gets

$$\frac{dT}{dt} = -\frac{\partial\tilde{H}}{\partial t} = -\frac{d\tilde{H}}{dt},$$

from which one obtains that

$$\begin{aligned}T(t) &= T(0) + \int_0^t \frac{dT(s)}{ds}ds = T(0) - \int_0^t \frac{d\tilde{H}(s)}{ds}ds \\ &= -\tilde{H}(0) - \int_0^t \frac{d\tilde{H}(s)}{ds}ds = -\tilde{H}(t).\end{aligned}$$

References

Celletti, A. (2002) Singularities, collisions and regularization theory, in *Singularities in Gravitational Systems*, D. Benest, C. Froeschlé eds., Springer-Verlag, Berlin, Heidelberg, 1–24

Celletti, A. (2002) The Levi–Civita, KS and radial–inversion regularizing transformations, in *Singularities in Gravitational Systems*, D. Benest, C. Froeschlé eds., Springer-Verlag, Berlin, Heidelberg, 25–48

Kustaanheimo, P. and Stiefel, E. (1965) Perturbation theory of Kepler motion based on spinor regularization, *J. Reine Angew. Math* **218,** 204–219

Stiefel, E. L. and ossler, M. R and Waldvogel, J. and Burdet, C. A. (1967) Methods of regularization for computing orbits in Celestial Mechanics, *NASA Contractor Report* CR-769, Washington

Stiefel, E. L and Scheifele, G. (1971) *Linear and Regular Celestial Mechanics*, Springer–Verlag, Berlin, Heidelberg, New York

Stiefel, E. (1972) A linear theory of the perturbed two-body problem (regularization), in *Recent advances in Dynamical Astronomy*, B. D. Tapley, V. Szebehely eds., NATO School proc., 3–20

Szebehely, V. (1967) *Theory of orbits,* Academic Press, New York and London

Waldvogel, J. (1972) Collision singularities in gravitational problems, in *Recent advances in Dynamical Astronomy*, B. D. Tapley, V. Szebehely eds., NATO School proc., 21–33

ORDER AND CHAOS IN SATELLITE ENCOUNTERS

Jörg Waldvogel
Swiss Federal Institute of Technology ETH, CH-8092 Zürich, Switzerland

Abstract In order to describe the motion of two weakly interacting satellites of a central body we suggest to use orbital elements based on the the linear theory of Kepler motion in Levi-Civita's regularizing coordinates. The basic model is the planar three-body problem with two small masses, a model in which both regular (e.g. quasi-periodic) as well as chaotic motion can occur.

This paper discusses the basics of this approach and illustrates it with a typical example. First, we will revisit Levi-Civita's regularization of the two-dimensional Kepler motion and introduce sets of orbital elements based on the differential equations of the harmonic oscillator. Then, the corresponding theory for the three-dimensional motion will be developed using a quaternion representation of Kustaanheimo-Stiefel (KS) regularization; we present it by means of an elegant new notation.

Keywords: Regularization, co-orbital satellites

1. Introduction

We begin by summarizing the equations of motion of the three-body problem with two small masses in the form of two weakly coupled Kepler motions, valid in two or three dimensions.

Let $m_k, x_k, (k = 0,1,2)$ be the masses and positions of the three bodies, where we assume $x_k \in \mathbf{R}^2$ or $x_k \in \mathbf{C}$ in the planar case and $x_k \in \mathbf{R}^3$ in the spatial case. We assume the center of mass to be at rest at the origin, $\sum_{k=0}^{2} m_k x_k = 0$, and the masses satisfy the hierarchy $m_j \ll m_0, (j = 1,2)$. The Newtonian equations of motion in inertial coordinates are

B.A. Steves et al. (eds.), Chaotic Worlds: from Order to Disorder in Gravitational N-Body Dynamical Systems, 231–251.

$$\begin{aligned}\ddot{x}_0 &= m_1 \frac{x_1 - x_0}{\|x_1 - x_0\|^3} + m_2 \frac{x_2 - x_0}{\|x_2 - x_0\|^3} \\ \ddot{x}_1 &= m_2 \frac{x_2 - x_1}{\|x_2 - x_1\|^3} + m_0 \frac{x_0 - x_1}{\|x_0 - x_1\|^3} \\ \ddot{x}_2 &= m_0 \frac{x_0 - x_2}{\|x_0 - x_2\|^3} + m_1 \frac{x_1 - x_2}{\|x_1 - x_2\|^3},\end{aligned} \tag{1}$$

where dots denote differentiation with respect to time t.

We introduce relative coordinates $r_j = x_j - x_0,\ j = 1, 2$, from which the inertial coordinates may be recovered via

$$x_0 = -\frac{1}{M} \sum_{j=1}^{2} m_j r_j\,, \quad M = m_0 + m_1 + m_2.$$

Subtracting the first equation of (1) from the second and third equation yields the equivalent system

$$\begin{aligned}\ddot{r}_1 + (m_0 + m_1) \frac{r_1}{\|r_1\|^3} &= m_2 \left(\frac{r_2 - r_1}{\|r_2 - r_1\|^3} - \frac{r_2}{\|r_2\|^3} \right) \\ \ddot{r}_2 + (m_0 + m_2) \frac{r_2}{\|r_2\|^3} &= m_1 \left(-\frac{r_2 - r_1}{\|r_2 - r_1\|^3} - \frac{r_1}{\|r_1\|^3} \right),\end{aligned} \tag{2}$$

which describes a system of two perturbed Kepler motions with weak coupling if the masses satisfy the above hierarchy and none of the distances $\|r_1\|,\ \|r_2\|,\ \|r_2 - r_1\|$ is small.

2. Levi-Civita Regularization of Perturbed Kepler motion

We first restrict ourselves to the two-dimensional case and take advantage of the fact that Levi-Civita's regularizing transformation (Levi-Civita 1920) has the agreeable property of transforming perturbed Kepler problems into perturbed harmonic oscillators, i.e. into perturbed *linear* problems. For a recent account of regularization theory see the article (Celletti 2002) and other contributions in the same volume.

We will use both vector notation $x = (x_1, x_2)^T \in \mathbf{R}^2$ and complex notation $\mathbf{x} = x_1 + i\, x_2 \in \mathbf{C}$ for convenience. Consider now the perturbed Kepler problem

$$\ddot{x} + \mu \frac{x}{\|x\|^3} = f(x, t), \tag{3}$$

where dots denote derivatives with respect to time t, and $f(x,t)$ is a small perturbation. The corresponding energy equation is obtained by integrating the dot product $\langle \dot{x}, (3) \rangle$ with respect to t:

$$\frac{1}{2} \|\dot{x}\|^2 - \frac{\mu}{r} = -h, \tag{4}$$

where $r := \|x\|$ is the distance of the moving particle from the origin and h is the energy satisfying the differential equation and initial condition

$$\frac{dh}{dt} = -\langle \dot{x}, f \rangle, \quad h(0) = \frac{\mu}{\|x(0)\|} - \frac{1}{2} \|\dot{x}(0)\|^2. \tag{5}$$

The **first step** of Levi-Civita's regularization consists of introducing the fictitious time τ by the differential relation $dt = r \cdot d\tau$ (differentiation with respect to τ will be denoted by primes). In view of the step to follow we write the result of transforming equation (3) in complex form, where $f = (f_1, f_2)^T$, $\mathbf{f} = f_1 + i\, f_2$:

$$r\, \mathbf{x}'' - r'\, \mathbf{x}' + \mu\, \mathbf{x} = r^3 \mathbf{f} \in \mathbf{C}. \tag{6}$$

The **second step** of Levi-Civita's regularization consists of representing the complex physical coordinate $\mathbf{x}$ as the square $\mathbf{u}^2$ of a complex variable $\mathbf{u} = u_1 + i\, u_2 \in \mathbf{C}$,

$$\mathbf{x} = \mathbf{u}^2, \tag{7}$$

i.e. the mapping from the parametric plane to the physical plane is chosen as a conformal squaring. This implies

$$r = |\mathbf{x}| = |\mathbf{u}|^2 = \mathbf{u}\,\bar{\mathbf{u}}, \tag{8}$$

and differentiation of the last two equations yields

$$\mathbf{x}' = 2\,\mathbf{u}\,\mathbf{u}', \quad \mathbf{x}'' = 2\left(\mathbf{u}\,\mathbf{u}'' + \mathbf{u}'^{\,2}\right) \in \mathbf{C}, \quad r' = \mathbf{u}'\,\bar{\mathbf{u}} + \mathbf{u}\,\bar{\mathbf{u}}'. \tag{9}$$

By substituting this into (6) we obtain

$$2\, r\, \mathbf{u}\,\mathbf{u}'' + \mathbf{u}^2 \left(\mu - 2\,|\mathbf{u}'|^2\right) = r^3\, \mathbf{f}, \tag{10}$$

where the two terms $2\, r\, \mathbf{u}'^2 = 2\, \mathbf{u}'\, \bar{\mathbf{u}}\, \mathbf{u}\, \mathbf{u}'$ have cancelled out.

Remark. Obtaining initial values $\mathbf{u}(0) = \sqrt{\mathbf{x}(0)}$ requires the computation of a complex square root. This can conveniently be accomplished by means of the formula

$$\sqrt{\mathbf{x}} = \frac{\mathbf{x} + |\mathbf{x}|}{\sqrt{2\,(|\mathbf{x}| + \mathrm{Re}\ \mathbf{x})}}, \tag{11}$$

which reflects the observation that the complex vector $\sqrt{\mathbf{x}}$ has the direction of the bisector between $\mathbf{x}$ and the real vector $|\mathbf{x}|$; it holds in the range $-\pi < \arg(\mathbf{x}) < \pi$. The alternate formula

$$\sqrt{\mathbf{x}} = \frac{\mathbf{x} - |\mathbf{x}|}{i\sqrt{2\,(|\mathbf{x}| - \mathrm{Re}\,\mathbf{x})}}$$

holds in $0 < \arg(\mathbf{x}) < 2\pi$ and agrees with (11) in the upper half plane.

The **third step** of Levi-Civita's regularization process produces linear differential equations for the unperturbed problem $\mathbf{f} = 0$ by combining equation (10) with the energy relation. By using $\dot{\mathbf{x}} = \frac{1}{r}\cdot 2\,\mathbf{u}\mathbf{u}'$ equation(4) becomes

$$\mu - 2\,|\mathbf{u}'|^{\,2} \;=\; r\,h; \tag{12}$$

therefore the perturbed Kepler problem (3) is equivalent with

$$\begin{aligned} 2\,\mathbf{u}'' + h\cdot\mathbf{u} &= r\,\bar{\mathbf{u}}\,\mathbf{f} \quad \text{where} \quad \mathbf{x} = \mathbf{u}^2 \in \mathbf{C} \\ h' &= -\langle x', f\rangle, \end{aligned} \tag{13}$$

as is seen by substituting (12) into (10) and dividing by $r\,\mathbf{u}$, using (8). Also, equation (5) for h has been added in order to obtain a complete system of differential equations for the dependent variables $\mathbf{u} \in \mathbf{C},\ h \in \mathbf{R}$.

The following cases are of particular interest:

1 $f = 0 \Longrightarrow\ h = h(0) = \text{const}$. Equation(12) describes a harmonic (linear) oscillator in two dimensions.

2 f has a potential V, $f = -\mathrm{grad}\,V \Longrightarrow h(x) = h(0) + V(x) - V(0)$. Equation (12) describes a perturbed harmonic oscillator with varying frequency.

3 $f = O(\varepsilon),\ \varepsilon \to 0 \Longrightarrow h(x) = h(0) + O(\varepsilon)$. Equation (12) describes a perturbed harmonic oscillator with slowly varying frequency.

3. Regular Elements

We will now take advantage of the linear structure of the unperturbed version $f = 0$ of equation(12). Consider, as a model problem, the perturbed harmonic oscillator

$$\frac{d^2u}{d\tau^2} + \omega^2\,u = F, \tag{14}$$

where F is small, and ω is slowly varying. First, we transform the perturbed oscillator (14) to constant frequency by introducing the new independent variable E according to the differential relation

$$dE = \omega \, d\tau, \quad \frac{d}{d\tau} = \omega \frac{d}{dE}, \quad \frac{d^2}{d\tau^2} = \omega^2 \frac{d^2}{dE^2} + \omega' \frac{d}{dE}, \tag{15}$$

where primes – in this section – denote derivatives with respect to E ($2E$ is the *eccentric anomaly* of the osculating Kepler motion). Equation (14) now becomes

$$\frac{d^2 u}{dE^2} + u = G \quad \text{with} \quad G = \frac{F - \omega' \, du/dE}{\omega^2}. \tag{16}$$

We now discuss two ways of introducing regular elements to equation (16):

3.1 Variation of the constant

With the notation $v := du/dE$ equation (16) may be written as the vector differential equation

$$\begin{pmatrix} u \\ v \end{pmatrix}' = A \begin{pmatrix} u \\ v \end{pmatrix} + \begin{pmatrix} 0 \\ G \end{pmatrix} \quad \text{with} \quad A = \begin{pmatrix} 0 & 1 \\ -1 & 0 \end{pmatrix}. \tag{17}$$

Departing from the matrix solution

$$U(E) = \begin{pmatrix} u(E) \\ v(E) \end{pmatrix} = \begin{pmatrix} \cos E & \sin E \\ -\sin E & \cos E \end{pmatrix} \tag{18}$$

satisfying the unperturbed equation $U' = AU$, the method of varying the constant consists of seeking a solution of (17) of the form

$$\begin{pmatrix} u(E) \\ v(E) \end{pmatrix} = U(E) \begin{pmatrix} \alpha(E) \\ \beta(E) \end{pmatrix}, \tag{19}$$

where $\alpha(E), \beta(E)$ are the (orbital) elements. Substituting this into (17) and solving for the derivatives of the elements yields

$$\begin{aligned} \frac{d\vec{\alpha}}{dE} &= -\vec{G} \cdot \sin E \\ \frac{d\vec{\beta}}{dE} &= \vec{G} \cdot \cos E. \end{aligned} \tag{20}$$

Here we have used vector symbols in order to indicate that equations (19) not only hold for scalars $\alpha, \beta, G \in \mathbf{R}$, but also for vectors $\vec{\alpha}, \vec{\beta}, \vec{G} \in \mathbf{R}^n$.

3.2 Singular-value decomposition

This set of elements is based on the original perturbation problem (14), now written for vectors $\vec{u}, \vec{F} \in \mathbf{R}^n$, here with $n = 2$,

$$\frac{d^2\vec{u}}{d\tau^2} + \omega^2\,\vec{u} = \vec{F}. \tag{21}$$

As it is often done, we define the *osculating orbit* at the fixed value $\tau = \tau_0$ as the orbit determined by equation (21) for $\tau \geq \tau_0$ if ω and $\vec{F}$ are fixed at the values $\omega_0 = \omega(\tau_0)$, $\vec{F_0} = \vec{F}(\tau_0)$ for all $\tau \geq \tau_0$. It is convenient to shift the origin by introducing the new coordinate $\vec{v}$ according to $\vec{u} = \vec{v} + \vec{F_0}/\omega_0^2$; $\vec{v}$ satisfies

$$\frac{d^2\vec{v}}{d\tau^2} + \omega_0^2\,\vec{v} = 0 \tag{22}$$

for $\tau \geq \tau_0$. Any four quantities uniquely characterizing the solution of (22) may be used as orbital elements, e.g. the initial values $\vec{v}_0 = \vec{v}(\tau_0)$, $\vec{v}\,'_0 = \vec{v}\,'(\tau_0)$ at $\tau = \tau_0$. With these initial values, the solution of (22) is

$$\vec{v}(\tau) = \vec{v}_0\,\cos(\omega_0\tilde{\tau}) + \frac{1}{\omega_0}\,\vec{v}\,'_0\,\sin(\omega_0\tilde{\tau}), \quad \tilde{\tau} = \tau - \tau_0\,, \tag{23}$$

or, by representing $\vec{v}(\tilde{\tau}) = (v_1(\tilde{\tau}),\, v_2(\tilde{\tau}))^T$ in components and using matrix notation:

$$\begin{pmatrix} v_1(\tilde{\tau}) \\ v_2(\tilde{\tau}) \end{pmatrix} = M \begin{pmatrix} \cos(\omega_0\tilde{\tau}) \\ \sin(\omega_0\tilde{\tau}) \end{pmatrix}, \quad M = \begin{pmatrix} v_{10} & v'_{10}/\omega_0 \\ v_{20} & v'_{20}/\omega_0 \end{pmatrix} \tag{24}$$

with $v_{10}, v_{20}, v'_{10}, v'_{20}$ being the components of $\vec{v}_0$ and $\vec{v}\,'_0$, respectively.

The osculating orbit (24) is an ellipse centered at $\vec{v} = 0$, or $\vec{u} = \vec{F_0}/\omega_0^2$. A more natural choice of orbital elements than $\vec{v}_0, \vec{v}\,'_0$ are four geometric parameters of the ellipse (24). We suggest to use the singular-value decomposition (SVD)

$$M = U\,S\,V^T \tag{25}$$

of the matrix M in (24), where U, V are orthogonal and S is diagonal,

$$U = \begin{pmatrix} \cos\varphi & -\sin\varphi \\ \sin\varphi & \cos\varphi \end{pmatrix}, \; V = \begin{pmatrix} \cos\psi & -\sin\psi \\ \sin\psi & \cos\psi \end{pmatrix}, \; S = \begin{pmatrix} \sigma_1 & 0 \\ 0 & \sigma_2 \end{pmatrix}, \tag{26}$$

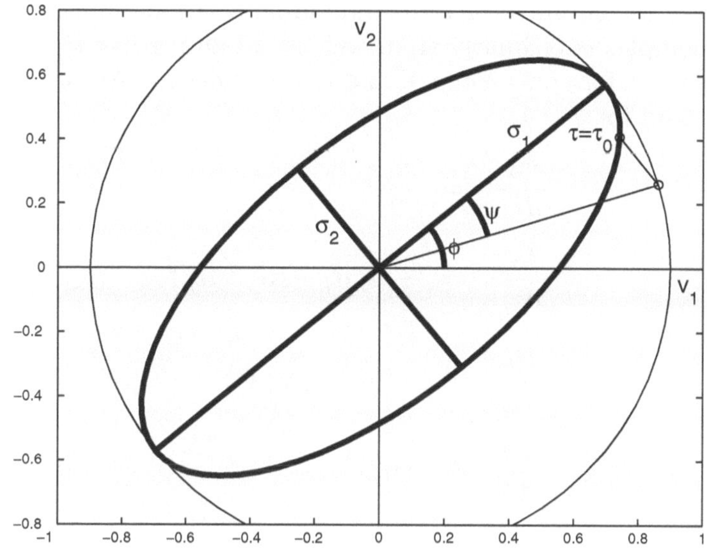

Figure 1. The harmonic elements $\sigma_1, \sigma_2, \varphi, \psi$. The axes are v_1, v_2.

with nonnegative singular values $\sigma_1 \geq \sigma_2 \geq 0$. The two quantities σ_1, σ_2 (the semi-axes of the osculating ellipse) and the two angles φ, ψ will be referred to as the *harmonic elements* of the perturbed oscillator (21). The geometric meaning of the angles φ, ψ is shown in Figure 1: φ is the angle of rotation of the axes of the ellipse with respect to fixed axes, ψ is related to the position of the moving point on the ellipse corresponding to $\tau = \tau_0$.

4. Weakly Coupled Harmonic Oscillators

In order to apply one of the proposed sets of elements for describing coorbital motion we formulate the equations of motion (1) of the weakly coupled Kepler motions in terms of the Levi-Civita coordinates of Section 2. In this way the unperturbed problem will be defined by *linear* differential equations. Using the symbols $\mu_j = m_0 + m_j$, $j = 1, 2$ as well as complex notation $\mathbf{r}_1, \mathbf{r}_2 \in \mathbf{C}$ and the abbreviations $\mathbf{f}_1, \mathbf{f}_2 \in \mathbf{C}$ for the right-hand sides, equation (1) reads as

$$\begin{aligned} \ddot{\mathbf{r}}_1 + \mu_1 \frac{\mathbf{r}_1}{|\mathbf{r}_1|^3} &= \mathbf{f}_1 \\ \ddot{\mathbf{r}}_2 + \mu_2 \frac{\mathbf{r}_2}{|\mathbf{r}_2|^3} &= \mathbf{f}_2. \end{aligned} \qquad (27)$$

For $j = 1, 2$ we will introduce the individual fictitious times τ_j and Levi-Civita's complex coordinates $\mathbf{u}_j$ as well as the derivatives $\mathbf{v}_j = 2\, d\mathbf{u}_j/d\tau_j$ (the factor 2 is for convenience), and the energies h_j. According to (12), (8) we obtain for $j = 1, 2$

$$\begin{aligned} \frac{d\,\mathbf{u}_j}{d\tau_j} &= \frac{\mathbf{v}_j}{2} \\ \frac{d\,\mathbf{v}_j}{d\tau_j} &= -\,h_j\,\mathbf{u}_j + |\mathbf{u}_j|^2\,\bar{\mathbf{u}}_j\,\mathbf{f}_j \\ \frac{d\,h_j}{d\tau_j} &= -\,\mathrm{Re}\,(\bar{\mathbf{u}}_j\,\bar{\mathbf{v}}_j\,\mathbf{f}_j) \\ \frac{d\,t}{d\tau_j} &= |\,\mathbf{u}_j\,|^2\,. \end{aligned} \tag{28}$$

The inconvenience of two individual fictitious times τ_1, τ_2 is easily circumvented by going back to physical time t as independent variable in both oscillators $j = 1, 2$ (by using the last equations of (27)):

$$\begin{aligned} \frac{d\,\mathbf{u}_j}{d\,t} &= \frac{\mathbf{v}_j}{2\,|\,\mathbf{u}_j\,|^2} \\ \frac{d\,\mathbf{v}_j}{d\,t} &= -\,\frac{h_j}{\bar{\mathbf{u}}_j} + \bar{\mathbf{u}}_j\,\mathbf{f}_j, \quad j = 1, 2 \\ \frac{d\,h_j}{d\,t} &= -\,\mathrm{Re}\left(\frac{\bar{\mathbf{v}}_j}{\mathbf{u}_j}\,\mathbf{f}_j\right). \end{aligned} \tag{29}$$

In the near-circular case we have $|\mathbf{u}_j| \approx$ const, and the osculating elements of Section 3.2 can be used.

Consider, as an example, the planar motion of two small satellites m_1, m_2 under the influence of the large central body m_0. We assume the satellites to be initially on nearly identical circular osculating orbits of radii $\varrho_1 \approx \varrho_2$ with velocities $s_j = \sqrt{M/\varrho_j}$, $j = 1, 2$, where $M = m_0 + m_1 + m_2$ is the total mass of the system. Assuming opposite initial postions of the satellites with respect to the central body yields the position and velocity vectors

$$r_1 = (\varrho_1, 0)^T, \quad r_2 = (-\varrho_2, 0)^T, \quad \dot{r}_1 = (0, s_1)^T, \quad \dot{r}_2 = (0, -s_2)^T\,,$$

to be used as initial conditions for the equations of motion (1). The choice $m_0 = 777$, $m_1 = 1$, $m_2 = 2$, $\varrho_1 = 1.01$, $\varrho_2 = 1.00$, which is the basis of the orbits shown in Figure 2, causes the considered system to display the typical behavior of coorbital motion. The mass ratios and

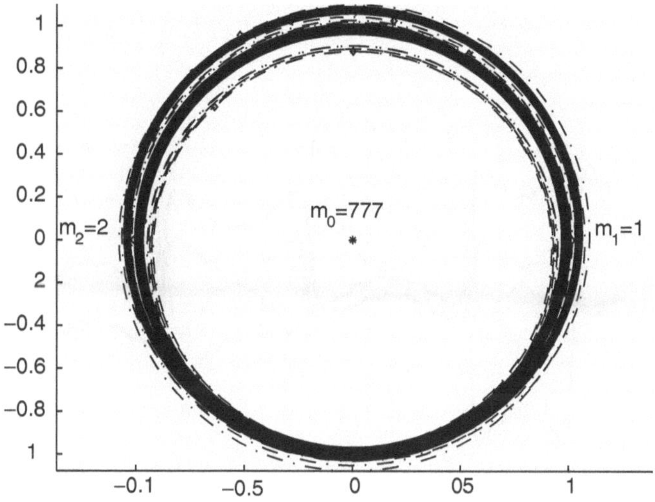

Figure 2. A few revolutions of the satellites $m_1 = 1$ (diamonds, dashdotted) and $m_2 = 2$ (circles, solid line) about the central body $m_0 = 777$, shown in heliocentric coordinates. Initial distances $\varrho_1 = 1.01, \varrho_2 = 1.00$, opposite starting points, circular initial velocities.

geometry are such that a few close encounters of the satellites with the well known orbit exchanges occur before the system suddenly breaks up.

Whereas Figure 2 hardly shows any structure, the behavior of the harmonic elements clearly reflects the dynamics of the orbits of the two satellites. In Figure 3 the semi-axes σ_1, σ_2 of the ellipses (24) associated with the two satellites are plotted versus time, the thin lines corresponding to the smaller satellite, m_1. The wiggles correspond to the near-Keplerian revolutions around the central body when the satellites are far apart; the motion is quite orderly. If the satellites have a close encounter the harmonic elements change dramatically, corresponding to the transition into new near-Keplerian orbits. This process repeats in a more or less regular way, with an increasing tendency towards chaos, however. When deviations from this pattern have sufficiently accumulated – here after 5 close encounters of the satellites – the orderly motion ceases: *Order and chaos in satellite encounters.* The reader is referred to the wide literature on this topic, beginning with Spirig and Waldvogel 1985 and Hénon and Petit 1986; a more extended list of references may be found in Waldvogel 1999.

The derivation of the perturbation equations for the harmonic elements and the development of a perturbation theory based on these elements will not be discussed here.

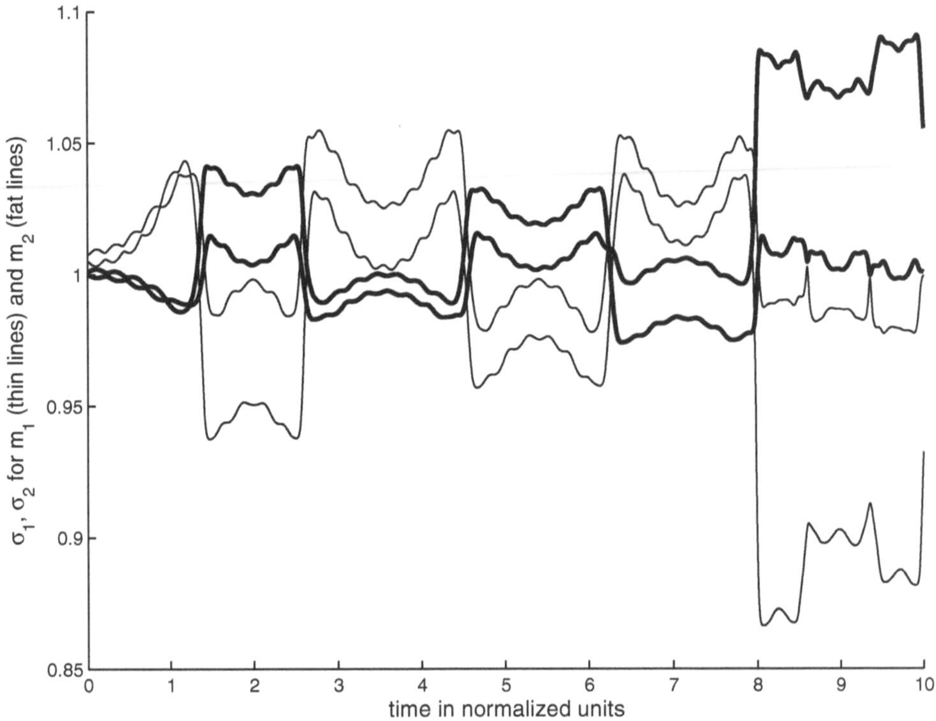

Figure 3. The harmonic elements σ_1, σ_2 of the orbits of Figure 2. The fat lines refer to the larger satellite $m_2 = 2$.

5. Quaternion Algebra

In the remaining sections we indicate how the ideas discussed above may be generalized to three-dimensional motion. The essential step is to replace Levi-Civita's regularization with the Kustaanheimo-Stiefel (KS) regularization, described in the original papers Kustaanheimo and Stiefel 1965 and Kustaanheimo 1964, and extensively discussed in the comprehensive text (Stiefel and Scheifele 1971). The relevant mapping from the 3-sphere onto the 2-sphere was discovered already in 1931 by Heinz Hopf (Hopf 1931) and is referred to in topology as the Hopf mapping.

Both the Levi-Civita and the Kustaanheimo-Stiefel regularization share the property of "linearizing" the equations of motion of the two-body problem. Quaternion algebra, introduced by W. R. Hamilton in 1856 (Hamilton 1856), was originally rejected (Stiefel and Scheifele 1971, p. 286) as a tool for regularizing the three-dimensional Kepler motion. Here we will present a new elegant way of extending the Levi-Civita regularization to three dimensions by means of quaternions. Similar techniques were used earlier by M. D. Vivarelli (Vivarelli 1994, Vrbik 1994, Vrbik 1995 and Vrbik 1998).

5.1 Basics

Quaternion algebra is a generalization of the algebra of complex numbers obtained by using three independent "imaginary" units i, j, k. As for the single imaginary unit i in the algebra of complex numbers, the rules

$$i^2 = j^2 = k^2 = -1$$

are postulated, together with the non-commutative multiplication rules

$$i\,j = -j\,i = k\,, \quad j\,k = -k\,j = i\,, \quad k\,i = -i\,k = j\,.$$

Given real numbers $u_l \in \mathbf{R}\,,\ l = 0, 1, 2, 3$, the object

$$\mathbf{u} = u_0 + i\,u_1 + j\,u_2 + k\,u_3 \tag{30}$$

is called a *quaternion* $\mathbf{u} \in \mathbf{U}$, where $\mathbf{U}$ denotes the set of all quaternions (in the remaining sections bold-face characters denote quaternions). The sum $iu_1 + ju_2 + ku_3$ is called the *quaternion part* of $\mathbf{u}$, whereas u_0 is naturally referred to as its real part. The above multiplication rules and vector space addition define the *quaternion algebra.* Multiplication is generally non-commutative; however, any quaternion commutes with a real:

$$c\,\mathbf{u} = \mathbf{u}\,c, \quad c \in \mathbf{R}, \ \mathbf{u} \in \mathbf{U}, \tag{31}$$

and for any three quaternions $\mathbf{u}$, $\mathbf{v}$, $\mathbf{w} \in \mathbf{U}$ the associative law holds:

$$(\mathbf{u}\,\mathbf{v})\,\mathbf{w} = \mathbf{u}\,(\mathbf{v}\,\mathbf{w}). \tag{32}$$

The quaternion $\mathbf{u}$ may naturally be associated with the corresponding vector $u = (u_0, u_1, u_2, u_3) \in \mathbf{R}^4$. For later reference we introduce notation for 3-vectors in two important particular cases: $\vec{u} = (u_1, u_2, u_3) \in \mathbf{R}^3$ for the vector associated with the *pure quaternion* $\mathbf{u} = i\,u_1 + j\,u_2 + k\,u_3$, and $\underline{u} = (u_0, u_1, u_2)$ for the vector associated with the quaternion with a vanishing fourth component, $\mathbf{u} = u_0 + i\,u_1 + j\,u_2$.

For convenience we also introduce the vector $\vec{\imath} = (i, j, k)$; the quaternion $\mathbf{u}$ may then formally be written as $\mathbf{u} = u_0 + \langle \vec{\imath}, \vec{u} \rangle$. For the two quaternion products of $\mathbf{u}$ with $\mathbf{v} = v_0 + \langle \vec{\imath}, \vec{v} \rangle$ we then obtain the concise expressions

$$\begin{aligned} \mathbf{u}\,\mathbf{v} &= u_0 v_0 - \langle \vec{u}, \vec{v} \rangle + \langle \vec{\imath},\, u_0\,\vec{v} + v_0\,\vec{u} + \vec{u} \times \vec{v}\,\rangle \\ \mathbf{v}\,\mathbf{u} &= u_0 v_0 - \langle \vec{u}, \vec{v} \rangle + \langle \vec{\imath},\, u_0\,\vec{v} + v_0\,\vec{u} - \vec{u} \times \vec{v}\,\rangle, \end{aligned} \tag{33}$$

where $\times$ denotes the vector product. Note that the non-commutativity shows only in the sign of the term with the vector product.

The *conjugate* $\bar{\mathbf{u}}$ of the quaternion $\mathbf{u}$ is defined as

$$\bar{\mathbf{u}} = u_0 - i\,u_1 - j\,u_2 - k\,u_3; \tag{34}$$

then the *modulus* $|\mathbf{u}|$ of $\mathbf{u}$ is obtained from

$$|\mathbf{u}|^2 = \mathbf{u}\,\bar{\mathbf{u}} = \bar{\mathbf{u}}\,\mathbf{u} = \sum_{l=0}^{3} u_l^2. \tag{35}$$

As transposition of a product of matrices, conjugation of a quaternion product reverses the order of its factors:

$$\overline{\mathbf{u}\,\mathbf{v}} = \bar{\mathbf{v}}\,\bar{\mathbf{u}}. \tag{36}$$

5.2 Rotations in Three Dimensions

This is a short digression in order to demonstrate the elegance of quaternion representations in three-dimensional geometry. Let $\vec{x}$ be a vector of the Euclidean 3-space $\mathbf{R}^3$, and consider the right-handed rotation about the unit vector $\vec{a} = (a_1, a_2, a_3)^T$, $|\vec{a}| = 1$ through the angle ω. One way of representing the mapping

$$\vec{x} \in \mathbf{R}^3 \mapsto \vec{y} = T\,\vec{x} \tag{37}$$

is to use Cayley's parametrization

$$T = \frac{I \cdot \cos\frac{\omega}{2} + S(\vec{a}) \cdot \sin\frac{\omega}{2}}{I \cdot \cos\frac{\omega}{2} - S(\vec{a}) \cdot \sin\frac{\omega}{2}} \tag{38}$$

of the orthogonal matrix T. Here I is the unit matrix, and matrix "division" may be interpreted as multiplication with the inverse of the denominator (if it exists) from the left *or* from the right. The skew symmetric matrix

$$S(\vec{a}) = \begin{pmatrix} 0 & -a_3 & a_2 \\ a_3 & 0 & -a_1 \\ -a_2 & a_1 & 0 \end{pmatrix}$$

is the *vector product matrix* associated with $\vec{a}$, i.e. for every column vector $\vec{x} \in \mathbf{R}^3$ we have $S(\vec{a})\,\vec{x} = \vec{a} \times \vec{x}$.

A *proof* of (38) may be obtained by considering the relation

$$\cos\frac{\omega}{2} \cdot (\vec{y} - \vec{x}) = \sin\frac{\omega}{2} \cdot \vec{a} \times (\vec{y} + \vec{x}), \tag{39}$$

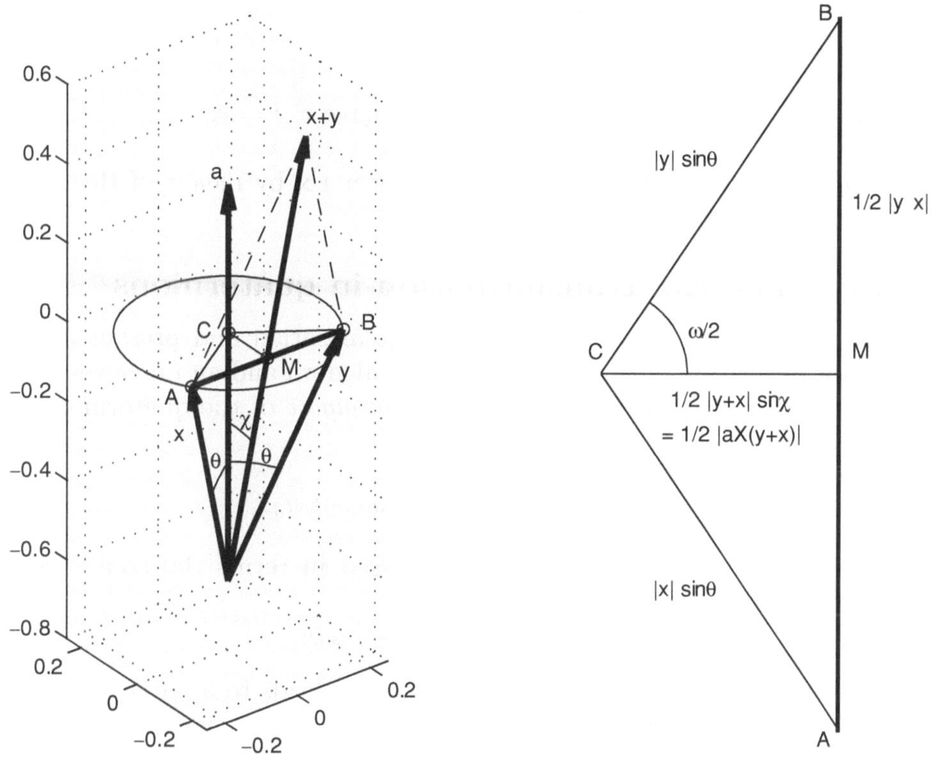

Figure 4. Geometric proof of equation (39) for the rotation about the axis $\vec{a}$, $|\vec{a}| = 1$ through the angle ω. equation (39) follows from the geometry of the triangle CMB.

which is equivalent with (37), (38) and may be obtained by multiplying (37) from the left by the denominator of (38). Equation (39), in turn, exactly reflects the geometry of the rotation under consideration, as is easily deduced from Figure 4, in particular from the triangle CMB.

The mapping (37) with T from (38) will now be written as a relation between the pure quaternions $\mathbf{x} = \langle \vec{\imath}, \vec{x} \rangle$, $\mathbf{y} = \langle \vec{\imath}, \vec{y} \rangle$.

Theorem 1. Let $\vec{a} \in \mathbf{R}^3$ be a unit vector and define the unit quaternion

$$\mathbf{r} := \cos\frac{\omega}{2} + \langle \vec{\imath}, \vec{a} \rangle \sin\frac{\omega}{2}.$$

Furthermore, let $\vec{x} \in \mathbf{R}^3$ be an arbitrary vector and $\mathbf{x} = \langle \vec{\imath}, \vec{x} \rangle$ the associated pure quaternion. Then the map

$$\mathbf{x} \mapsto \mathbf{y} = \mathbf{r}\,\mathbf{x}\,\mathbf{r}^{-1}$$

describes the right-handed rotation of $\vec{x}$ about the axis $\vec{a}$ through the angle ω.

Remark. Since $\mathbf{r}$ is a unit quaternion we have $\mathbf{r}^{-1} = \bar{\mathbf{r}}$.

Proof. Show the equivalent statement $\mathbf{y}\,\mathbf{r} = \mathbf{r}\,\mathbf{x}$ by means of the multiplication rules (32) and equation (39).

6. The KS Transformation in quaternions

In this section we will revisit KS regularization and present a new, elegant derivation of it, using quaternion algebra and an unconventional "conjugate" $\mathbf{u}^\star$ referred to as the *star conjugate* of the quaternion $\mathbf{u} = u_0 + i\,u_1 + j\,u_2 + k\,u_3$:

$$\mathbf{u}^\star := u_0 + i\,u_1 + j\,u_2 - k\,u_3. \tag{40}$$

The star conjugate of $\mathbf{u}$ may be expressed in terms the conventional conjugate $\bar{\mathbf{u}}$ as

$$\mathbf{u}^\star = k\,\bar{\mathbf{u}}\,k^{-1} = -k\,\bar{\mathbf{u}}\,k;$$

however, it turns out that the definition (40) leads to a particularly elegant treatment of KS regularization. The following elementary properties are easily verified:

$$\begin{aligned} (\mathbf{u}^\star)^\star &= \mathbf{u}. \\ |\,\mathbf{u}^\star|^2 &= |\mathbf{u}|^2 \\ (\mathbf{u}\,\mathbf{v})^\star &= \mathbf{v}^\star\,\mathbf{u}^\star \end{aligned} \tag{41}$$

Consider now the mapping

$$\mathbf{u} \in \mathbf{U} \longmapsto \mathbf{x} = \mathbf{u}\,\mathbf{u}^\star. \tag{42}$$

Star conjugation immediately yields $\mathbf{x}^\star = (\mathbf{u}^\star)^\star\,\mathbf{u}^\star = \mathbf{x}$; hence $\mathbf{x}$ is a quaternion of the form $\mathbf{x} = x_0 + i\,x_1 + j\,x_2$ which may be associated with the vector $\underline{x} = (x_0, x_1, x_2) \in \mathbf{R}^3$. From $\mathbf{u} = u_0 + i\,u_1 + j\,u_2 + k\,u_3$ we obtain

$$\begin{aligned} x_0 &= u_0^2 - u_1^2 - u_2^2 + u_3^2 \\ x_1 &= 2(u_0\,u_1 - u_2\,u_3) \\ x_2 &= 2(u_0\,u_2 + u_1\,u_3), \end{aligned} \tag{43}$$

which is exactly the KS transformation in its classical form or – up to a permutation of the indices – the Hopf map. Therefore we have

Theorem 2: The KS transformation which maps $u = (u_0, u_1, u_2, u_3) \in \mathbf{R}^4$ to $\underline{x} = (x_0, x_1, x_2) \in \mathbf{R}^3$ is given by the quaternion relation

$$\mathbf{x} = \mathbf{u}\,\mathbf{u}^\star,$$

where $\mathbf{u} = u_0 + i\,u_1 + j\,u_2 + k\,u_3$, $\mathbf{x} = x_0 + i\,x_1 + j\,x_2$.

Corollary: The norms of the vectors u and $\underline{x}$ satisfy $\|\underline{x}\| = \|u\|^2$.

Proof: By appropriately combining the two conjugations and using the rules (31), (32), (35), (36), (40) we obtain

$$\|\underline{x}\|^2 = \mathbf{x}\,\bar{\mathbf{x}} = \mathbf{u}\,(\mathbf{u}^\star\,\bar{\mathbf{u}}^\star)\,\bar{\mathbf{u}} = |\mathbf{u}^\star|^2\,|\mathbf{u}\,|^2 = |\mathbf{u}\,|^4 = \|u\|^4,$$

from where the statement follows.

As a side step, we will briefly discuss two topics not directly related to our primary objective, perturbation theory. Both the inverse map and the Birkhoff Transformation in three dimensions allow for an elegant treatment in terms of quaternions.

6.1 The Inverse Map

Being given a quaternion $\mathbf{x} = x_0 + i\,x_1 + j\,x_2$ with vanishing fourth component, $x_3 = 0$, we want to find all quaternions $\mathbf{u}$ such that $\mathbf{u}\,\mathbf{u}^\star = \mathbf{x}$. We propose the following solution in two steps:

First step: Find a particular solution $\mathbf{u} = \mathbf{v} = \mathbf{v}^\star = v_0 + i\,v_1 + j\,v_2$ which has also a vanishing fourth component. Since $\mathbf{v}\,\mathbf{v}^\star = \mathbf{v}^2$ we may use equation (11), which was developed for the complex square root, also for the square root of a quaternion:

$$\mathbf{v} = \frac{\mathbf{x} + |\mathbf{x}|}{\sqrt{2\,(|\mathbf{x}| + x_0)}}.$$

Clearly, $\mathbf{v}$ has a vanishing fourth component.

Second step: The entire family of solutions (geometrically a circle in $\mathbf{R}^4$), parametrized by the angle φ, is given by

$$\mathbf{u} = \mathbf{v}\cdot e^{k\,\varphi} = \mathbf{v}\,(\cos\varphi + k\sin\varphi).$$

Proof. $\mathbf{u}\,\mathbf{u}^\star = \mathbf{v}\,e^{k\,\varphi}\,e^{-k\,\varphi}\,\mathbf{v}^\star = \mathbf{v}\,\mathbf{v}^\star = \mathbf{x}.$

6.2 The Birkhoff Transformation

This regularizing transformation was proposed in 1915 by George David Birkhoff (Birkhoff 1915) in order to regularize all singularities of the planar restricted three-body problem with a single transformation. In 1965 E. Stiefel and this author (Stiefel and Waldvogel 1965) published a generalization of Birkhoff's transformation to three dimensions, using the KS transformation. Later these ideas resulted in the publications Waldvogel 1967a and Waldvogel 1967b.

Here we will first revisit the classical Birkhoff Transformation (the same conformal map is known in aerodynamics as the Joukowsky transformation) and represent it as the composition of three conformal mappings; this will then readily generalize to the spatial situation by means of quaternions.

Consider a rotating physical plane parametrized by the complex variable $\mathbf{y} \in \mathbf{C}$; for convenience we assume the fixed primaries of the restricted three-body problem to be situated at the points A, C given by the complex posititons $\mathbf{y} = -1$ and $\mathbf{y} = 1$, respectively (see Figure 5). The complex variable of the parametric plane will be denoted by $\mathbf{v}$ and will be normalized in such a way that the primaries are mapped to $\mathbf{v} = -1$ or $\mathbf{v} = 1$, respectively.

The key observation is that Levi-Civita's conformal map (7), $\mathbf{u} \mapsto \mathbf{x} = \mathbf{u}^2$, not only regularizes collisions at $\mathbf{x} = 0$ but also analogous singularities at $\mathbf{x} = \infty$. This is seen by closing the complex planes to become Riemann spheres (by adding the point at infinity) and using inversions $\mathbf{x} = 1/\tilde{\mathbf{x}}$, $\mathbf{u} = 1/\tilde{\mathbf{u}}$.

Taking advantage of this fact, we first map the $\mathbf{v}$-sphere to an auxiliary $\mathbf{u}$-sphere by the Möbius transformation

$$\mathbf{v} \longmapsto \mathbf{u} = \frac{\mathbf{v}+1}{\mathbf{v}-1} = 1 + \frac{2}{\mathbf{v}-1}, \tag{44}$$

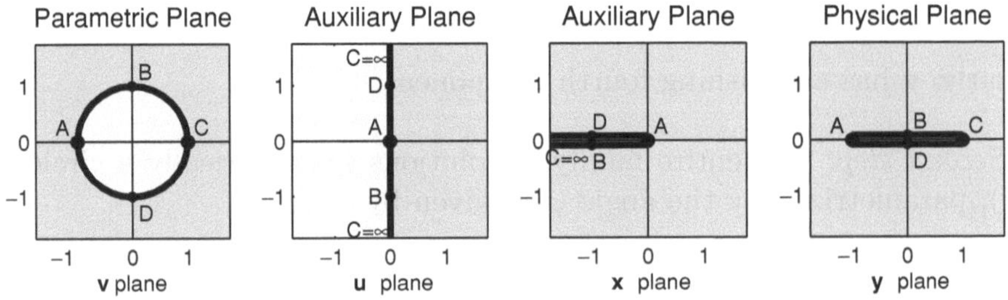

$$\mathbf{v} \mapsto \mathbf{u} = \frac{\mathbf{v}+1}{\mathbf{v}-1} \qquad \mathbf{u} \mapsto \mathbf{x} = \mathbf{u}^2 \qquad \mathbf{x} \mapsto \mathbf{y} = \frac{\mathbf{x}+1}{\mathbf{x}-1}$$

Figure 5. The sequence of conformal maps generating the Birkhoff transformation.

which takes the primaries A, C to the points $\mathbf{u} = 0$, $\mathbf{u} = \infty$, respectively. The Levi-Civita map (7) will leave these points invariant while regularizing collisions at A or C. Finally, the Möbius transformation

$$\mathbf{x} \longmapsto \mathbf{y} = \frac{\mathbf{x}+1}{\mathbf{x}-1} = 1 + \frac{2}{\mathbf{x}-1} \tag{45}$$

maps A, C to $\mathbf{y} = -1$ and $\mathbf{y} = 1$, respectively. The composition of the maps (44), (7), (45) yields

$$\mathbf{y} = \frac{\left(\dfrac{\mathbf{v}+1}{\mathbf{v}-1}\right)^2 + 1}{\left(\dfrac{\mathbf{v}+1}{\mathbf{v}-1}\right)^2 - 1} \quad \text{or} \quad \mathbf{y} = \frac{1}{2}\left(\mathbf{v} + \frac{1}{\mathbf{v}}\right), \tag{46}$$

the well known map used by Birkhoff.

In the spatial case we choose $\mathbf{v}, \mathbf{u}, \mathbf{x}, \mathbf{y} \in \mathbf{U}$ to be quaternions, $\mathbf{x} = \mathbf{x}^\star$, $\mathbf{y} = \mathbf{y}^\star$ being quaternions with vanishing fourth components, associated with 3-vectors $\underline{x}$, $\underline{y}$. Then the mappings (44), (45), being shifted inversions in 4 or 3 dimensions, are both conformal maps, in fact the only conformal maps existing in those dimensions, except for the translations, magnifications, and rotations. Composing these with the KS or Hopf map (42), $\mathbf{u} \mapsto \mathbf{x} = \mathbf{u}\,\mathbf{u}^\star$, yields

$$\mathbf{y} = 1 \;+\; (\mathbf{v}^\star - 1)\,(\mathbf{v} + \mathbf{v}^\star)^{-1}\,(\mathbf{v} - 1) \tag{47}$$

after a few lines of careful noncommutative algebra. This is easily split up into components by means of the inversion formula $1/\mathbf{v} = \bar{\mathbf{v}}/|\mathbf{v}|^2$; it agrees with the results given in [47] up to the sign of v_3. Both transformations regularize; the discrepancy is due to the different definition of the orientation in the inversions.

7. Perturbation theory in Three Dimensions

In order to regularize the perturbed three-dimensional Kepler motion by means of the KS transformation it is necessary to look at the properties of the map (42) under differentiation.

7.1 Differentiation

The transformation (42) is a mapping from $\mathbf{R}^4$ to $\mathbf{R}^3$; it therefore leaves one degree of freedom in the parametric space undetermined. In KS theory (Kustaanheimo and Stiefel 1965, Stiefel and Scheifele 1971), this freedom is taken advantage of by trying to inherit as much as possible of the conformality properties of the Levi-Civita map, but other approaches exist (Vrbik 1998). By imposing the "bilinear relation"

$$u_3\, du_0 - u_2\, du_1 + u_1\, du_2 - u_0\, du_3 = 0 \tag{48}$$

between the vector $u = (u_0, u_1, u_2, u_3)$ and its differential du on orbits the tangential map of (42) becomes a linear map with an orthogonal (but non-normalized) matrix.

This property has a simple consequence on the differentiation of the quaternion representation (42) of the KS transformation. Considering the noncommutative multiplication of quaternions, the differential of equation (42) becomes

$$d\mathbf{x} = d\,\mathbf{u} \cdot \mathbf{u}^\star + \mathbf{u} \cdot d\,\mathbf{u}^\star, \tag{49}$$

whereas (48) takes the form of a commutator relation,

$$\frac{1}{2}\,(-d\,\mathbf{u} \cdot \mathbf{u}^\star + \mathbf{u} \cdot d\,\mathbf{u}^\star) = 0. \tag{50}$$

Combining (49) with the relation (50) yields the elegant result

$$d\mathbf{x} = 2\,\mathbf{u} \cdot d\,\mathbf{u}^\star, \tag{51}$$

i.e. the bilinear relation (48) of KS theory is equivalent with the requirement that the tangential map of $\mathbf{u} \mapsto \mathbf{u}\,\mathbf{u}^\star$ behaves like in a commutative algebra.

7.2 KS Regularization

The procedure of Section 2 for regularizing the planar case now carries over almost identically to the spatial case; care must be taken to preserve the order of the factors in quaternion products. Changing the order is only permitted if one of the factors is real. Let $\mathbf{x} = x_0 + i\,x_1 + j\,x_2 \in \mathbf{U}$ be the quaternion associated with the vector $\underline{x} = (x_0, x_1, x_2)$; then the perturbed Kepler problem (3) is given by

$$\ddot{\mathbf{x}} + \mu\,\frac{\mathbf{x}}{\mathbf{r^3}} = \mathbf{f}(\mathbf{x}, t) \in \mathbf{U}, \quad r = |\mathbf{x}|, \tag{52}$$

where $\mathbf{f}(\mathbf{x}, t) = f_0(\mathbf{x}, t) + i\,f_1(\mathbf{x}, t) + j\,f_2(\mathbf{x}, t) = \mathbf{f}^\star(\mathbf{x}, t)$ is the quaternion associated with the perturbation $\underline{f}(\underline{x}, t) \in \mathbf{R}^3$. The energy equation (4) becomes

$$\frac{1}{2}\,|\dot{\mathbf{x}}|^2 - \frac{\mu}{r} = -h, \tag{53}$$

whereas the result of the **first step**, i.e. introducing τ by $dt = r \cdot d\tau$ exactly agrees with equation (6),

$$r\,\mathbf{x}'' - r'\,\mathbf{x}' + \mu\,\mathbf{x} = r^3\,\mathbf{f} \ \in \mathbf{U}. \tag{54}$$

The relations (7), (8), (9) needed in the **second step** read as

$$\mathbf{x} = \mathbf{u}\,\mathbf{u}^\star, \qquad r = \mathbf{u}\,\bar{\mathbf{u}} \tag{55}$$

and

$$\mathbf{x}' = 2\,\mathbf{u}\,\mathbf{u}^{\star\prime}, \qquad \mathbf{x}'' = 2\,\mathbf{u}\,\mathbf{u}^{\star\prime\prime} + 2\,\mathbf{u}'\,\mathbf{u}^{\star\prime}, \qquad r' = \mathbf{u}'\,\bar{\mathbf{u}} + \mathbf{u}\,\bar{\mathbf{u}}'. \tag{56}$$

The energy equation (53) needed in the **third step** becomes

$$\frac{1}{2\,r^2}\,\left|\mathbf{x}'\right|^2 - \frac{\mu}{r} = -h \qquad \text{or} \qquad \frac{4}{2\,r^2}\,\mathbf{u}\,(\,\mathbf{u}^{\star\prime}\,\bar{\mathbf{u}}^{\star\prime}\,)\,\bar{\mathbf{u}} - \frac{\mu}{r} = -h$$

which results in the relation

$$\mu - 2\,\left|\mathbf{u}'\right|^{\,2} \;=\; r\,h, \tag{57}$$

in complete agreement with (12) found for the planar case.

Substitution of (55) and (56) into (54) yields the lengthy formula

$$(\mathbf{u}\,\bar{\mathbf{u}})\,(2\,\mathbf{u}\,\mathbf{u}^{\star\prime\prime} + 2\,\mathbf{u}'\,\mathbf{u}^{\star\prime}) \;-\; (\mathbf{u}'\,\bar{\mathbf{u}} + \mathbf{u}\,\bar{\mathbf{u}}')\cdot 2\,\mathbf{u}\,\mathbf{u}^{\star\prime} \;+\; \mu\,\mathbf{u}\,\mathbf{u}^\star \;=\; r^3\,\mathbf{f}, \tag{58}$$

which is considerably simplified by observing that the second and third term – after applying the distributive law – compensate:

$$2\,(\mathbf{u}\,\bar{\mathbf{u}})\,\mathbf{u}'\,\mathbf{u}^{\star\prime} \;-\; 2\,\mathbf{u}'\,(\bar{\mathbf{u}}\,\mathbf{u})\,\mathbf{u}^{\star\prime} \;=\; 0.$$

Furthermore, the fourth term of (58) reduces to

$$-2\,\mathbf{u}\,\bar{\mathbf{u}}'\cdot\mathbf{u}\,\mathbf{u}^{\star\prime} \;=\; -\,2\,\mathbf{u}\,(\bar{\mathbf{u}}'\,\mathbf{u}')\,\mathbf{u}^\star \;=\; -\,2\,\mathbf{u}\,|\,\mathbf{u}'|^2\,\mathbf{u}^\star \;=\; (r\,h - \mu)\,\mathbf{u}\,\mathbf{u}^\star$$

by using (57). Therefore, (58) becomes

$$2\,r\,\mathbf{u}\,\mathbf{u}^{\star\prime\prime} \;+\; r\,h\,\mathbf{u}\,\mathbf{u}^\star \;=\; r^2\,\mathbf{u}\,\bar{\mathbf{u}}\,\mathbf{f},$$

and, finally, left-division by $r\,\mathbf{u}$ and star-conjugation yields

$$2\,\mathbf{u}'' \;+\; h\,\mathbf{u} \;=\; r\,\mathbf{f}\,\bar{\mathbf{u}}^\star, \tag{59}$$

a differential equation in perfect agreement with (12) for the planar case; however, it takes more than an educated guess to get the correct right-hand side.

7.3 Osculating Elements

The considerations of Section 3.2., i.e. the introducing osculating *harmonic* elements by means of the singular-value decomposition, readily carries over to the spatial case. Consider, as in equation (22), the vector-valued harmonic oscillator

$$\frac{d^2\vec{v}}{d\tau^2} + \omega_0^2\,\vec{v} = 0, \tag{60}$$

with initial values $\vec{v}(\tau_0) = \vec{v}_0$, $\vec{v}\,'(\tau_0) = \vec{v}\,'_0$ at some time $\tau = \tau_0$, now with $\vec{v} \in \mathbf{R}^4$. As in equation (24), we write the solution in matrix form as

$$\begin{pmatrix} v_0(\tau) \\ v_1(\tau) \\ v_2(\tau) \\ v_3(\tau) \end{pmatrix} = M \begin{pmatrix} \cos(\omega_0(\tau-\tau_0)) \\ \sin(\omega_0(\tau-\tau_0)) \end{pmatrix}, \quad M = \begin{pmatrix} v_{00} & v'_{00}/\omega_0 \\ v_{10} & v'_{10}/\omega_0 \\ v_{20} & v'_{20}/\omega_0 \\ v_{30} & v'_{30}/\omega_0 \end{pmatrix} \tag{61}$$

with $v_{00}, v_{10}, v_{20}, v_{30}, v'_{00}, v'_{10}, v'_{20}, v'_{30}$ being the components of $\vec{v}_0$ and $\vec{v}\,'_0$, respectively. The SVD of M becomes

$$M = U\,S\,V^T \tag{62}$$

with an orthogonal matrix $U \in \mathbf{R}^{4\times 4}$ having 5 essential degrees of freedom and matrices

$$S = \begin{pmatrix} \sigma_1 & 0 \\ 0 & \sigma_2 \\ 0 & 0 \\ 0 & 0 \end{pmatrix}, \qquad V = \begin{pmatrix} \cos\psi & -\sin\psi \\ \sin\psi & \cos\psi \end{pmatrix}$$

with 3 more degrees of freedom, totally 8, as expected.

In the preceding text we have presented a unified theory of regularization of the perturbed Kepler motion. Quaternion algebra allows for an elegant treatment of the spatial case in a way completely analogous to the way the planar case is traditionally handled by means of complex numbers. As a consequence of the linearity of the regularized equations of the perturbed Kepler motion, the problem of satellite encounters reduces to a linear perturbation problem, the problem of coupled harmonic oscillators. Orbital elements based on the oscillators may lead to a simpified discussion of ordered and chaotic behavior in repeated satellite encounters. This has been demonstrated by means of an instructive example.

References

George David Birkhoff (1915). The restricted problem of three bodies. *Rendiconti del Circolo Matematico di Palermo* **39**, 1.

Alessandra Celletti (2002). The Levi-Civita, KS and radial-inversion regularizing transformations. In: D. Benest and C. Frœschlé (eds.): *Singularities in Gravitational Systems. Applications to Chaotic Transport in the Solar System.* Lecture Notes in Physics, Springer 25-48.

William Rowan Hamilton (1856). Memorandum respending a new system of Roots of Unity. *Philosophical Magazine* **12**, 4, 446.

Michel Hénon and Jean-Marc Petit (1986). Series expansions for encounter-type solutions of Hill's problem. *Celest. Mech.* **38**, 67-100.

Heinz Hopf(1931). Über die Abbildung der dreidimensionalen Sphäre auf die Kugelfläche. *Math. Ann.* **104**, (1931). Reprinted in Selecta Heinz Hopf, p. 38-63, Springer-Verlag Berlin Heidelberg New York, 1964.

Paul Kustaanheimo (1964). Spinor regularization of the Kepler motion. *Ann. Univ. Turku*, Ser. AI **73**.

Paul Kustaanheimo and Eduard L. Stiefel (1965). Perturbation theory of Kepler motion based on spinor regularization. *J. Reine Angew. Math.* **218** (1965), 204-219.

Franz Spirig and Jörg Waldvogel (1985). The three-body problem with two small masses: A singular-perturbation approach to the problem of Saturn's coorbiting satellites. In: V. Szebehely (ed.), *Stability of the Solar System and its Minor Natural and Artificial Bodies*, Reidel, 53-63.

Eduard L. Stiefel et Jörg Waldvogel (1965). Généralisation de la régularisation de Birkhoff pour le mouvement du mobile dans l'espace à trois dimensions. *C.R. Acad. Sc. Paris* **260**, 805.

Eduard L. Stiefel and Gerhard Scheifele (1971). *Linear and Regular Celestial Mechanics.* Springer-Verlag Berlin Heidelberg New York, 301 pp.

Tullio Levi-Civita (1920). Sur la régularisation du problème des trois corps. *Acta Math.* **42**, 99-144.

Maria Dina Vivarelli (1994). The KS transformation revisited. *Meccanica* **29**, 15-26.

Jan Vrbik (1994). Celestial Mechanics via quaternions. *Can. J. Phys.* **72**, 141-146.

Jan Vrbik (1995). Perturbed Kepler problem in quaternionic form. *J. Phys. A* **28**, 193-198.

Jan Vrbik (1998). Novel analysis of tadpole and horseshoe orbits. *Celestial Mechanics and Dynamical Astronomy* **69**, 283-291.

Jörg Waldvogel (1967a). Die Verallgemeinerung der Birkhoff-Regularisierung für das räumliche Dreikörperproblem. *Bulletin Astronomique*, Série 3, Tome II, Fasc. 2 (1967), 295-341.

Jörg Waldvogel (1967b). The restricted elliptic three-body problem. In: E. Stiefel et. al., *Methods of Regularization for Computing Orbits in Celestial Mechanics*, NASA Contractor Report NASA CR **769**, 88-115.

Jörg Waldvogel (1999). Long-term evolution of coorbital motion. In: B.A. Steves and A.E. Roy (eds.), *The Dynamics of Small Bodies in the Solar System: A Major Key to Solar System Studies*, NATO ASI Series C, Plenum, 257-276.

IV

APPLICATIONS

REGULAR MOTIONS IN EXTRA-SOLAR PLANETARY SYSTEMS

S. Ferraz-Mello, T.A. Micchtchenko
Instituto de Astronomia, Geofísica e Ciências Atmosféricas, Universidade de São Paulo, Brasil
sylvio@usp.br

C. Beaugé
Observatório Astronómico, Universidad Nacional de Córdoba, Argentina
beauge@oac.unicor.edu

Abstract This paper is a review of the dynamics of a system of planets. It includes the study of averaged equations in both non-resonant and resonant systems and shows the great deal of situations in which the angle between the two semi-major axes oscillates around a constant value. It introduces the Hamiltonian equations of the N-planet problem and Poincaré's reduction of them to $3N$ degrees of freedom with a detailed discussion of the non-osculating "canonical" heliocentric Keplerian elements that should be used with Poincaré relative canonical variables. It also includes Beaugé's approximation to expand the disturbing function in the exoplanetary case where masses and eccentricities are large. The paper is concluded with a discussion of systems captured into resonance and their evolution to symmetric and asymmetric stationary solutions with Apsidal corotation.

Keywords: extra solar planets

1. Introduction

The discovery of the first extra-solar planet in orbit around a main-sequence star was announced in 1995. Since then, the number of known Extra-Solar Planets has not ceased to grow. As observations accumulate, planetary systems with 2 and 3 planets are being discovered. More than

B.A. Steves et al. (eds.), Chaotic Worlds: from Order to Disorder in Gravitational N-Body Dynamical Systems, 255–288.

ten such systems are, presently, known[1]. As the discoveries are recent and many of the discovered planets are at the edge of observational capabilities, the uncertainties on their orbital elements and masses are large. It is worth recalling that one of the 2-planet systems previously announced, HD 83443, vanished from the lists after new observations failed to show the radial velocity variations previously identified with a second planet. Another important example of the current uncertainties is the "jump" suffered by the determined eccentricity of HD 82943b. For a long time, it was listed as ~ 0.4, while a new determination using observations over a long span of time gives only 0.18. In a similar manner, the mass of HD 82943c became twice bigger than believed before (Mayor et al., 2004). These discrepancies should be enough to show us how hazardous is the task of drawing conclusions from the present data, and that we should avoid conclusions critically dependent on the available data.

In the current state of the art, we are only capable of discovering big planets with not too large periods. Therefore, the planets so far discovered are big and most of them have orbits close to their parent star. Another characteristic is the large eccentricities of many of them. Even if large eccentricities favour discovery, this characteristic is not only due to observational bias and needs an explanation. (See Perryman, 2000 for a review of the existing hypotheses). Large eccentricities are considered to be the result of early migration processes. It is generally believed that the planets did not form at their present observed locations, but were driven by a migration process due to tidal interaction of the planets with the discs where they were formed (see Papaloizou, 2003). Whether this orbital drift is still at work or not is a matter of debate, although it is more plausible to assume that it stopped after the end of the planetary formation stage. These early processes were also responsible for having driven the (surviving) systems to very stable conditions in which orbit periapses appear close to alignment or anti-alignment. This condition is observed in several systems.

Periapses alignment (or anti-alignment) may occur in resonant and non-resonant systems alike. In resonant systems, they are the natural states after the system is trapped into a mean-motion resonance (see Section 6). Conversely, in non-resonant systems they are a consequence of the angular momentum variations during resonance crossings without capture (Ferraz-Mello et al., in preparation). However, and independently of how they reached this condition, an important consequence of

[1]For an up-to-date list see the web page "Extra-Solar Planets Encyclopaedia", by J.Schneider at www.obspm.fr/planets and links therein.

this type of configuration is that they constitute a stabilizing mechanism for planetary orbits, especially if they have large eccentricities.

Four extra-solar systems seem to satisfy the resonance condition: Gliese 876, HD 82943, 55 Cnc and 47 UMa. The first two have planets with periods in a 2/1 commensurability, the third in a 3/1, and the last close to a 7/3. With regard to Gliese 876, numerical simulations (Laughlin and Chambers, 2001; Lee and Peale, 2002) seem to indicate that these bodies are actually deeply trapped in an Apsidal corotation (see Section 6.1): They exhibit a libration of both resonant angles $\sigma_i = 2\lambda_2 - \lambda_1 - \varpi_i$, and also an alignment of their major axes. Apsidal corotation seems to be the natural issue of a capture in resonance in the case of two planets with initially low eccentricities (Ferraz-Mello et al., 2003; *cf.* this paper, Section 6). The alignment (or anti-alignment) of periapses has not yet been confirmed in the case of the other planetary systems above mentioned.

The most conspicuous non-resonant system showing nearly aligned periapses is υ Andromedae. This system has been the object of many numerical and analytical studies (for references, see Michtchenko and Malhotra, 2004). The orbit of the planets *c* and *d* in this system are such that the distance between their periapses oscillates about zero with half-amplitude $\sim$ 60 degrees and period $\sim$ 7260 years.

2. Hamiltonian Equations of the N-planet problem

This section considers the Hamiltonian formulation of the problem of N planets orbiting a star in an arbitrary configuration. This is a well-known problem in Celestial Mechanics. However, the vast majority of papers in Celestial Mechanics deal with the so-called restricted 3-body problems in which only 2 bodies have finite masses. Therefore, some basic topics of the general problem need to be remembered.

2.1 Barycentric Hamiltonian Equations

The barycentric Hamiltonian equations of the N+1 body problem are obtained using the basic principles of mechanics. Let m_i $(i = 0, 1, \cdots, N)$ be their masses. If we denote as $\mathbf{X}_i$ the position vectors of the $N+1$ bodies with respect to an inertial system, and $\mathbf{\Pi}_i = m_i\dot{\mathbf{X}}_i$ their linear momenta, these variables are canonical and the Hamiltonian of the system is nothing but the sum of their kinetic and potential energies:

$$\tilde{H} = T + U = \frac{1}{2}\sum_{i=0}^{N} \frac{\mathbf{\Pi}_i^2}{m_i} - G\sum_{i=0}^{N}\sum_{j=i+1}^{N} \frac{m_i m_j}{\Delta_{ij}}, \qquad (1)$$

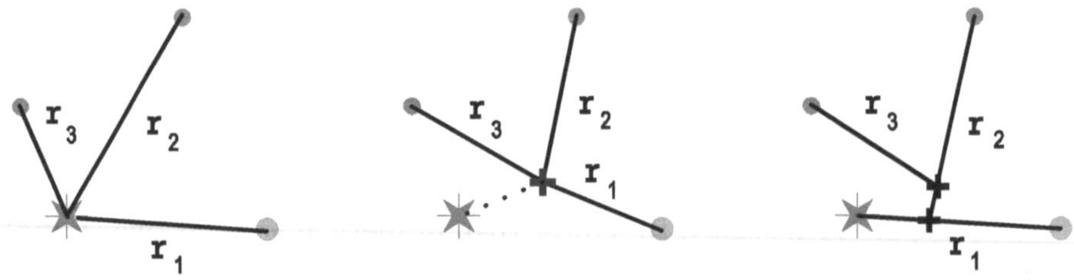

Figure 1. Main systems of coordinates: heliocentric (left), barycentric (center) and Jacobi's (right).

where G is the gravitational constant and $\Delta_{ij} = |\mathbf{X}_i - \mathbf{X}_j|$. This system has, however, $3(N+1)$ degrees of freedom – i.e., 6 equations more than the usual Laplace-Lagrange formulation of the heliocentric equations of motion. The system can be reduced to $3N$ degrees of freedom through the convenient use of the trivial conservation laws concerning the inertial motion of the barycenter. There are two sets of variables used to reduce to $3N$ the number of degrees of freedom of the above system. The most popular reduction, due to Jacobi, is widely used in the study of the general three-body problem and of planetary and stellar systems. A less popular reduction is due to Poincaré; it was first published in 1897, but Poincaré himself did not use it because of difficulties related with the definition of the associated Keplerian elements (see Poincaré, 1905, and next section). It appeared in the literature from time to time and started being more frequently used around the eighties (Yuasa and Hori, 1979; Hori, 1985; Laskar, 1990). Hagihara (1970) notes that it was discovered by Cauchy.

2.2 Poincaré's Reduction to $3N$ degrees of freedom

In Poincaré's reduction, the variables are the components of the heliocentric position vectors $\mathbf{X}_i - \mathbf{X}_0$ and the momenta are the same linear momenta $\mathbf{\Pi}_i$ of the barycentric formulation. Hence,

$$\mathbf{r}_i = \mathbf{X}_i - \mathbf{X}_0, \qquad \mathbf{p}_i = \mathbf{\Pi}_i, \tag{2}$$

$(i = 1, 2, \cdots, N)$. The given system has $N+1$ bodies and we thus need to introduce one more pair of (vector) variables. Let them be

$$\mathbf{r}_0 = \mathbf{X}_0, \qquad \mathbf{p}_0 = \sum_{i=0}^{N} \mathbf{\Pi}_i. \tag{3}$$

A trivial calculation shows that the variables $\mathbf{r}_i, \mathbf{p}_i$ ($i = 0, 1, \cdots, N$) are canonical. Let us, now, write the Hamiltonian in terms of the new variables. The transformations of T and U give, respectively,

$$T = \frac{1}{2}\sum_{i=1}^{N}\frac{p_i^2}{m_i} + \frac{1}{2}\sum_{i=1}^{N}\frac{p_i^2}{m_0} + \frac{1}{2}\frac{p_0^2}{m_0} - \sum_{i=1}^{N}\frac{\mathbf{p}_0\cdot\mathbf{p}_i}{m_0} + \sum_{i=1}^{N}\sum_{j=i+1}^{N}\frac{\mathbf{p}_i\cdot\mathbf{p}_j}{m_0}, \quad (4)$$

and

$$U = -G\sum_{i=1}^{N}\frac{m_0 m_i}{r_i} - G\sum_{i=1}^{N}\sum_{j=i+1}^{N}\frac{m_i m_j}{\Delta_{ij}}, \quad (5)$$

where $p_i = |\mathbf{p}_i|$ and $r_i = |\mathbf{r}_i| = |\Delta_{0i}|$.

The reduction of the system is immediate. We note, beforehand, that the variable $\mathbf{r}_0$ is ignorable. Consequently, $\mathbf{p}_0$ is a constant that, by construction, we set equal to zero. The resulting equations may be separated into two parts:

A. The first pair of equations, corresponding to the subscript 0, is:

$$\dot{\mathbf{p}}_0 = 0 \qquad\qquad \dot{\mathbf{r}}_0 = \mathrm{grad}_{\mathbf{p}_0}\tilde{H}. \quad (6)$$

We note that the second of equations (6) gives

$$\dot{\mathbf{r}}_0 = \frac{\mathbf{p}_0}{m_0} - \sum_{i=1}^{N}\frac{\mathbf{p}_i}{m_0}. \quad (7)$$

B. The canonical equations in the variables $\mathbf{r}_i$, $\mathbf{p}_i$, ($i \neq 0$) are given by the reduced Hamiltonian

$$H = \tilde{H} - \frac{1}{2}\frac{p_0^2}{m_0} + \sum_{i=1}^{N}\frac{\mathbf{p}_0\cdot\mathbf{p}_i}{m_0}. \quad (8)$$

This subsystem has $3N$ degrees of freedom and is separated from the previous one, since $\mathbf{p}_0$ is constant. (We did assume $\mathbf{p}_0 = 0$.)

The Hamiltonian of the reduced system is $H = H_0 + H_1$ where

$$H_0 = \sum_{i=1}^{N}\left(\frac{1}{2}\frac{p_i^2}{\beta_i} - \frac{\mu_i\beta_i}{r_i}\right), \quad (9)$$

$$H_1 = \sum_{i=1}^{i=N}\sum_{j=i+1}^{j=N}\left(-\frac{Gm_i m_j}{\Delta_{ij}} + \frac{\mathbf{p}_i\cdot\mathbf{p}_j}{m_0}\right), \quad (10)$$

and

$$\mu_i = G(m_0 + m_i), \qquad \beta_i = \frac{m_0 m_i}{m_0 + m_i}. \tag{11}$$

We note that H_0 is of the order of the planetary masses m_i while H_1 is of order two with respect to these masses. Then H_0 may be seen as the new expression for the undisturbed energy while H_1 is the potential energy of the interaction between the planets.

It is worth noting that each term

$$F_i = \frac{1}{2}\frac{p_i^2}{\beta_i} - \frac{\mu_i \beta_i}{r_i}, \tag{12}$$

is the Hamiltonian of a two-body problem in which the mass-point m_i is moving around the mass point m_0. In fact, from the Hamiltonian given by equation (12), it is easy to obtain the second-order differential equation

$$\ddot{\mathbf{r}}_i = -\mu_i \frac{\mathbf{r}_i}{r_i^3} = -G(m_0 + m_i)\frac{\mathbf{r}_i}{r_i^3}. \tag{13}$$

One of the canonical equations spanned by F_i is

$$\dot{\mathbf{r}}_i = \frac{\mathbf{p}_i}{\beta_i}. \tag{14}$$

This equation apparently contradicts the statements made that $\mathbf{r}_i$ is the heliocentric radius vector and $\mathbf{p}_i$ is the barycentric linear momentum. However, this only means that the variation of $\mathbf{r}_i$ in the reference Keplerian motion is not the actual relative velocity of the ith body but $\mathbf{p}_i/\beta_i$. This means that, at variance with other formulations, the Keplerian motions defined by equations (12) are not tangent to the actual motions. To distinguish them from "heliocentric osculating", when necessary, we will use the term "heliocentric canonical".

2.3 Action-angle variables. Delaunay elements

The solution of H_0 is a set of N Keplerian motions whose generic Hamiltonian is F_i. The purpose of this and the forthcoming section is to obtain the Keplerian elements and the Delaunay variables corresponding to the relative coordinates introduced before, which must be used when a canonical perturbation theory is constructed using H_0 as an "unperturbed" approximation. For that sake, we have to solve the corresponding Hamilton-Jacobi equation and construct the action-angle variables of the given problem. We only give here the more important steps characterizing the variables appearing in the definitions of their action-angle variables and in the associated Delaunay elements. To accomplish it, the study of the planar case is enough and is preferable since the rotations necessary when the spatial case is considered, although trivial, introduce many new equations. All conceptual questions appear in the planar case, which has the advantage of making the calculations much easier, thus allowing the crucial points to be clearly identified. Once the conceptual problems are solved in the planar case, the usual three-dimensional equations can be easily adapted to give the elements we are looking for. In the plane, the Hamiltonian is separable in polar coordinates. To introduce these variables let us remember that, in the reference Keplerian motion, $\mathbf{p} = \beta\dot{\mathbf{r}}$. (For the sake of simplicity, we omit the subscript i in the forthcoming equations). Then

$$\mathbf{p} = \beta\left(\dot{r}\mathbf{a} + r\dot{\psi}\mathbf{b}\right), \tag{15}$$

where $\mathbf{a}, \mathbf{b}$ are the right-handed set of unit vectors at $\mathbf{r}$ in the positive directions of the increments of r, ψ. $\dot{r}, \dot{\psi}$ are the time derivatives of r, ψ in the reference Keplerian motion. The kinetic energy term is, then,

$$T = \frac{\beta}{2}(\dot{r}^2 + r^2\dot{\psi}^2), \tag{16}$$

or, introducing the momenta $p_r = \partial T/\partial\dot{r}$ and $p_\psi = \partial T/\partial\dot{\psi}$, we obtain

$$T = \frac{1}{2\beta}\left(p_r^2 + \frac{p_\psi^2}{r^2}\right). \tag{17}$$

The potential energy term is given by

$$U(r) = -\frac{\mu\beta}{r}, \tag{18}$$

and the resulting Hamilton-Jacobi equation is the classical one of the two-body problem with β instead of m and μ instead of $G(M + m)$:

$$F = \frac{1}{2\beta}\left(p_r^2 + \frac{p_\psi^2}{r^2}\right) - \frac{\mu\beta}{r}. \tag{19}$$

The solution of this equation is well known and does not need to be reproduced here with all the details. This equation is separable into:

$$p_r = \sqrt{2\beta(E + \frac{\mu\beta}{r}) - \frac{C^2}{r^2}}, \tag{20}$$

$$p_\psi = C. \tag{21}$$

C, E are integration constants ($E = F$ is the "energy" and $C = \mathbf{r} \times \mathbf{p}$ is the "angular momentum"; the quotation marks are necessary because of the particular definitions of $\mathbf{r}$ and $\mathbf{p}$ in the considered formulation).

The actions associated with the given Hamiltonian are

$$J_r = \frac{1}{2\pi} \oint p_r dr, \qquad J_\psi = \frac{1}{2\pi} \oint p_\psi d\psi, \tag{22}$$

whose integrations give

$$J_r = -C + \mu\beta\sqrt{\frac{\beta}{-2E}}, \qquad J_\psi = C. \tag{23}$$

The Delaunay actions are:

$$\begin{aligned} L &= J_r + J_\psi = \beta\sqrt{\mu a} \\ G &= J_\psi \qquad\;\; = \beta\sqrt{\mu a}\sqrt{1-e^2} \end{aligned}, \tag{24}$$

where a and e are two constants introduced in the integration giving the action J_r:

- The mean distance (or semi-major axis)

$$a \overset{\text{def}}{=} -\frac{\mu\beta}{2E} \tag{25}$$

- The eccentricity

$$e \overset{\text{def}}{=} \sqrt{1 + \frac{2EC^2}{\mu^2\beta^3}}. \tag{26}$$

Since, in general, the planets do not move in the same plane, we have to introduce the inclinations I of their planes of motion with respect to a fixed reference plane and add the third Delaunay action $H = \beta\sqrt{\mu a}\sqrt{1-e^2}\cos I$. The Delaunay angles $\ell, \omega = \psi - v$ (and Ω) are obtained in the usual way.

2.4 Canonical heliocentric elements

For each planet, we may transform $\beta, \mu, \mathbf{r}, \mathbf{p}$ into the elements a, e, λ, ϖ employing the same transformations used to define the ordinary osculating heliocentric elements $a_{\text{osc}}, e_{\text{osc}}, \lambda_{\text{osc}}, \omega_{\text{osc}}$ of the two-body problem as functions of $m, G(M+m), \mathbf{r}, m\dot{\mathbf{r}}$. However, the equations giving the osculating heliocentric elements depend on m only through μ. In order to use always the same routines, the above equations may be transformed.

We substitute, in equations (25) and (26), E and C by their definitions $E = F$ and $C = \mathbf{r} \times \mathbf{p}$. We obtain the well-known equations

$$a = \frac{\mu r}{2\mu - rw^2}, \tag{27}$$

$$e = \sqrt{\left(1 - \frac{r}{a}\right)^2 + \frac{(\mathbf{r}.\mathbf{w})^2}{\mu a}}, \tag{28}$$

where we have used the velocity of the reference Keplerian motion

$$\mathbf{w} = \frac{\mathbf{p}}{\beta}, \tag{29}$$

instead of the actual planetary velocity, and $w = |\mathbf{w}|$. The Keplerian motion corresponding to H_0 in Poincaré 's relative canonical coordinates may be obtained with the ordinary routines substituting the heliocentric velocities by

$$\mathbf{w} = \frac{m}{\beta}\mathbf{V}, \tag{30}$$

where $\mathbf{V}$ is the absolute (i.e. barycentric) velocity.

The angles are obtained with the usual equations. In the planar problem, the true longitude (φ) is given by the angle formed by the radius vector with the first axis of the reference system (to be obtained through $\arctan y/x$ where x, y are the components of $\mathbf{r}$). In the spatial problem, some rotations are necessary beforehand. The anomalies may also be easily obtained, starting with the eccentric anomaly (u), which is given by

$$u = \arctan\left(\sqrt{\frac{a}{\mu}}\frac{\mathbf{r}.\mathbf{w}}{a - r}\right). \tag{31}$$

The true (v) and mean (ℓ) anomalies are obtained by means of classical 2-body equations. The other angles to determine are the longitude of periapsis ($\omega = \varphi - v$) and the mean longitude ($\lambda = \ell + \omega$).

The elements of the reference Keplerian orbit at the time t are a, e, ω, λ. Since the parameter λ is variable, it is convenient to replace it by the so-called "mean longitude at the epoch" (λ_0), which is the value of λ at a standard "epoch" t_0:

$$\lambda = \lambda_0 + n(t - t_0), \tag{32}$$

where $n = \sqrt{\mu/a^3}$ is the mean-motion in the reference orbit.

2.5 The Conservation of the angular momentum

If the only forces acting on the $N+1$ bodies are their point-mass gravitational attractions, the angular momentum is conserved:

$$\mathcal{L} = \sum_{i=0}^{N} m_i \mathbf{X}_i \times \dot{\mathbf{X}}_i \tag{33}$$

Since $\sum_0^N m_i \mathbf{X}_i = \sum_0^N m_i \dot{\mathbf{X}}_i = 0$, the above equation gives

$$\mathcal{L} = \sum_{i=1}^{N} \mathbf{r}_i \times \mathbf{p}_i, \tag{34}$$

that is

$$\mathcal{L} = \sum_{i=1}^{N} \beta_i \sqrt{\mu_i a_i (1 - e_i^2)} \cdot \mathbf{k}_i, \tag{35}$$

where $\mathbf{k}_i$ are the unit vectors normal to the Keplerian planes. This is an exact conservation law. In this equation a_i and e_i are not the usual heliocentric osculating elements but the canonical heliocentric elements defined by equations (27) – (28) where $\mathbf{w}_i$ are the absolute velocities corrected by the factors m_i/β_i.

The conservation law given by equation (34) is also true if Jacobian coordinates are used. However, the expression

$$\widehat{\mathcal{L}} = \sum_{i=1}^{N} m_i \sqrt{\mu_i a_i (1 - e_i^2)} \cdot \mathbf{k}_i \tag{36}$$

where a_i and e_i are the heliocentric osculating elements (Keplerian elements defined by equations (27) – (28) with the heliocentric velocities $\mathbf{v}_i$ instead of $\mathbf{w}_i$) often found in the literature, is not a true conservation law. One may easily see that:

$$\widehat{\mathcal{L}} = \mathcal{L} - \sum_{i=1}^{N} m_i \mathbf{X}_0 \times \dot{\mathbf{X}}_0, \tag{37}$$

showing that the quantity $\widehat{\mathcal{L}}$ has in fact a variation of order $\mathcal{O}(m_i^2)$.

2.6 Two-body Expansions

For the sake of future calculations, let us recall some series expansions of the two-body problem. These expansions are helpful in the task of writing computer codes for automatic expansion of H_1 and hold in all systems of elements founded on unperturbed Keplerian motions.

The first result to be recalled concerns the Fourier expansion of some functions of the radius vector and true anomaly. They are the convergent series

$$
\begin{aligned}
\left(\frac{r}{a}\right)^n \cos(kf) &= \sum_{j=0}^{\infty}(X_j^{n,k} + X_{-j}^{n,k}) \cos(j\ell), \\
\left(\frac{r}{a}\right)^n \sin(kf) &= \sum_{j=0}^{\infty}(X_j^{n,k} - X_{-j}^{n,k}) \sin(j\ell),
\end{aligned} \tag{38}
$$

where the superscript n may be either positive or negative. The coefficients $X_j^{n,k}$ are the Hansen coefficients (see Tisserand, 1960; Kaula, 1962). Hansen coefficients are functions of the eccentricity. They may be developed into power series of the eccentricities:

$$
X_j^{n,k} = e^{|k-j|} \sum_{s=0}^{\infty} Y_{s+u_1,s+u_2}^{n,k} e^{2s}, \tag{39}
$$

($u_1 = \max(0, j-k)$ and $u_2 = \max(0, k-j)$) where the $Y_{s+u_1,s+u_2}^{n,k}$ are the Newcomb operators. Newcomb operators obey some simple recurrence relations, which allow them to be easily calculated for all values of the indices (see Brouwer & Clemence, 1961).

Introducing equation (39) into equation (37), we obtain, after some algebra,

$$
\begin{aligned}
\left(\frac{r}{a}\right)^n \cos(kf) &= \sum_{i=0}^{\infty} \sum_{m=-\infty}^{\infty} B_{n,k,i,m} e^i \cos(m\ell), \\
\left(\frac{r}{a}\right)^n \sin(kf) &= \sum_{i=0}^{\infty} \sum_{m=-\infty}^{\infty} C_{n,k,i,m} e^i \sin(m\ell),
\end{aligned} \tag{40}
$$

where $B_{n,k,i,m}$ and $C_{n,k,i,m}$ are constant coefficients expressed as functions of Newcomb operators. These coefficients, first calculated by Leverrier, do not depend on the orbital parameters and may be calculated once and for all. They have some interesting properties. The most important of them is the d'Alembert property: $B_{n,k,i,m} = C_{n,k,i,m} = 0$ when $|m| < i$ or when $|m| - i$ is odd.

The latest expansions are power series in e and their convergence depend on the singularities of the analytic function $u = u(e, \ell)$, which are at $|e| = 0.6627434\cdots$. This is the convergence radius of the given series (see Wintner, 1941).

3. Expansion of the Disturbing Function

The Hamiltonian equations in relative coordinates may be used to study the planetary motions. In analytical studies, once new variables have been introduced, the next step is to write H_1 in terms of the Keplerian elements. A well-known approach to this problem is the classical Laplacian expansion of H_1 into a Fourier series in the angles and a power series in the eccentricities, which introduces the functions of the semi-major axes known as "Laplace coefficients". Another expansion sometimes found in the literature uses the expansion of $\frac{1}{\Delta}$ in Legendre polynomials of the ratio of the distances of the two planets to the central star. These expansions work well in their domains of validity. The Laplacian expansion is a good approximation if the orbital eccentricities are small. However, the radius of convergence of the expansion decreases (see Ferraz-Mello, 1994) with the increase of the ratio α of the two semi-major axes. For $\alpha \sim 0.6$ the series is no longer convergent for eccentricities as small as ~ 0.2. The expansions with Legendre polynomials are more stringent: they may only be used in the study of well hierarchized systems where the ratio of the distances of the perturbed and perturbing bodies to the central body remain small forever. This is the case of the lunar theory, in which the motion of the Moon around the Earth is disturbed by the Sun. Otherwise, the convergence of the expansion in Legendre polynomials is very slow and its use in planetary problems accounts for many wrong results. We present, in this lecture, an improvement of the technique first developed by Beaugé (1996). This "expansion" is valid in large domains of the phase space excluding a domain around the singularities associated to collisions between the two bodies. In Beaugé's approximation, the number of terms necessary to represent H_1 depends on the magnitude of H_1 in the domain to be studied: near the minimum of $|H_1|$, a few terms are enough to have a good representation. This number increases quickly as we approach orbits that may come close to a collision. At variance with Beaugé's early expansion, the present one (Beaugé & Michtchenko, 2003) has no explicit restrictions with regard to eccentricities and inclinations.

3.1 Beaugé's approximation. The parameter δ

The big problem in the expansion of H_1 comes from the terms having Δ in the denominator. In heliocentric coordinates, we can write:

$$\frac{1}{\Delta} = \left(r_1^2 + r_2^2 - 2r_1 r_2 \cos S\right)^{-1/2}, \tag{41}$$

where S is the angle between both bodies as seen from the central mass. Introducing the ratio $\varrho = r_1/r_2$, equation (41) becomes

$$\frac{r_2}{\Delta} = \left(1 + \varrho^2 - 2\varrho\cos S\right)^{-1/2}. \tag{42}$$

Instead of expanding this function in Fourier series in S (the Laplace approach) or power series in ϱ (Legendre polynomials), a best-fit approach is used. We write

$$\frac{r_2}{\Delta} = \left(1 + x\right)^{-1/2}. \tag{43}$$

where

$$x = \varrho^2 - 2\varrho\cos S, \tag{44}$$

and represent the function $(1+x)^{-1/2}$ by a polynomial of order N in x:

$$(1+x)^{-1/2} \simeq \sum_{n=0}^{N} b_n x^n, \tag{45}$$

whose coefficients b_n are determined numerically through a linear regression.

The variable x is a measure of the proximity of the initial condition to the singularity in $\frac{1}{\Delta}$. It is equal to -1 at the singularity, and takes values larger than this for every point $(\varrho, \cos S)$ away from the collision curve (see Figure 3). We note that the values of ϱ and S are not separately significant; only the distance from the singularity is important.

The numerical fit is performed using values of $x > -1 + \delta$, where δ is a positive parameter close to zero. The smaller its value, the better the approximation to the real function near the singularity. However, when δ is small, the number N of terms to be considered in the representation of $(1+x)^{-1/2}$ to guarantee adequate precision for all values of the independent variable is necessarily large.

Figure 2 shows the relative error of equation (45) for $N = 30$ and two values of δ. We can see that for most of the interval of x, the fit with $\delta = 0.1$ yields a much higher precision than the fit with $\delta = 0.01$. In the fit with $\delta = 0.1$, the errors are of the order of 10^{-6}, that is, about 3 orders of magnitude lower than in the other case. Conversely, as $x \to -1$, the fit with $\delta = 0.01$ is more precise. Larger values of N will diminish the error in both cases, but at the cost of increasing the number of terms enormously.

In the general case, the motion of the two bodies is unconstrained and the distance between the two planets is a minimum in a symmetric

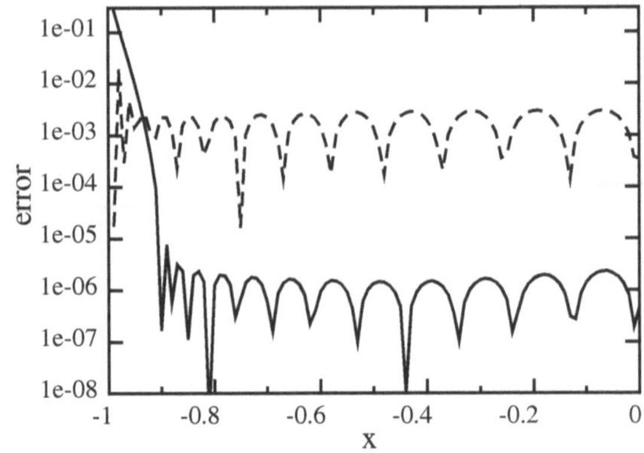

Figure 2. Relative error of the approximation of $1/\sqrt{1+x}$ given by equation (45) as a function of x, for two values of δ. The continuous line shows the case $\delta = 0.1$ and the broken line shows $\delta = 0.01$. In both examples, $N = 30$.

conjunction with the outer planet at periapsis and the inner planet at apoapsis. In this case, we have to choose $\delta < (1-\xi)^2$ where $\xi = \alpha(1+e_1)(1-e_2)^{-1}$. Beaugé's technique no longer requires that the ratio of the distances of the two planets is small, but it requires $\xi < 1$. However, when the planets are in resonant motion, the method is valid even for crossing orbits because the resonance does not allow the planets to come close to one another. The limits of x when the motion of the two planets is constrained by a 2:1 commensurability ($\alpha = 0.63$) are shown in Figure 3 in the particular case where $e_2 = 0$.

The geometry of the curves in Figure 3 follows very closely (but not identically) the topology of the phase portrait of the 2:1 resonant restricted three-body problem averaged over short-period terms. The maximum value of $x_{\min}$ lies at $e_1 = 0.8, \sigma_1 = 0$ (on the horizontal axis) and corresponds to the minimum of $|H_1|$. This point is very close to the corotation Stationary solution of the 2:1 asteroidal resonance ($e_1 = 0.73$ when $e_2 \to 0$; see Ferraz-Mello et al., 1993). Similarly, the minimum value of $x_{\min}$ (equal to -1) corresponds to the singularities of H_1. There is no direct relationship between the eccentricity and $x_{\min}$. An orbit with a large eccentricity near the corotation center may have a larger value of $x_{\min}$, while an almost circular orbit with a lower eccentricity may reach values very close to the limit $x = -1$. It is worth recalling that several extra-solar planet pairs observed in resonant configuration lie near corotation centers where $x_{\min}$ is large and good Beaugé's approximations may be obtained with small N.

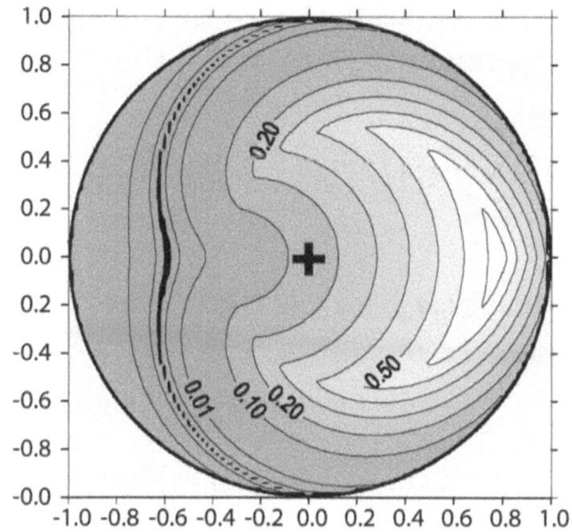

Figure 3. Limits of validity of Beaugé's approximations for planets in the 2:1 resonance with $e_2 = 0$ for different values of δ. The thick black line on the left-hand side is the locus of the points where $x_{\min} = -1$ (collision curve). The non-labeled curves adjacent to it correspond to $\delta = 0.001$. Horizontal axis: $e_1 \cos\sigma_1$; Vertical axis: $e_1 \sin\sigma_1$. ($\sigma_1 = 2\lambda_2 - \lambda_1 - \varpi_1$).

3.2 The Direct Part

To transform the above approximation into a function having the form needed in a theory, many transformations have to be carried out. Introducing the explicit expression for x into equation (45), it becomes

$$\frac{r_2}{\Delta} \simeq \sum_{k=0}^{N} \sum_{j=0}^{n} c_k (-2)^j \begin{pmatrix} k \\ j \end{pmatrix} \varrho^{2k-j} \cos^j S, \tag{46}$$

where the c_k are constant coefficients, easily obtainable in terms of the b_k. From now on, we will restrict ourselves to coplanar orbits. Changing from powers of the cosines to multiples of the argument, and introducing the planar approximation $S = f_1 - f_2 + \Delta\varpi$, we can rewrite it as:

$$\frac{a_2}{\Delta} \simeq \sum_{k=0}^{N} \sum_{i=0}^{N-k} 2A_{k,i}\alpha^m \left(\frac{r_1}{a_1}\right)^m \left(\frac{r_2}{a_2}\right)^{-m-1} \cos k(f_1 - f_2 + \Delta\varpi), \tag{47}$$

where $m = 2i + k$.

At last, introducing equation (39) into the expression of the direct part of the disturbing function, and reordering the terms, we get:

$$\frac{a_2}{\Delta} \simeq \sum_{j,k=0}^{\infty} \sum_{m,n=-\infty}^{\infty} \sum_{l=0}^{N} \sum_{i=0}^{N-l} A_{l,i} D_{2i+l,j,k,m,n} \alpha^{2i+l} e_1^i e_2^j \cos(m\ell_1 - n\ell_2 + l\Delta\varpi), \tag{48}$$

where the coefficients $D_{2i+l,j,k,m,n}$ are given by:

$$\begin{aligned} D_{2i+l,j,k,m,n} &= \frac{1}{2\gamma_m\gamma_n}(B_{2i+l,l,j,|m|} + \text{sign}(m) C_{2i+l,l,j,|m|}) \\ &\quad \times (B_{-2i-l-1,l,k,|n|} + \text{sign}(n) C_{-2i-l-1,l,k,|n|}), \end{aligned} \tag{49}$$

and γ_m is a simple bi-valuated function defined as:

$$\gamma_m = \begin{cases} 1/2 & \text{if} \quad m = 0 \\ 1 & \text{if} \quad m > 0 \end{cases} \tag{50}$$

Equation (48) multiplied by the factor $\frac{Gm_1m_2}{a_2}$ gives the term of the direct part corresponding to the given pair of planets.

3.3 The Indirect Part

In Poincaré heliocentric relative coordinates, the indirect part of H_1 is (see equation 10):

$$T_1 = \sum_{i=1}^{N} \sum_{j=i+1}^{N} \frac{\mathbf{p}_i \mathbf{p}_j}{m_0}. \tag{51}$$

The linear momenta $\mathbf{p}_i$ may be obtained from the derivatives of the vector radii $\mathbf{r}_i(t), \mathbf{r}_j(t)$ in the Keplerian reference orbit (see equation 14). Then,

$$T_1 = \sum_{i=1}^{N} \sum_{j=i+1}^{N} \beta_i \beta_j \frac{\dot{\mathbf{r}}_i(t) \dot{\mathbf{r}}_j(t)}{m_0}, \tag{52}$$

or

$$T_1 = \frac{\beta_1\beta_2}{m_0} n_1 n_2 \left[\frac{\partial x_1}{\partial \ell_1} \frac{\partial x_2}{\partial \ell_2} + \frac{\partial y_1}{\partial \ell_1} \frac{\partial y_2}{\partial \ell_2} \right], \tag{53}$$

where ℓ_i are the mean anomalies and n_i the mean motions. x_i and y_i are the components of $\mathbf{r}_i$ and are given by $x_i = r_i \cos(f_i + \varpi_i)$ and $y_i = r_i \sin(f_i + \varpi_i)$.

In the sequence, we substitute the mean motions by the values issued from Kepler's third law and put into evidence the same factor used at the end of the previous section. Hence

$$T_1 = \frac{Gm_1m_2}{a_2}\mathcal{A}\alpha^{-1/2}\left[\frac{\partial}{\partial \ell_1}\left(\frac{x_1}{a_1}\right)\frac{\partial}{\partial \ell_2}\left(\frac{x_2}{a_2}\right) + \frac{\partial}{\partial \ell_1}\left(\frac{y_1}{a_1}\right)\frac{\partial}{\partial \ell_2}\left(\frac{y_2}{a_2}\right)\right]. \tag{54}$$

$\mathcal{A} = \sqrt{\frac{\beta_1\beta_2}{m_1m_2}} \approx 1 - \frac{m_1+m_2}{2m_0}$ is taken hereafter to equal 1, introducing an error of third order in the planetary masses. Using the expansions given in Section 5.2.6, it follows that

$$\begin{aligned} \frac{x_1}{a_1} &= \sum_{i=0}^{\infty}\sum_{j=-\infty}^{\infty} I_{i,j}e^i \cos(j\ell_1 + \varpi_1), \\ \frac{y_1}{a_1} &= \sum_{i=0}^{\infty}\sum_{j=-\infty}^{\infty} I_{i,j}e^i \sin(j\ell_1 + \varpi_1), \end{aligned} \tag{55}$$

where

$$I_{i,j} = \frac{1}{2\gamma_j}\Big(B_{1,1,i,|j|} + \mathrm{sign}(j)C_{1,1,i,|j|}\Big). \tag{56}$$

After differentiating these equations with respect to the mean anomalies, and substituting in T_1, we obtain

$$T_1 = \frac{Gm_1m_2}{a_2\alpha^{1/2}}\sum_{j,k=0}^{\infty}\sum_{m,n=-\infty}^{\infty} mnI_{j,m}I_{k,n}e_1^je_2^k\cos(m\ell_1 - n\ell_2 + \Delta\varpi). \tag{57}$$

Notice that, except for the dependence on α, this series is formally similar to that giving the direct part of F_1. To complete the similarity, we can substitute for the factor $\alpha^{-1/2}$ with a power series expansion in the neighborhood of the exact resonant value and write it as

$$\alpha^{-1/2} = \sum_{i=0}^{2N}\overline{A}_i\alpha^i, \tag{58}$$

where the $\overline{A}_i$ are constant coefficients. With this change, T_1 now reads:

$$T_1 = \frac{Gm_1m_2}{a_2}\sum_{i=0}^{2N}\sum_{j,k=0}^{\infty}\sum_{m,n=-\infty}^{\infty}\overline{A}_i mnI_{j,m}I_{k,n}\alpha^ie_1^je_2^k\cos(m\ell_1 - n\ell_2 + \Delta\varpi), \tag{59}$$

which is the final expression for the indirect part of the disturbing potential.

3.4 The Disturbing Function

Since both parts are formally similar, we can unify both expressions and obtain a single series for the disturbing function of the planetary three-body problem in heliocentric relative coordinates:

$$H_1 = \frac{Gm_1m_2}{a_2} \sum_{j,k=0}^{\infty} \sum_{m,n=-\infty}^{\infty} \sum_{l=0}^{N} \sum_{i=0}^{2N} R_{i,j,k,m,n,l} \alpha^i e_1^j e_2^k \cos(m\ell_1 - n\ell_2 + l\Delta\varpi), \tag{60}$$

where $\alpha = \frac{a_1}{a_2}$ and

$$R_{i,j,k,m,n,l} = A_{l,(i-l)/2} D_{i,j,k,m,n} - \delta_{l,0} \overline{A}_i mn I_{j,m} I_{k,n}, \tag{61}$$

where $\delta_{l,0}$ is Kronecker's delta function . Note that these coefficients are constant for all initial conditions, and therefore need only be determined once. (For more details, see Beaugé & Michtchenko, 2003).

It is important to note that each term in H_1 depends on the mean anomalies ℓ_i and on the difference of the periapses longitudes $\Delta\varpi$. This means that if the arguments are written in terms of longitudes λ_i, ϖ_i only, they become $\kappa_1\lambda_1 + \kappa_2\lambda_2 + \kappa_3\varpi_1 + \kappa_4\varpi_2$ with $\sum \kappa_i = 0$. That is, H_1 is invariant to rotations of the reference frame.

4. Secular Dynamics of 2 Planets

The study of the secular dynamics is the study of the secular part of the Hamiltonian, obtained after an averaging over the mean longitudes. We will restrict ourselves in this text to the case of only two planets. To first-order in the masses, the averaged Hamiltonian is the mean value of H:

$$< H > = \frac{1}{4\pi^2} \int_0^{2\pi} \int_0^{2\pi} H d\lambda_1 d\lambda_2, \tag{62}$$

or $< H > = -\sum_{i=1}^{2} \frac{\mu_i^2 \beta_i^3}{2L_i^2} - R_{\rm sec}(L_i, I_i, \Delta\varpi)$, where we have introduced the Delaunay variables $L_i, I_i = L_i - G_i$ defined in Section 5.2.3. Because of the invariance of H_1 with respect to rotations, once the λ_i are averaged out, only terms with arguments $\kappa_3\varpi_1 + \kappa_4\varpi_2$ with $\kappa_3 = -\kappa_4$ can remain in $R_{\rm sec}$. That is, $< H >$ depends on only one angle, the difference $\Delta\varpi$. This means that the averaged equations have three ignorable angles – i.e., three first integrals (conservation laws). They are

$L_1 =$ const.

$L_2 =$ const.

$K_2 = I_1 + I_2 =$ const.

The third of these integrals,

$$K_2 = I_1 + I_2 = L_1(1 - \sqrt{1 - e_1^2}) + L_2(1 - \sqrt{1 - e_2^2}). \tag{63}$$

was called angular momentum Deficit by Laskar (2000). It is a combination of the conservation of the angular momentum ($G_1 + G_2$=const., in

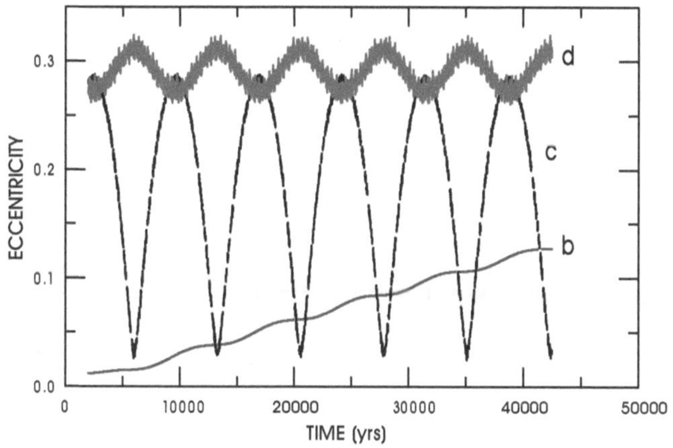

Figure 4. Secular variation of the osculating eccentricities of the υAnd planets. The eccentricities of the two largest planets have variations in anti-phase as necessary to have AMD $\approx$ const.

the planar case) and the secular invariance of the L_i. The I_i are positive quantities increasing from 0 when $e_i = 0$ to L_i when $e_i = 1$ – that is, $0 < I_i < L_i$. Therefore, in a system formed by only two planets, the I_i shall vary in contrary directions and so the eccentricities shall vary: when one eccentricity increases, the other decreases (see Figure 4).

This conservation law has some algebraic consequences. Assuming that $a_1 < a_2$, we have the following possibilities:

- $K_2 < L_1 < L_2$

  (1) I_1 and I_2 cannot reach their maximum values L_1 and L_2, resp.;

  (2) $e_1 < 1$ and $e_2 < 1$ (for all t);

- $L_1 < L_2 < K_2$

  (1) The AMD does not bound the eccentricities (both can reach $e = 1$);

  (2) $e_1 > 0$ and $e_2 > 0$ (for all t).

- $L_1 < K_2 < L_2$

  (1) I_1 can reach its maximum value L_1;

  (2) I_2 cannot reach its maximum value L_2 ($I_2 < L_2 - L_1$);

  (3) The AMD does not bound e_1 (it can reach $e_1 = 1$);

(4) The AMD bounds e_2 ($e_2 < 1$ for all t);

(5) $I_2 > K_2 - L_1 > 0$;

(6) $e_2 > 0$ (for all t).

The conservation law is also found in N-planet systems. In the coplanar case, the angular momentum deficit is $K_N = I_1 + I_2 + \cdots + I_N = \sum_1^N L_i(1-\sqrt{1-e_i^2})$. It is worth emphasizing that this conservation law of the averaged system is not a rigorous one like the angular momentum conservation discussed in Section 2.5. It is approximate and valid strictly only as far as the hypotheses done to average the system are satisfied and the semi-major axes remain approximately constant.

The equations of motion derived from $< H >$ are

$$\dot{I}_1 = -\frac{\partial < H >}{\partial \Delta\varpi}, \qquad \Delta\dot{\varpi} = \frac{\partial < H >}{\partial I_1}. \tag{64}$$

This system has only one degree of freedom and is integrable.

4.1 The Mode I and Mode II Periodic Motions

The exact solution of equations (64) is not easy to obtain analytically, but some insight can be gained by examining their equilibrium points (which correspond to periodic solutions of the two-degrees-of-freedom Hamiltonian $< H >$). They are defined by

$$\dot{I}_1 = 0, \qquad \Delta\dot{\varpi} = 0. \tag{65}$$

For non-singular I_1 ($I_1 \neq 0$ and $I_1 \neq K_2$), we have two trivial solutions: $\Delta\varpi = 0$ and $\Delta\varpi = \pi$. These solutions are often referred as Mode I ($\Delta\varpi = 0$) and Mode II ($\Delta\varpi = \pi$). In mode I, the lines of apses of the two planets are aligned having the periapses on the same side; In mode II, the situation is similar but the two periapses are in opposite directions (the periapses are anti-aligned). Ordinary motions are oscillations around these fixed points.

Solutions of the above equations corresponding to the masses, semi-major axes and energy level of planets c and d of υAnd are shown in Figure 5. They are presented in two different planes. One in which the coordinates are $e_1 \cos\Delta\varpi$, $e_1 \sin\Delta\varpi$ (e_1 is the eccentricity of υAnd c) and another, not independent, in which the coordinates are $e_2 \cos\Delta\varpi$, $e_2 \sin\Delta\varpi$ (e_2 is the eccentricity of υAnd d).

On each figure, we can see the two fixed points above called Mode I and Mode II. In the left-hand phase plane, corresponding to the eccentricity of planet c, motions around the Mode I fixed point dominate; the Mode

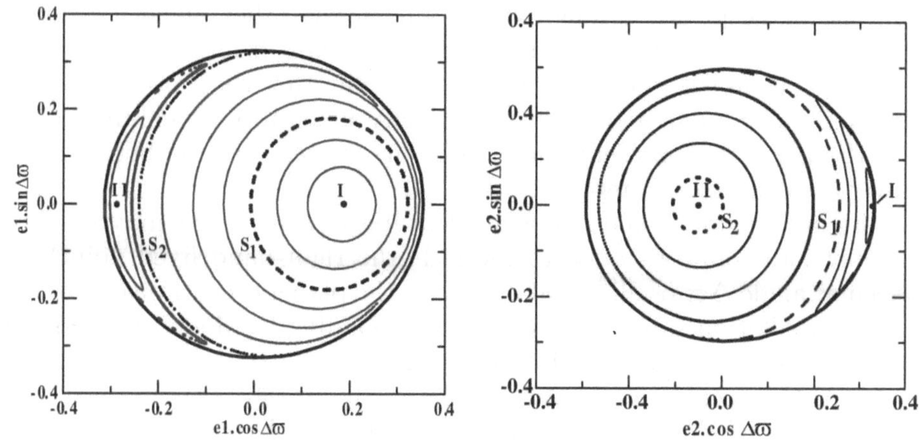

Figure 5. Secular variations of a system of planets with the same masses and semi-major axes as planets c and d of υAnd. The outer curves show the boundary of the energy manifold. The dashed curves S_1 and S_2 represent motions through the singularities of equations (65) (see text). The actual motion of planets υAnd d and υAnd c (see Figure 4) is an oscillation around the Mode I fixed point.

II fixed point lies near the left-hand boundary of the energy surface. In the right-hand figure, corresponding to planet d's eccentricity, the situation is reversed and the flow is dominated by motions around the Mode II fixed point which lies near the center. (For a discussion on the periodic orbits corresponding to the fixed points, see Michtchenko & Ferraz-Mello, 2001; Michtchenko and Malhotra, 2004.)

It is important to note that even though the motion of angle $\Delta\varpi$ may be either an oscillation (about 0 or 180°) or a circulation, there is no separatrix associated with an unstable infinite-period solution separating these motions. To better understand this feature, we plot by dashed lines two special solutions on each figure. These solutions are associated with the singularities in equations (65), which take place at $I_1 = 0$ and $I_1 = K_2$ (that is, $I_2 = 0$). One of these solutions, presented by the curve S_1, was calculated with initial condition $I_1 \simeq 0$ and is seen as a smooth curve passing through the origin on the left-hand figure. At variance, S_2, calculated with initial condition $I_2 \simeq 0$ is seen as the 'false' separatrix between the domains of the motion around the two different fixed points. An analogous situation is seen in the right-hand figure where, now, S_2 is a smooth curve passing through the origin and S_1 separates the domains of the motion around the two different fixed points. The motion along these separatrix-like curves is such that, when the representative point in one plane passes through the origin, in the other plane it is at the boundary of the separatrix-like curve and jumps from one boundary to another. However, such a jump does not mean that the motion is passing through a singularity. It is just the result of the topological inadequacy of the plane to represent these solutions; they would be better drawn over a sphere (see Pauwels, 1983).

Figure 5 shows some important features of the secular motion of two planets. In solutions close to Mode I (the right-hand fixed point), the secular angle $\Delta\varpi$ oscillates about 0 and the planet eccentricities undergo small oscillations about the value corresponding to Mode I equilibrium. In a similar way, in the solutions close to Mode II (the left-hand fixed point), the secular angle $\Delta\varpi$ oscillates about 180°. At mid-way from Mode I to Mode II, there is a large region of the phase space, corresponding to solutions where the motion of the secular angle $\Delta\varpi$ is a prograde circulation.

The motions around Mode I and Mode II are two opposite stable ways for the planetary system to be aligned. In Mode II, the closest approaches between the planets occur when υAnd c is at apoapsis and υAnd d at periapsis, simultaneously. This situation can never occur in Mode I.

Figure 5 is akin to surfaces of section of the two-degrees-of-freedom system. The curves in each plane are defined by initial conditions on the plane plus one condition out of the plane (the other eccentricity, or, equivalently, K_2), which is adjusted in such a way that all curves correspond to the same energy. Therefore, it is not a phase portrait. (Phase portraits of the one-degree-of-freedom Hamiltonian are sets of trajectories of different energies. See the phase portraits of $< H >$ in Pauwels, 1983.) This choice makes Figure 5 more suitable for comparison to similar plots obtained for 2-planet resonant systems (Michtchenko & Ferraz-Mello, 2001; Callegari Jr. et al., 2004).

5. Resonant Dynamics

In the previous section, the Hamiltonian was averaged over the two mean longitudes λ_1 and λ_2. This procedure is not valid if the two planets have commensurable periods, since, in this case, λ_1 and λ_2 are no longer independent: $\frac{p+q}{p}$ resonance $\Longleftrightarrow \frac{T_2}{T_1} \simeq \frac{p+q}{p}$. The averaging over the two longitudes will kill all terms depending on the longitudes including those depending on the critical combination $(p+q)\lambda_2 - p\lambda_1$. However, these terms play a major role in the dynamics of the two planets and should remain in $< H >$. To preserve them, we define, before the averaging, the following set of planar canonical variables:

$$\begin{array}{rl} \lambda_1 & J_1 = L_1 + s(I_1 + I_2) \\ \lambda_2 & J_2 = L_2 - (1+s)(I_1 + I_2) \\ (1+s)\lambda_2 - s\lambda_1 - \varpi_1 = \sigma_1 & I_1 = L_1 - G_1 \\ (1+s)\lambda_2 - s\lambda_1 - \varpi_2 = \sigma_2 & I_2 = L_2 - G_2 \end{array}, \qquad (66)$$

where $s = p/q$. The two angular variables σ_i are the critical angles. With the angles thus introduced, the generic argument $m\ell_1 - n\ell_2 + l\Delta\varpi$ of the disturbing function becomes $(m-l)\sigma_1 - (n-l)\sigma_2 + [m(1+s) - ns](\lambda_1 - \lambda_2)$. Note that, because of the invariance of H_1 to rotations, the mean longitude only appears through the mean synodic longitude $\lambda_1 - \lambda_2$. It is easy to see that the "action" conjugate to the missing angle is the total angular momentum $\mathcal{L} = G_1 + G_2 = (L_1 - I_1) + (L_2 - I_2) = J_1 + J_2$. The averaging over the mean longitudes (or over the mean synodic longitude) can, now, be done and the critical angles will be preserved inside σ_1 and σ_2.

After the averaging, $< H > = -\sum_{i=1}^{2} \frac{\mu_i^2 \beta_i^3}{2L_i^2} - R_{\rm res}$, where $R_{\rm res} = \frac{Gm_1m_2}{a_2} \sum_{i,j,k,m',n'} C_{[\ldots]} \alpha^i e_1^j e_2^k \cos[m'q\sigma_1 + n'(\sigma_2 - \sigma_1)]$. The momenta whose conjugate angles no longer appear in $< H >$ are first integrals (only 2, now):

$$\begin{aligned} J_1 &= \text{const.} \\ J_2 &= \text{const.} \end{aligned} \qquad (67)$$

$J_1 + J_2 = G_1 + G_2$ is the angular momentum, whose conservation in the system before the averaging was discussed. It is worth emphasizing the fact that the L_i (i.e. the semi-major axes) are no longer invariant.

The two integrals above may be combined to give

$$(1+s)L_1 + sL_2 = \text{const.} \qquad (68)$$

(Sessin and Ferraz-Mello, 1984). This integral of the resonant dynamics means that a_1 and a_2 vary in anti-phase. When one of the semi-axis increases, the other necessarily decreases.

The above variables may also be combined to give:

$$\text{AMD} = I_1 + I_2 = \text{const.} + \frac{L_1}{s}. \qquad (69)$$

The AMD also is no longer invariant, but its variation is small and thus limitations of the eccentricities similar to that observed in the secular motion (but different) exist.

5.1 Resonant Stationary solutions. Apsidal corotation

The averaged system is, now, an irreducible two-degrees-of-freedom system. An important feature of this system is the existence of stationary solutions (Beaugé et al., 2003; Ferraz-Mello et al. 2003; Lee and Peale, 2003). These solutions are defined by the equations

$$\frac{dI_i}{dt} = \frac{\partial < H >}{\partial \sigma_i} = \frac{\partial R_{\rm res}}{\partial \sigma_i} = 0, \qquad \frac{d\sigma_i}{dt} = \frac{\partial < H >}{\partial I_i} = 0. \quad (70)$$

They are such that I_i and σ_i are constant (except for the short period terms eliminated by the averaging and for contributions of higher orders). Constant I_i's mean semi-major axes and eccentricities constant in these solutions; σ_1 and σ_2 constant mean that $\Delta\varpi = \sigma_1 - \sigma_2$ is constant – that is, the periapses are moving with same velocities so that their mutual separation do not vary. This frozen relative state in resonant systems is known as *Apsidal corotation.*

Equations (70) may be studied separately. The first equation says that the stationary solutions lie at the extrema of the function $R_{\rm res}$ with respect to the variables σ_i. These extrema depend only on the ratio of the masses of the two planets and on the eccentricities (constants in the stationary solution). The factor $\frac{Gm_1m_2}{a_2}$ does not affect the results.

Figures 6 are contour plots of the function $R_{\rm res}$ for given e_1, e_2 and α (taken at $\alpha = a_1/a_2 \simeq 0.63$, value corresponding to the resonance 2/1). For the sake of an easier interpretation, we used the angles $\sigma_1, \Delta\varpi = \sigma_2 - \sigma_1$, instead of σ_1, σ_2. The extrema seen in these figures may correspond to stable stationary solutions or not. $< H >$ is a function of four variables and only two variables are considered in these figures. Therefore, what appears as an extremum in this picture is not necessarily one extremum in the full phase space. The stable solutions considered in this paper are those corresponding to the centers in the white areas. However, one should be aware that they are not the only stable stationary solutions in this problem (see Hadjidemetriou and Psychoyos, 2003).

The first two plots correspond to low e_1 ($e_1 = 0.02$). For small e_2 ($e_2 = 0.02$) the extremum corresponding to stable solutions is such that $\Delta\varpi = \pi$ ($\sigma_1 = 0, \sigma_2 = \pi$). In this solution, the periapses are anti-aligned. When e_2 is larger ($e_2 = 0.04$ in the right-hand plot), the extremum seen in the left-hand plot becomes a saddle point and a bifurcation gives rise to two extrema symmetric with respect to the saddle. These extrema correspond to asymmetric stationary solutions where $\Delta\varpi = \sigma_2 - \sigma_1$

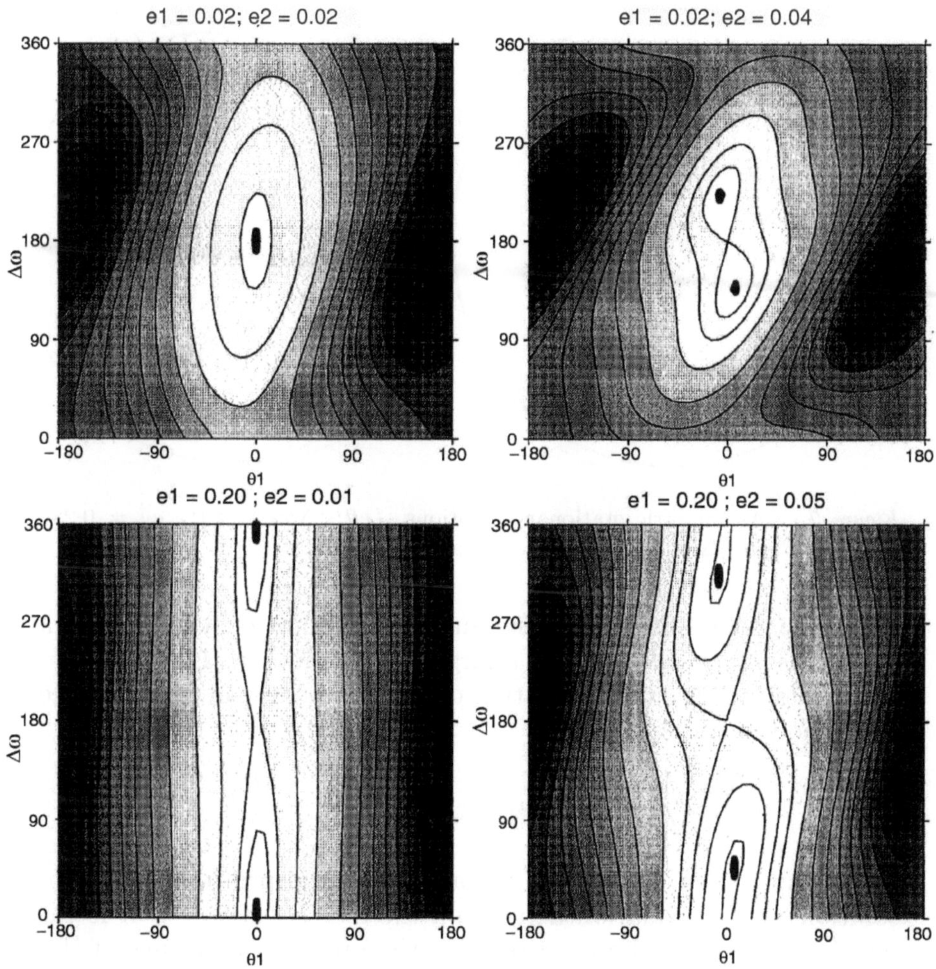

Figure 6. Contour plots of $R_{\rm res}$ in the 2/1 resonance for four given pairs of eccentricity values. Abcissas: $\sigma_1 = 2\lambda_2 - \lambda_1 - \varpi_1$; Ordinates: $\Delta\varpi = \varpi_1 - \varpi_2$.

remains constant but with a value not necessarily equal to zero or π or commensurable with π. The second row of plots correspond to high e_1 ($e_1 = 0.2$). For small e_2 ($e_2 = 0.01$) the extremum corresponding to stable solutions is such that $\Delta\varpi = 0$ ($\sigma_1 = \sigma_2 = 0$). In this solution, the periapses are aligned. When e_2 is larger ($e_2 = 0.05$ in the right-hand plot), the same phenomenon seen in the first row occurs: the extremum seen for low e_2 gives rise to two extrema symmetric with respect to the saddle. As in the previous case, these extrema correspond to asymmetric

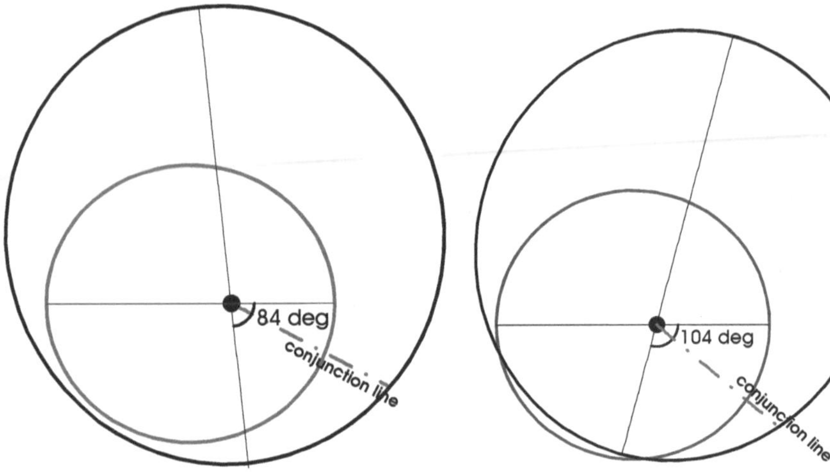

Figure 7. Asymmetric stationary solutions. *Left:* $|\Delta\varpi| = 84°$, $e_1 = 0.286$ and $e_2 = 0.3$. *Right:* $|\Delta\varpi| = 104°$, $e_1 = 0.17$ and $e_2 = 0.38$.

stationary solutions. These asymmetric solutions, depending on the eccentricities, may be found on a large set of points $\sigma_1, \Delta\varpi$.

To complete the determination of the stationary solution, we need to solve the remaining equation:

$$\frac{\partial < H >}{\partial I_i} = s\mu_1 n_1 - (1+s)\mu_2 n_2 + \frac{\partial R_{\rm res}}{\partial I_i} = 0. \tag{71}$$

(See Beaugé et al. 2003). At variance with the previous equation, the solutions of this equation depend on the masses. However, it is easy to see that they depend almost only on the ratio of the masses of the two planets. Indeed, the n_i are constants in the stationary solutions and the commensurability relation at the resonance $\frac{p+q}{p}$ is $sn_1 - (1+s)n_2 = 0$; Hence, since $\mu_i = G(m_0 + m_i)$, $s\mu_1 n_1 - (1+s)\mu_2 n_2 = sGm_1 n_1 - (1+s)Gm_2 n_2$, that is, this part of the equation is homogeneous of degree 1 in the masses. In the remaining part, the derivatives of $R_{\rm res}$ with respect to I_i change the dependence on the masses. We remember that $I_i = m_i(1+\frac{m_i}{m_0})^{-1/2}\sqrt{Gm_0 a_i}(1-\sqrt{1-e_i^2})$. Thus, the coefficient in front of the summation in $R_{\rm res}$ becomes linear in the planet masses after the derivative with respect to I_i. Therefore, equation (71) has the form $A_1 m_1 + A_2 m_2 = 0$, whose solutions do not depend on the masses themselves but only on the mass ratio m_2/m_1. This is not a rigorous statement. In fact, the semi-major axes and eccentricities are functions of I_i that include also the factor $m_0 + m_i$. This means that A_1 and A_2 are independent on the masses only in a first approximation. Even if their

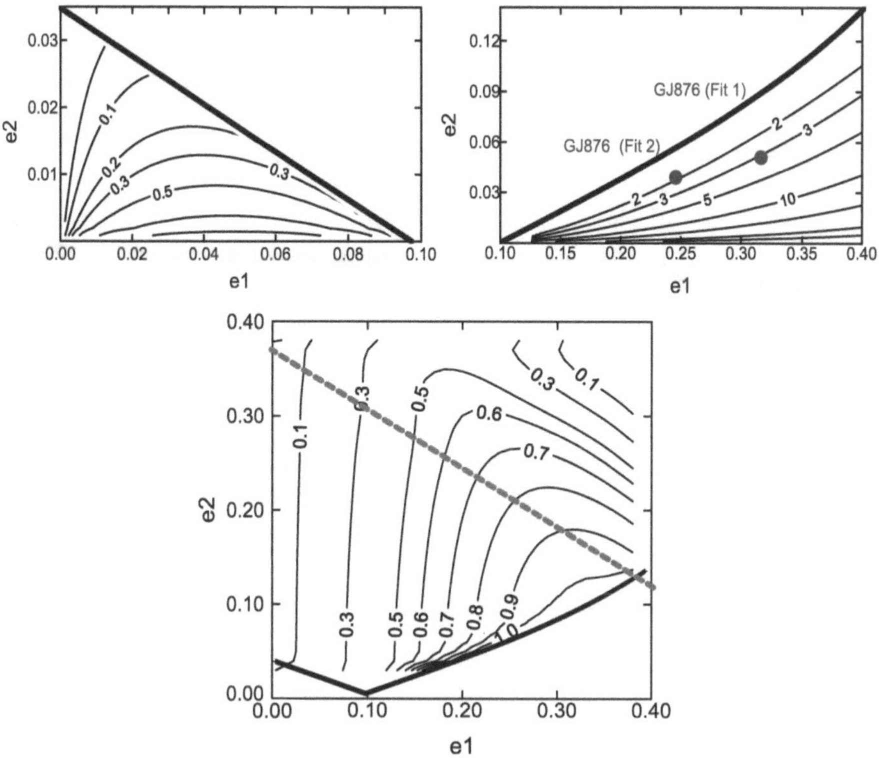

Figure 8. Loci of the stationary corotation solutions of the 2/1 resonance for several mass ratios m_2/m_1. Top figures correspond to the symmetric solutions of the two left-hand side plots in Figure 6. The points corresponding to two early determinations of the elements of Gliese 876 are shown in one of these plots. The bottom figure corresponds to the asymmetric solutions of the two right-hand side plots in Figure 6. The line across these curves shows the values of the eccentricities for which $0.63(1 + e_1) = (1 - e_2)$. In all panels, the thick line shows the boundary between the domains of symmetric and asymmetric solutions.

variation with the masses is small for the range of masses of the considered planets, this variation exists and will affect the solutions in case of large ratios m_i/m_0. Beaugé *et al.* (2003) have shown that the stationary orbits obtained in this section exist for planet masses less than $\sim 10^{-2}$ of the star mass.

The above equations were used to find Apsidal corotation solutions in the case of planets in 2/1 and 3/1 mean-motion resonances. The relationship between eccentricities and mass ratios in some of these solutions is shown in Figure 8. The top panels correspond to symmetric solutions. In the left-hand side panel, the periapses are anti-aligned. This is the case of the two innermost Galilean satellites of Jupiter: Io and Europa. In the right-hand side panel, the periapses are aligned. This is the case of the two planets in orbit around the star Gliese 876. The thick lines in the two top panels show the boundary above which symmetric solutions no longer exist. At the thick line, the solutions bifurcate into pairs of asymmetric solutions. The relationship between eccentricities and mass ratios in the domain of asymmetric solutions is shown in the bottom panel. It is worth noting that the mass ratio m_1/m_2 in the bottom panel is always smaller than a limit close to 1.0. This situation is often called the "exterior case" since it corresponds to having the smaller body in an orbit exterior to that of the more massive one. Asymmetric Apsidal corotations are known in the exterior asteroidal case (Beaugé, 1994). Asymmetric periodic solutions in the restricted three-body problem were first shown to exist by Message (1958). We may also mention a similar behavior, in deep resonance, of the Laplacian critical angle of the Galilean satellites of Jupiter: $\lambda_1 - 3\lambda_2 + 2\lambda_3$ (Greenberg, 1987).

6. Capture into Resonance

In this section, we present the results of a series of numerical simulations of the dynamical evolution of fictitious pairs of planets under the action of a non-conservative perturbation that adds angular momentum and energy to the orbit of the innermost planet. The planets are small (some 10^{-5} of the central body mass) and the mass ratio is $m_2/m_1 = 0.538$ (i.e., the so-called exterior case). The actual calculations were done with satellites instead of planets, but the physical nature of the system does not matter in the following discussion. The physics and used methodology are in Ferraz-Mello et al. (2003).

The initial distances to the star are just behind the 2/1 resonance: $\alpha = a_1/a_2 = 0.612$. When the semi-major axis of m_1 increases, a_1 increases and the mean-motion resonance ($\alpha = 0.63$) between m_1 and m_2 is reached. Capture then can take place. The probability of capture depends on the rate of variation of a_1 – if the rate is high, the orbit crosses the resonance without capture, one phenomenon very well studied in the case of one massless particle. Other factors influencing the probability of capture are the orbital eccentricities – capture is more probable when orbital eccentricities are small (Dermott et al., 1988; Gomes, 1995). In our calculations, initial eccentricities were lower than 0.001 and the physical parameters were adjusted to have slow resonance approximation. Figure 9 shows the evolution of the semi-major axes.

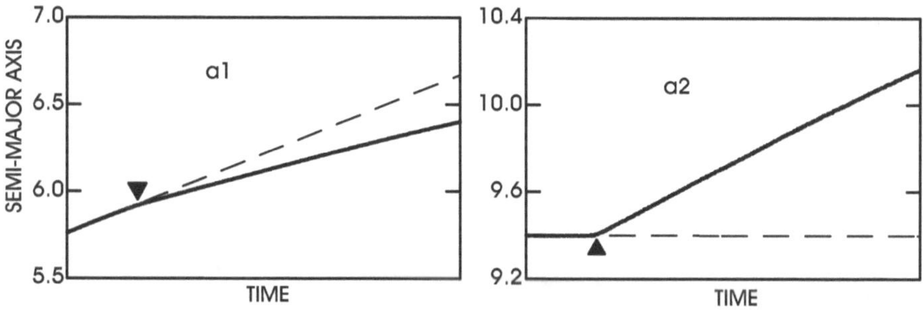

Figure 9. Evolution of the semi-major axes before and after the capture into resonance. Triangles mark the moment of the capture. Dashed lines extrapolate the evolution before the capture and show the change in slope of the evolution lines. (arbitrary units).

6.1 Capture into Apsidal corotation

The system evolves with the innermost orbit receding from the central body (because of the non-conservative forces acting on m_1) up to the moment where the system is captured into a resonance. a_2 is almost constant. When the 2/1-resonance is reached, the system is trapped by the resonance. As known since Laplace, after the capture, m_1 continuously transfers one fraction of the energy that it is getting from the non-conservative source to m_2, so that a_2 also increases. One may note from Figure 9 that, after the capture into the resonance, a_1 increases at a smaller pace than before the capture. The increase of the semi-major axes is such that the ratio a_1/a_2 remains constant.

Figures 10 show the variation of the eccentricities, critical angles $\sigma_i = 2\lambda_2 - \lambda_1 - \varpi_i$ and $\Delta\varpi$ in the same time interval as the previous figures. They show that, after capture, the two critical angles become trapped in the neighborhood of 0 and π, respectively and, consequently, the angle $\Delta\varpi$ is trapped in the neighborhood of π. The capture into a symmetric Apsidal corotation with anti-aligned periapses is thus simultaneous with the capture into the resonance.

6.2 Evolution after Capture

Figure 10 also shows that, after some time, σ_2 jumps from π to 0 and the Apsidal corotation becomes one with aligned periapses . This change is not the result of a discontinuous process. The left-hand side plot shows that the change happens when the eccentricity e_2 is zero. Thus, we may describe the process by a momentary circularization of the orbit such that, when it becomes an ellipse again, the periapses is not

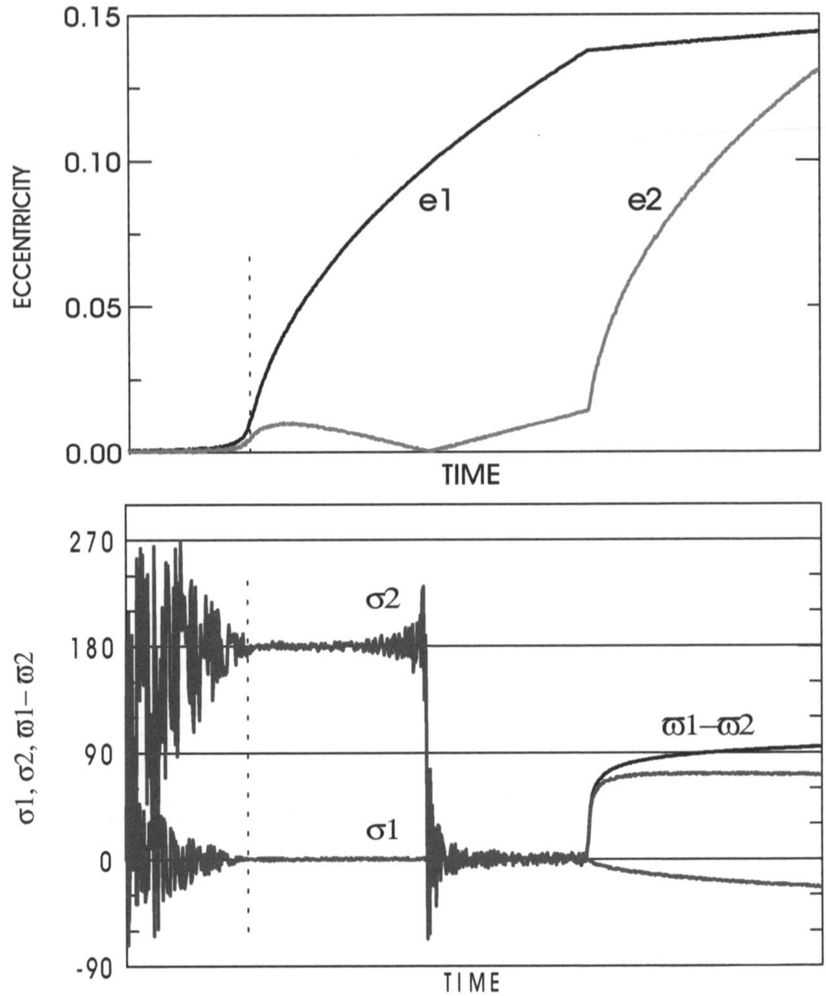

Figure 10. Variation of the eccentricities, critical angles (σ_i) and $\Delta\varpi$ in the same time interval as Figure 9. The vertical dotted lines show the moment of the capture. $\Delta\varpi$ is only shown in the final part since it does not differ significantly from σ_2 in the time interval between the capture into resonance and the bifurcation.

at the same side as before. The large transients shown by the variation of the angle σ_2 are just due to the sensibility of the angle ϖ_2 to small changes when $e_2 \sim 0$. The Apsidal corotation with aligned periapses does not last long. The figures show that the angles depart from zero and the Apsidal corotation becomes asymmetric. At this moment, there is a discontinuity in the rates of variation of the eccentricities – the elbows seen in the curves $e_i(t)$).

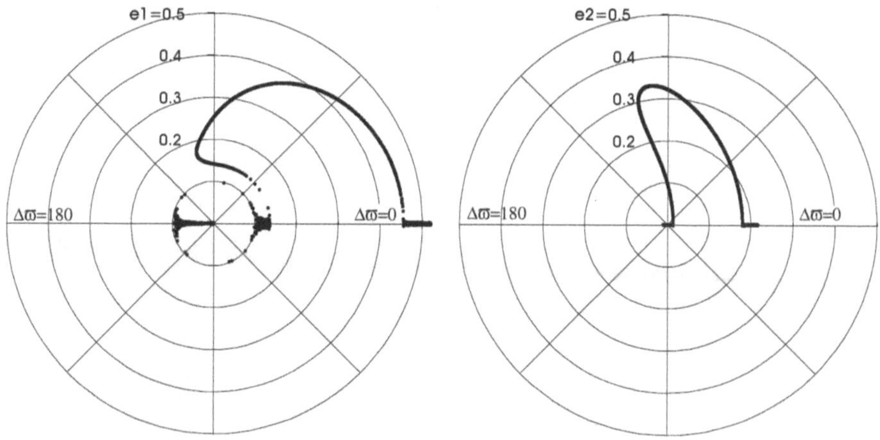

Figure 11. Polar plots in the planes $e_i \exp i\Delta\varpi$.

Figure 11 shows the evolution over a time interval almost 10 times longer. The first point to stress here is that such a time span is likely beyond physical significance. Each panel of Figure 11 combines the variations of e_1 (resp. e_2) and $\Delta\varpi$ in a same plot in polar coordinates in which the radius vector is the eccentricity and the polar angle is $\Delta\varpi$. We itemize the important points to be noted:

- The eccentricity e_1 increases monotonically. When it reaches $\sim$ 0.46, the asymmetric Apsidal corotation changes back to a symmetric configuration with aligned periapses.

- After the bifurcation, the solution followed lies in the upper half-plane. This is not the only possibility. The motion could have followed a mirror path in the lower half-plane. The probability of following one or another branch is the same.

- The phenomenon leading to the transformation from anti-aligned to aligned periapses is clearly seen in the right-hand side panel, where the trajectory is seen crossing the origin of the plot.

It is interesting to note that this picture has a counterpart in the study of periodic orbits of the 3-body problem. The study of symmetric periodic solutions shows the existence of two separated stable branches with aligned periapses ; these two branches are tied with continuity by a branch of unstable periodic orbits (Hadjidemetriou, 2002). The unstable branch corresponds to the saddles shown in the fourth panel of Figure 6. In Figure 11 it would appear as a right shortcut on the horizontal axis tying the initial and final segments of stable solutions with aligned periapses.

7. Conclusions

The contents of this paper include, with variable emphasis, the topics of a series of lectures whose main title was "Routes to Order: Capture into resonance". This was indeed the subject of the last section above. The study of this subject has, however, shown that – unlike the restricted three-body problem – capture into resonance drives the system immediately to stationary solutions known as "Apsidal corotations". The whole theory of these solutions was also included in the paper from the beginning – that is, from the formulation of the Hamiltonian equations of the planetary motions and the expansion of the disturbing function in the high-eccentricity planetary three-body problem. The secular theory of non-resonant systems was also given. Motions with aligned or anti-aligned periapses, resonant or not, resulting from non-conservative processes (tidal interactions with the disc) in the early phases of the life of the system, seem to be frequent in extra-solar planetary systems.

References

Beaugé, C. (1994), Asymmetric librations in exterior resonances. *Cel. Mech. & Dynam. Astron.* **60**, 225-248.

Beaugé, C. (1996), On a global expansion of the disturbing function in the planar elliptic three-body problem. *Cel. Mech. & Dynam. Astron.* **64**, 313-349.

Beaugé, C. and Michtchenko, T.A. (2003), Modelling the high-eccentricity planetary three-body problem. Application to the GJ876 planetary system. *Month. Not. Roy. Astron. Soc.* **341**, 760-770.

Beaugé, C., Ferraz-Mello, S. and Michtchenko, T.A. (2003), Extrasolar planets in mean-motion resonance: Apses alignment and other stationary solutions. *Astroph. J.* **593**, 1124-1133. (astro-ph 0219577).

Brouwer, D. and Clemence, G.M. (1961), *Methods of Celestial Mechanics.* Academic Press Inc., New York.

Callegari Jr., N., Michtchenko, T. and Ferraz-Mello, S. (2004), Dynamics of two planets in 2:1 mean-motion resonance. *Cel. Mech. & Dynam. Astron.* **89** (in press).

Dermott, S., Malhotra, R. and Murray, C.D. (1988), Dynamics of the Uranian and Saturnian satellite systems: A chaotic route to melting Miranda. *Icarus* **76**, 295-334.

Ferraz Mello, S., Tsuchida, M. and Klafke, J.C. (1993), On symmetrical planetary corotations. *Cel. Mech. & Dynam. Astron.* **55**, 25-45.

Ferraz-Mello, S. (1994), The convergence domain of the Laplacian expansion of the disturbing function. *Cel. Mech. & Dynam. Astron.* **58**, 37-52.

Ferraz-Mello, S., Beaugé, C. and Michtchenko, T.A. (2003), Evolution of migrating planet pairs in resonance. *Cel. Mech. & Dynam. Astron.* **87**, 99-112 (astro-ph 0301252).

Gomes, R.S. (1995), resonance trapping and evolution of particles subject to Poyinting-Robertson drag: Adiabatic and non-adiabatic approaches. *Cel. Mech. & Dynam. Astron.* **61**, 97-113.

Greenberg, R. (19..) Galilean satellites: Evolutionary paths in deep resonance. *Icarus* **70**, 334-347.

Hagihara, Y. (1970), *Celestial Mechanics*, vol. I, MIT Press.

Hadjidemetriou, J. (2002). Resonant periodic motion and the stability of extrasolar planetary systems. *Cel. Mech. & Dynam. Astron.* **83**, 141-154.

Hadjidemetriou, J. D. and Psychoyos, D. (2003), Dynamics of extrasolar planetary systems: 2/1 resonant motion. In *Galaxies and chaos* (G. Contopoulos and N. Voglis, eds.). Lectures Notes in Physics, Springer, 412-432.

Hori, G.-I. (1985), Mutual perturbations of 1:1 commensurable small bodies with the use of the canonical relative coordinates. Part I. In *Resonances in the Motion of Planets, Satellites and Asteroids* (S. Ferraz-Mello & W. Sessin, eds.), IAG-USP, 53-66.

Kaula, W.M. (1962), Development of the lunar and solar disturbing functions for a close satellite. *Astron. J.*, **67**, 300-303.

Laskar, J. (1990), Systèmes de variables et éléments. In *Les Méthodes Modernes de la Mécanique Céleste* (D. Benest & Cl. Froeschlé, eds), Ed. Frontières, pp. 63-87.

Laskar, J. (2000), On the spacing of planetary systems. *Phys. Rev. Let.* **84**, 3240-3243.

Laughlin, G. and Chambers, J. (2001), Short-term dynamical interactions among extrasolar planets. *Astroph. J.* **551**, L109-L113.

Lee, M. H. and Peale, S. J. (2002), Dynamics and origin of the 2:1 orbital resonances of the GJ 876 planets. *Astroph. J.* **567**, 596-609.

Lee, M. H. and Peale, S. J. (2003), Extrasolar planets and mean-motion resonances. In *Scientific Frontiers in Research on Extrasolar Planets* (D. Deming and S. Seager, eds.) *ASP Conf. Series* (in press).

Lissauer, J. L. and Rivera, E. J. (2001), Stability analysis of the planetary system orbiting υ Andromedae. II. Simulations using new Lick Observatory fits. *Astrophys. J.* **554**, 1141-1150.

Mayor, M., Udry, S., Naef, D., Pepe, F., Queloz, D., Santos, N.C. and Burnet, M. (2004), The CORALIE survey for southern extra-solar planets. XII. Orbital solutions for 16 Extra-Solar Planets discovered with CORALIE. *Astron. Astrophys.* (in press).

Message, P.J. (1958), The search for asymmetric periodic orbits in the restricted problem of three bodies. *Astron. J.* **63**, 443-448.

Michtchenko, T. and Ferraz-Mello, S. (2001), Modeling the 5:2 mean-motion resonance in the Jupiter-Saturn planetary system. *Icarus* **149**, 357-374.

Michtchenko, T. and Malhotra, R. (2004), secular dynamics of the three-Body problem: Application to the υ Andromedae planetary system. *Icarus* (in press).

Papaloizou, J.C.B. (2003), Disc-planet interactions: Migration and resonances in extrasolar planetary systems. *Cel. Mech. & Dynam. Astron.* **87**, 53-83.

Pauwels, T. (1983), Secular orbit-orbit resonance between two satellites with non-zero masses. *Celest. Mech.* **30**, 229-247.

Perryman, M. (2000), Extra-Solar Planets. *Rep. on Progress in Phys.* **63**, 1209-1272.

Poincaré, H. (1897), Sur une forme nouvelle des équations du problème des trois corps. *Bull. Astron.* **14**, 53-67.

Poincaré, H. (1905), *Leçons de Mécanique Céleste*, Gauthier-Villars, Paris, vol. I

Rivera, E. J. and Lissauer, J. L. (2000), Stability analysis of the planetary system orbiting υ Andromedae. *Astrophys. J.* **530**, 454-463.

Sessin, W. and Ferraz-Mello, S. (1984), Motion of two planets with periods commensurable in the ratio 2:1. Solutions of the Hori auxiliary system. *Celest. Mech.* **32**, 307-332.

Tisserand, F. (1960), *Traité de Mécanique Céleste*, vol. I, Gauthier-Villars, Paris, 2nd. ed., chap. XV.

Wintner, A. (1941), *The Analytical Foundations of Celestial Mechanics*, Univ. Press, Princeton.

Yuasa, M. and Hori, G.-I. (1979), New approach to the planetary theory, *IAU Symp.* **81**, 69-72.

TIDES ON EUROPA

Linking Celestial Mechanics, geology, and geophysics

Richard Greenberg
University of Arizona
greenberg@lpl.arizona.edu

Keywords: Europa, Celestial Mechanics, Tides

1. Introduction

Gravitational N-body dynamical systems usually have only the current positions and velocities of the bodies as the observational evidence of their past history. The traces of their past behavior exist only as conceptual, theoretically computed orbits. However, the Laplace orbital resonance among the galilean satellites actually leaves a visible record of its behavior in a surprising and seemingly unlikely form: Linear traces in distinctive geometric patterns clearly visible on the surface of the satellite Europa are a direct recording of the effect of the orbital resonance.

These lineaments are cycloid-shaped features, unique to Europa, but ubiquitous on it. Examples are shown in Figures 1-4. The Cycloids are cracks in the surface ice. Typically, these chains of arcs run for a thousand km or more (nearly a radian) across the surface, with each arc having a length ~100 km. In many cases, in fact in most of the visible examples, the cracks are lined by a pair of ridges on either side of the crack. Figure 1 shows the Voyager spacecraft image in which these remarkable features became evident. Compared with the later images from the Galileo spacecraft, Figure 1 is relatively low resolution, but the ridges allow us to see the geometry of the Cycloids quite clearly. Figures 2-4 are Galileo images that show several examples with details of their structure. Global lineament patterns can be seen in Figure 5, where cycloid-like patterns and arcs are recognizable on a global scale.

B.A. Steves et al. (eds.), Chaotic Worlds: from Order to Disorder in Gravitational N-Body Dynamical Systems, 289–305.

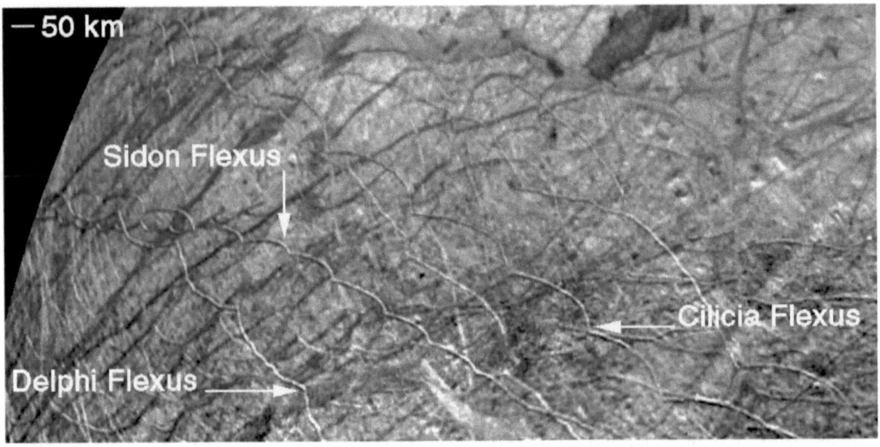

Figure 1. A Voyager image of a far southern region on Europa, showing many distinctive cycloidal ridges, which mark the paths of cracks in Europa's crust. These linear trajectories are the result of the Laplace orbital resonance among the Galilean satellites. Three of the Cycloids have IAU assigned names, as shown. Cycloids are chains of arcs, each typically ~100 km long, often with a dozen or so arcs in each chain.

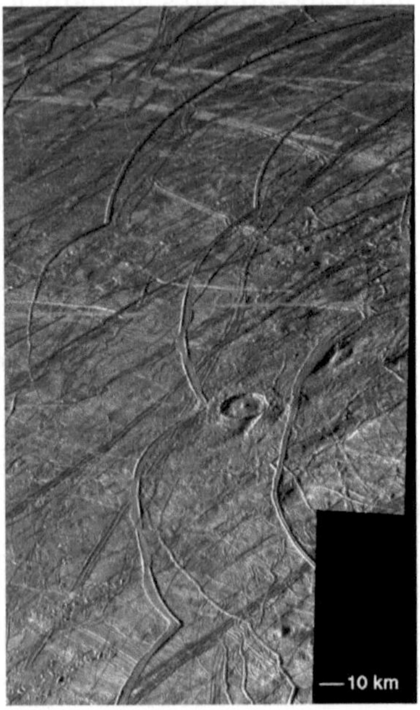

Figure 2. A Galileo image of cycloid arcs that run roughly north-south in the northern hemisphere. By chance, a rare crater is also seen in this region. That the ridges are double can be seen in this image. Also, the Cycloids' appearance is somewhat complicated here by a small amount of dilation (opening of the crack) along a couple of arcs.

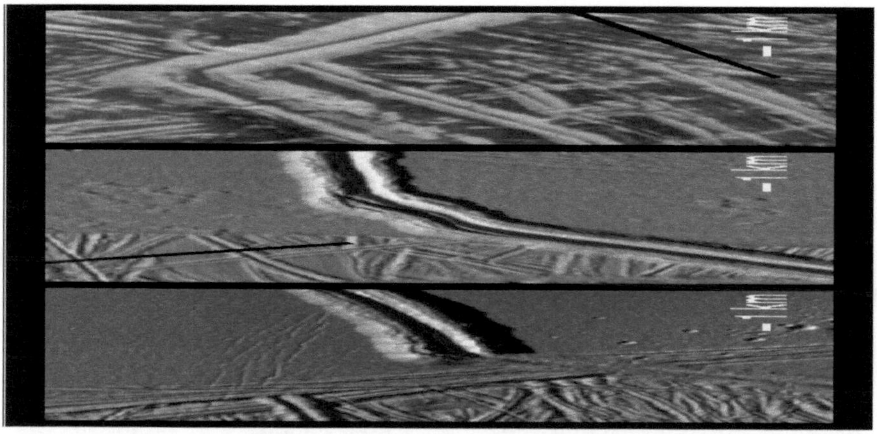

Figure 3. These high-resolution Galileo images show the pointy cusps of some Cycloids, where the double ridges make sharp turns. The crack in the crust lies between the ridges in each ridge-pair. The two examples to the left are details of cusps that appear in Figure 1.

Figure 4. Some Cycloids have been modified by considerable dilation as the crack opened. The dark band that runs from the bottom center to the upper left is an example of a cycloidal crack that has dilated. This example is in a region called the "Wedges", because it contains many dilational bands with this same wedged-shaped geometry. In fact, most of the dilational bands in the wedges originated as cycloidal cracks.

How can a phenomenon of orbital dynamics and celestial mechanics, such as the Laplace resonance, result in distinctive tectonic patterns on a planetary body? The linking process is tides. The orbital resonance forces orbital eccentricities. Thus, as a satellite orbits Jupiter, even if its

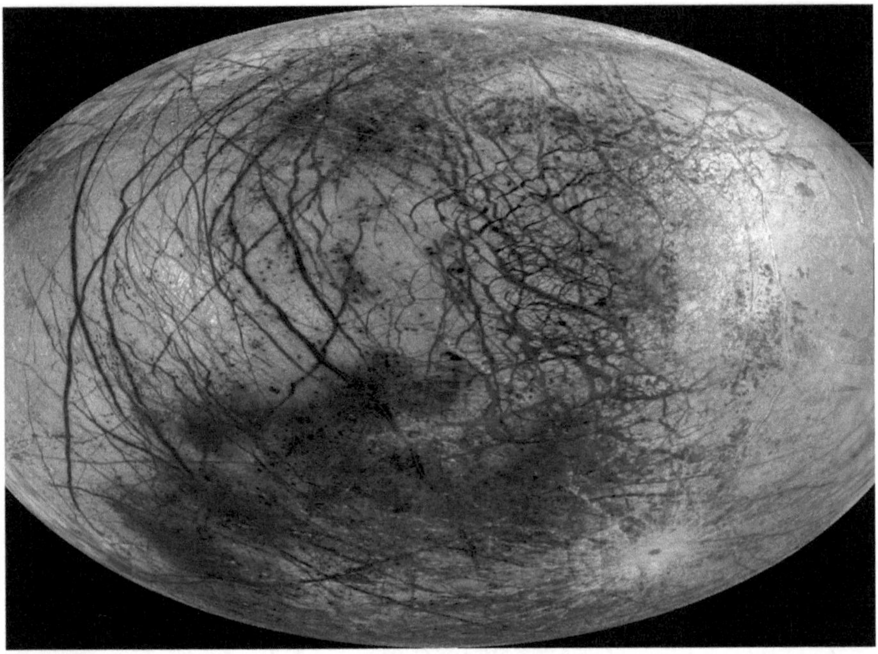

Figure 5. A global view of Europa. The Wedges region lies just south of the equator in this hemisphere. A rare crater, called Pwyll, is seen at the lower left at the center of its system of bright rays. Pwyll (as well as a distinctive X-shaped lineament intersection near a splotch called Conamara chaos about an equal distance north of the equator) lies at longitude 270°W, which by definition is the center of the trailing hemisphere as Europa moves around in its orbit. (The direction of Jupiter defines the prime meridian, longitude 0°). The anti-Jove region (longitude 180°W) appears in this image just to the right of the Wedges. Cycloid patterns, or arcuate segments of Cycloids, can be recognized in many places over the surface. Identification of the location of Figure 4 in this global context is left as an exercise for the reader.

rotation is synchronous or nearly synchronous, the tide-raising potential will vary periodically over each orbital period. As a result, the shape of the satellite is distorted in a regular, periodic cycle. In turn, the crust of the satellite is stressed, and this regular stressing results in many of the tectonic patterns that are visible on Europa, including the Cycloids.

The tides driven by the orbital resonance have other important geological and geophysical effects as well on Europa. Because the body is regularly worked, the material is heated by friction. This heating seems to be adequate to maintain a liquid water ocean just below the icy surface. The existence of the ocean allows the tidal distortion of the surface to be ~30 times greater than if the entire satellite were solid (as discussed in Section 3 below), greatly enhancing the tectonic and thermal effects of the tides. Moreover, modest local concentrations of heat produced thermal effects that have marked the surface. Both stress and heat affect the crust in ways that allow the ocean to be directly linked to the surface, through cracks and probable melt-through sites.

The tides also have important effects on the dynamics (both rotation and orbit) of Europa. Because the orbit is eccentric, tidal torques tend to drive spin to a state that is slightly faster than synchronous (Greenberg and Weidenschilling 1984). This effect in turn can increase the tidal stress on the surface substantially, although it acts on a much longer timescale (~10^5 years) than the 3 1/2 day orbital period. The tides also effect the long-term evolution of the orbits of the satellites, so that the Laplace resonance itself probably changes over ~10^7 years or more.

A global view of Europa (e.g. Figure 5) shows the results of many of these effects. In this view we see the disk of Europa, about 1560 km in radius. The surface has long been known to be composed predominantly of water ice (Pilcher et al. 1972). Dark markings are due to impurities, probably brought up from the ocean, and they are only visible because the contrast in this image has been stretched. Otherwise Europa would appear as a smooth white sphere.

Gravity studies, based on tracking spacecraft orbits, have revealed that the density of the outer 100-150 km is about 1 gm/cm^3 (Anderson et al. 1997). Evidently, the H_2O extends from the surface down to that depth. The outermost portion of this thick layer of water is a "lithosphere", meaning that it is cold, brittle, elastic ice. The lithosphere is probably ~1 km thick, although that value is somewhat uncertain, and also depends on the timescale for deformation; the transition depth from elastic to viscous ice depends on the rate of strain as well as on the temperature. The viscous portion of the ice crust lies below the lithosphere, and extends to a depth of several km below the surface. A belief that the ice crust is 20 km thick or greater has been widely promoted, but the evidence cited for that model (Pappalardo et al., 1998) has not survived quantitative study (Greenberg et al. 2003b). If the ice were that thick, the ocean would be isolated from the surface.

Tidal theory applied to observations of tectonic and thermal features suggests that the ice is significantly thinner, so that the ocean can connect with the surface through cracks and melt sites. Tidal stress can probably open cracks no deeper than a few km, so we infer that the ice is probably less than 10 km thick.

The surface of Europa is continually resurfaced by the tidal tectonic and thermal effects. The rapidity of resurfacing is immediately apparent from Figure 5. The surfaces of most airless bodies, like our Moon, are heavily cratered, recording billions of years of bombardment by small bodies. In contrast, Europa has almost no craters. One example is Pwyll, with its spectacular icy rays sprayed across the surface. Another is Cilix, a small bright spot near the equator toward the right in Figure 5, surrounded by a dark halo. Cilix lies near the anti-jovian point, opposite the direction of Jupiter. If Europa's surface were very old, it would be densely covered by craters, but these examples are rare. Estimates of the expected bombardment rate, mostly by comets, indicate that the surface is no older than about 50 million years (Zahnle *et al.* 2003). Evidently the tidal processes that have created the features that we do see on the surface have been very active, continually refreshing the surface and erasing what was there before.

Even at the low resolution of a global view like Figure 5, we can see the results of the two major types of resurfacing. Tectonics create the dark lines and thermal effects produce the dark splotches. At higher resolution, the lines are usually either ridge systems that form along cracks or dilational bands, as seen in Figures 1-4, although not all are in the shape of Cycloids. The splotches at higher resolution have morphological character that is called "chaotic terrain". Chaotic terrain is characterized by disruption of the older surface, which is replaced by a lumpy matrix, often with displaced rafts of the older surface. One example, part of Conamara chaos, is shown in Figure 6. The geology of Europa is not only controlled by an orbital resonance, but chaos is involved as well, albeit a different kind of chaos than is usually discussed in celestial mechanics.

Both tectonic cracks and thermally produced Chaotic terrain are caused by tides, and both involve linkage of the underlying ocean to the surface. The darkening associated with each of these formation processes probably involves substances from the ocean, which makes the features visible even at the low resolution of global images like Figure 5. The cracks are caused by tidal stress and the chaos is probably caused by melt-through of patches of crust by tidal heat.

A comprehensive discussion of the character of the tidally driven tectonic and thermal features of Europa was presented at the NATO Advanced Study Institute in Cortina. Reviews of specific topics of particular interest to celestial mechanicians include a summary of evidence regarding Europa's rotation (Greenberg *et al.* 2002b) and computations of tidal stress (Greenberg *et al.* 2003a). In the next section, I summarize a few of the key points regarding how tides have controlled the observed properties and features of Europa and what they tell us about its character and structure. For a more complete and integrated review of these topics, see Greenberg and Geissler (2002) or Greenberg *et al.* (2002a), which contain references to the supporting literature.

2. Tides and observed features

2.1 Cycloids

The mechanism by which tides create the cycloidal crack patterns was discovered by B.R. Tufts and G.V. Hoppa (Hoppa *et al.* 1999c). Due to the orbital eccentricity, which is pumped and maintained by the Laplace resonance, the tidal stress on Europas ice crust changes periodically with each orbit. We call this variation the diurnal tide, because the 85 hr orbital period is approximately the length of a Europan day, and is comparable to the length of a day on Earth.

At any location, the tensile stress on the thin shell may increase during the day to the point where it exceeds the strength of the ice. A crack will begin to propagate across the surface. As time goes on and as the tip of the crack moves, the direction of the local stress will change, causing the cracks trajectory to curve, forming one arc of the cycloid. Eventually, after several hours, the diurnal tidal stress decreases to the point where the crack stops propagating. Then, several hours later, after the stress has again increased, the crack begins to propagate again, but in a different direction from when it had stopped. In this way a pointed cusp is formed, and the next arc begins to be traced by the advancing crack.

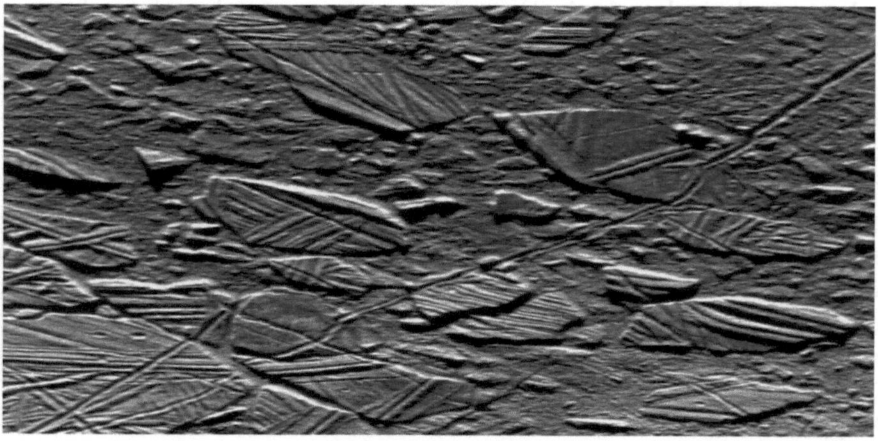

Figure 6. A portion of Conamara chaos, which lies just south of the lineament intersection mentioned in the caption for Figure 5. Chaotic terrain is characterized by a disrupted older surface, replaced by a lumpy matrix, often with displaced rafts of the older surface.

The process continues for several europan days, with one arc segment being produced each day.

In that way, a crack propagating across Europa can record the stress, which originates with the action of the Laplace resonance.

2.2 Ridge Formation

Once a crack is formed, whether or not it takes on a cycloidal shape, the stress cycle of the diurnal tide will continue to work the crack, opening and closing it on a daily basis. If the crack penetrates to liquid water, while it is open, water will rush up through it to the float line near the surface. At the top, the liquid will begin to freeze, but within hours the walls of the crack will begin to close, crushing the new ice and squeezing some of it to the surface, contributing to ridge-building. As the crack reopens due to the diurnal cycle, some material may fall back, so that a double ridge forms along both sides of the crack. This process would repeat on a daily cycle, building substantial ridges up to $\sim$ 100 m high in a few thousand years.

This tidal tectonic process was described by Greenberg *et al.* (1998). A key point is that the magnitude of the tidal tensile stress is probably not adequate to open cracks to a depth greater than a few km, because the overburden compression would be too great at depth. Therefore for this ridge-building process to work, the ice must be fairly thin.

2.3 Strike-slip displacement

Strike-slip displacement, in which one crustal plate shears past another, is relatively rare in the solar system except in a few places. On Earth, with its global tectonics, strike-slip between plates is well known. The San Andreas fault in California is a famous example. Similar strike-slip is ubiquitous on Europa. Faults there include some longer than the San Andreas, with Strike-slip displacement of tens of km. Strike-slip displacement on Europa is probably driven by diurnal tides, and thus by the Laplace resonance. Over the course of a day, the tidal stress follows a sequence that can open a crack, shear it, then close it. Then, stress that would reverse the shear is resisted while the crack is closed. This process repeats on a daily basis. In this process, which is analogous to walking, one plate of crust can shear past another, with Strike-slip displacement visible along the boundary. This process has been described in detail by Hoppa et al. (1999b).

This theoretical process can be used to predict the sense of shear that would be expected to develop along any crack, depending on its location and azimuthal orientation. Surveys of strike slip seem to agree with these predictions (Hoppa *et al.* 2000, Sarid *et al.* 2002). For example, faults in the far north shear in one direction ("left-lateral" on geological jargon), while those in the far south are all "right lateral", just as predicted by the theory.

Using this tool, Sarid *et al.* (2002) were able to show that most Strike-slip displacement features in the leading hemisphere had formed further north than their current positions, while those in the trailing hemisphere had formed further south. This result has been used to infer that polar wander has occurred. That is to say that the icy shell of Europa has slipped around as a whole, moving the former pole locations to places tens of degrees away from the spin axis. Strike-slip also provides additional evidence that cracks from the surface penetrate all the way down to liquid water. This result argues against the notion that the ice is so thick that the ocean is isolated from the surface.

2.4 Non-synchronous rotation

The lag in the tidal response to Non-synchronous rotation causes a torque that tends to make the rotation synchronous, but that effect is counteracted in the case of an eccentric orbit. In the latter case, the lag in the diurnal tide yields a torque that varies over each orbit. The torque, averaged over each orbit, tends to drive the rotation to a rate slightly different from synchronous. That rate depends on details of the tidal response, but according to theory is expected to be only slightly faster than synchronous (Greenberg and Weidenschilling 1984).

Even given Europa's orbital eccentricity e, the rotation might be synchronous if a non-spherically symmetric, frozen-in density distribution (like that of the Earth's Moon) were locked to the direction of Jupiter. Given that Europa is substantially heated by tidal friction, it may not be able to support such a frozen-in asymmetry. It is also conceivable that the silicate interior is locked to the direction of Jupiter by a mass asymmetry, while the ice crust, uncoupled from the silicate by an intervening liquid water layer, rotates non-synchronously due to the tidal torque.

However, even a completely solid Europa would rotate non-synchronously as long as mass asymmetries are small enough. Non-synchronous rotation does not in itself imply the existence of an ocean. However, both the existence of an ocean and Non-synchronous rotation are made possible by the substantial tidal heating.

Observational evidence places some constraints on the actual rotation rate. A complete summary is provided by Greenberg *et al.* (2002), but the key points are the following. First, a comparison of Europa's orientation during the Voyager 2 encounter with that observed 18 yr later by Galileo

(Hoppa *et al.* 1999a), showed no detectable deviation from synchroneity. Any deviation must be so small that the period is >12,000 yr, relative to the direction of Jupiter. Nevertheless, Geissler *et al.* (1998) found systematic changes in the orientation of regional scale lineaments, based on a cross-cutting sequence, indicating that terrain had moved west-to-east through the theoretical tidal stress field, consistent with some (albeit slow) Non-synchronous rotation. As the tectonic effects of tidal stress began to be better understood, various features were found to have likely formed further west than their current positions, including strike-slip faults and cycloidal crack patterns. From the latter, Hoppa *et al.* (2001) inferred a non-synchronous period of (very roughly) ~50,000 yr. More recently however, we have reviewed the evidence from azimuthal orientations, and have found that it may not be significant. Sarid *et al.* (2004) surveyed the rich record of cross-cutting features in the southern leading hemisphere and found no clear indication of Non-synchronous rotation. This issue is important, but at the current time is unresolved.

2.5 Life

The evidence discussed above, as well as other evidence reviewed in Greenberg and Geissler (2002) or Greenberg *et al.* (2002a), indicates that Europas icy crust is thin enough for the ocean to be linked to the surface through cracks and thermal melt-through sites. This linkage enhances the plausibility that there might be life on Europa, because it would allow organisms in the liquid to access oxidants and other vital substances from the surface, while remaining protected from the harsh radiation environment. The cracks in the ice might be important habitable niches, as well as the ocean itself.

With these linkages, the biosphere of Europa would extend to near the surface, where it could easily be sampled by spacecraft landers. At the same time, this exposure means that the habitable zone of Europa would be vulnerable to contamination by any organisms that might inadvertently be transported to the surface. These astrobiological issues are discussed in Greenberg and Geissler (2002) or Greenberg *et al.* (2002a).

3. Tidal Amplitude

All of the activity on Europa is driven by tides, so it is worth considering what we know about the amplitude of these tides. According to Tidal theory, Europa's figure is distorted by the tide-raising potential of Jupiter according to

$$H = h_2(Re(M_j/M_e)(R_e/a)^3)(3/2cos^2\vartheta - 1/2) \quad (1)$$

where H is the height of the surface (relative to the radius of a sphere of equal volume) at a location an angular distance ϑ from the direction of Jupiter, R_e is the satellite's radius, M_j and M_e are the mass of Jupiter and Europa respectively, and a is the distance from Jupiter. The coefficient h_2 is one of the tidal "Love numbers", which account for the properties of the material of which Europa is composed. A second Love number k_2 is the coefficient for the formula that represents the tidal distortion of Europas gravitational field.

The classical textbook formula, derived by Love (1944), for h_2 is

$$h_2 = \frac{5/2}{1 + 19\mu/2\varrho g R_e} \tag{2}$$

where μ is the rigidity of the material, ϱ is the density, and g is the surface gravity. This formula assumes that the body is of uniform density and incompressible. If the body is fluid or fully relaxed to conform to the tidal potential, $h_2 = 5/2$. For Europa, the quantity in brackets has a value of 800 m, so with $h_2 = 5/2$, the height H of this fixed tide would be 2 km at the sub- and anti-Jove points. (Actually, if the tidal potential were only due to the gravitation pull of Jupiter, such a relaxed Europa would have h2 = 1. However, the tidal potential includes not only the direct tidal stretching by Jupiter, but also the gravitational field due to the distortion of the satellite itself. This latter effect enhances the tides, which allows h_2 to be substantially greater than 1.)

For a real moon, the density of the satellite increases with depth and the material is compressible. For a reasonable non-uniform density distribution for Europa with a thick layer of liquid water, h_2 is probably $\sim$1.2, about half that for the uniform, low-rigidity, incompressible case. In that case, according to equation 1, H would be $\sim$1 km at the sub- and anti-Jove points.

Now consider the effects of an eccentric orbit. In this case, even with synchronous rotation, the magnitude and orientation of the tidal distortion changes throughout each orbital period. The tide-raising Gravitational potential is at a maximum at pericenter and a minimum at apocenter. The height of the diurnal tidal variation is 3e times the value of H from equation 1, a result derived from the change in H due to variation of the distance from Jupiter over each orbit. For Europa with $e = 0.01$, the amplitude is 30 m, i.e. the tidal height at the sub- and anti-Jove points at pericenter (or apocenter) is 30 m higher (or lower) than the surface of the fixed tide given by equation 1.

Computation of the amplitude of tides for a realistic layered model is challenging. There is no simple formula available for evaluating the Love numbers h_2 and k_2. We have to rely on results from proprietary computer codes (e.g. Yoder and Sjogren 1996, Moore and Schubert 2000). A graphical representation of such results is shown in Figure 7a, which was presented by NASA's Europa Orbiter Science Definition Team (1999). This graph shows that varying the assumptions about the physical state and thickness of the several layers (illustrated in Figure 7b) that compose Europa can significantly affect the values of the Love numbers. The value of such a plot, and the reason it was created, is that it allows space mission planners to determine what might be learned about Europa by measuring the tidal distortion of Europas shape, for example by a laser altimeter on an orbiting spacecraft, which can constrain h_2, and the tidal distortion of Europas gravitational field, by tracking the spacecrafts orbit, which can constrain k_2.

Unfortunately, the labeling on Figure 7 from the NASA report was rather obscure and misleading, so it is difficult to understand the meaning, the trends, or the sensitivities of the values of the Love numbers relative to the models of Europas interior. For example, the labeling on the two points furthest to the right (both near $h_2 = 1.3$) suggests that the only difference between them is a change in the character of the rocky mantle, which (as we shall see) is not at all the case. Moreover the report contained no quantitative information regarding the assumptions about the layers in Figure 7b. We can create a much more meaningful and useful version of Figure 7 if we go back to the source of these results, which is primarily a set of calculations carried out and tabulated by Yoder and Sjogren (1996). The calculations were based on the following assumptions about the interior layers, as represented by Figure 7b:

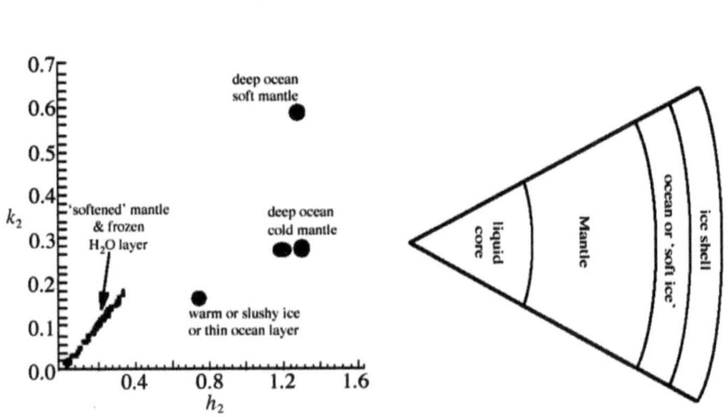

Figure 7. (a) The values of the Love numbers for various assumptions regarding the interior structure of Europa, as plotted by NASA's Europa Orbiter Science Definition Team (1999). (b) The layers assumed in the models of the interior upon which (a) was based.

The core has a radius of 600 km, and consisting of liquid iron with a density of 5 gm/cm^3, for all cases considered.

The mantle was assumed (it appears) to be made of rock and have a thickness of 800 km. Various cases were considered with the character of the rock ranging from solid and rigid to completely melted.

An "ocean or soft ice"layer (as labeled in Figure 7b) lies above the mantle. In all cases, this layer is composed of H_2O, However, cases are considered where it is assumed to have some rigidity, as well as cases where it is liquid. The variation from liquid to solid is represented by changes in the sound speed, which Yoder and Sjogren used as a proxy for the rigidity; It ranges from 2 km/sec for solid ice down to 0.001 km/sec for essentially liquid water.

An ice shell lies above the ocean. In all cases considered by Yoder and Sjogren, the total thickness of the H_2O layers (the ice shell, plus the ocean or soft ice layer) was 105 km. However, the thickness of the ice shell is not the same for all cases.

Figure 8 shows the results reported by Yoder and Sjogren, but in a way that elucidates the trends and is more accurately labeled than in the NASA report. First, we identify the important end-member cases that were not marked on Figure 7a. If all the layers were solid, the Love number values would lie at the spot marked by the small black spot at the lower left in Figure 8. If all the layers were fluid, then h_2 and k_2 would be 2.05 and 1.05 respectively (Moore and Schubert 2000). Also, recall that if the same mass were uniformly distributed, rather than in layers, and were fluid and incompressible, then h_2 and k_2 would be 2.5 and 1.5, respectively, according to the classical result by Love (1944).

Next consider the several cases computed by Yoder and Sjogren in which the ice shell is 25 km thick and the ocean (80 km thick) ranges in rigidity between solid ice (the end member at the lower left in Figure 8) and completely melted. As the ocean is increasingly fluid, the values follow the shallow-sloped (lower) line toward the right. Figure 8 includes the large dot along this line which represents a case intermediate between the solid and the fluid ocean, which the NASA report labeled "warm or slushy ice". The dot at the upper-right end of this line represents the fluid ocean limit, which the NASA report labeled "deep ocean, cold mantle". That label is accurate, but confusing, because the ocean is equally deep (80 km) and the mantle equally cold all along this line. What varies is the sound speed (equivalent to rigidity) of the water in the ocean layer, not its depth. Of course, down at the high-rigidity limit (completely frozen water), the word "ocean" would become less appropriate.

We also have information about how the Love numbers would change if we started from any point along that line and increasingly softened the

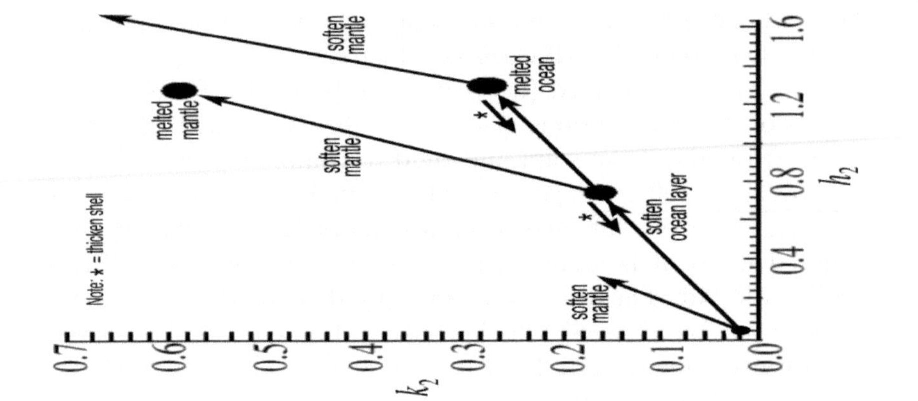

Figure 8. A new version of Figure 7a, in which the trends that result from changing the parameters of the multi-layer model in Figure 7b are more explicit. The case where the rock mantle and the entire 105 km thick layer of H_2O are both solid is shown by the small dot at the lower left. Cases in which the rock mantle and the outer 25 km ice shell remain solid lie along the shallow sloped line, with the 80 km of H_2O under the shell becoming progressively "softer", up to the extreme limit of a liquid ocean. Starting at several points along the latter line (the solid mantle locus), if the mantle is then progressively softened, the Love number values move upward on the steeper paths. Along each of these paths, the thickness of the ice shell remains 25 km, and the rigidity of the lower H_2O layer remains constant (solid along the left path, liquid along the right path, and intermediate rigidity along the middle path). The path at the upper right continues off-scale up to about (h_2 = 2.05, k_2 = 1.05), where the core, mantle, and ocean are all liquid. Starting at any point along the lower path (the solid mantle locus), if the ice shell is thickened while the total H_2O thickness remains 105 km and everything else stays constant, the values move back down the path as shown by the short arrows (marked with $*$). The values of the Love numbers are not sensitive to changes in the ice thickness value below 25 km.

mantle. First, the NASA report shows that if we started with the completely solid body and softened the mantle, the Love numbers would follow the steep line with $h_2 = 2k_2$. Second, a sequence of results tabulated by Yoder and Sjogren shows the change if we start with the intermediate-rigidity ocean layer and soften the mantle, all the way up to the case of a fully melted mantle. The later point is the one that the NASA report labeled confusingly as "deep ocean, soft mantle". In fact, at that point the mantle is fully melted and the ocean is no deeper than in the other cases shown. However, recall that at this point the ocean is characterized by intermediate rigidity (i.e. it is warm ice or slush). Finally, if we start at the point with a solid mantle and a completely melted ocean (1.29,0.28), and then melt the mantle, the Love number values must move up toward the limit of the fully melted satellite (2.05,1.05), as indicated by the steep line at the right in Figure 8.

We can summarize the major trend lines shown in Figure 8 as follows: All along the lower sloping line, the core is liquid iron and 600 km in radius, the mantle is solid rock and 800 km thick, the ice shell is solid and 25 km thick, and the 80 km ocean layer ranges from solid (lower left) to liquid (upper right). Along each of the steeper lines that branch upward, the only parameter that changes is the rigidity of the mantle, which is increasingly fluid toward the upper right.

Of considerable interest is the effect of changing the thickness of the ice shell that overlies the ocean. The evidence discussed above suggests that the ice is thin enough for the ocean to be linked to the surface. If we could measure the Love numbers with sufficient precision, we might be able to determine, or constrain the thickness of the shell independently. Yoder and Sjogren considered a couple of cases with the shell thickness of 65 km, but otherwise similar to the two cases shown by the larger dots in Figure 8 along the solid mantle locus (lower line). In both cases, the effect was roughly equivalent to increasing the rigidity of the ocean somewhat, i.e. moving the values of the Love numbers to the lower left along the shallow line as indicated by the short arrows in Figure 8.

While there would be a measurable difference in the Love numbers between the case of a 25 km ice shell and a 65 km ice shell, there is little change once the ice is thinner that about 25 km. This result was noted in the NASA report, and it (as well as most of these results) was corroborated by calculations by Moore and Schubert (2000).

Nevertheless, it would be useful to be able to explore the variations in the Love numbers over a much wider range of interior models than we have been able to glean from the few results of propriety codes that are available. In principle, multi-layered models can be evaluated analytically. Work on this problem is under way (e.g. Hurford and Greenberg 2002, Hurford *et al.* 2002). As a first step toward extending the derivation of the Love numbers to more general cases, T. Hurford, S. Frey and I have been reexamining the original derivation. Loves solution did not only include the case of an incompressible body (equation 2), but the uniform compressible case as well. Love evaluated his formulae for cases similar to the Earth. We have extended the evaluation over a broad region of parameter space. Some surprising singularities appear in the solution. These results are described and discussed in the next chapter (by Hurford *et al.*).

4. Conclusion

Europa has emerged as one of the most interesting objects in the solar system for a variety of reasons. It has some of the most active and dynamic geology in the solar system. Its ocean offers the exciting prospect of a habitable setting for extraterrestrial life. This chapter touches only briefly on some highlights, with reference to more detailed and complete surveys in the literature. Most remarkably, all of the characteristics that make Europa so exciting are the results of orbital dynamics, specifically the Laplace resonance, and all of them are made possible by the action of tides. Celestial mechanics has now become central to understanding planetary geology and geophysics, as well as the new discipline of astrobiology.

References

Anderson, J. D., Lau, E. L., Sjogren, W. L., Schubert, G., and Moore, W. B. (1997). Europa's differentiated internal structure: Inferences from two Galileo encounters. *Science*, 276:1236–1239.

Europa Science Definition Team (1998). *Report of the Europa Science Definition Team.* NASA.

Geissler, P. E., Greenberg, R., Hoppa, G., Helfenstein, P., McEwen, A., Pappalardo, R., Tufts, R., Ockert-Bell, M., Sullivan, R., and Greeley, R. (1998). Evidence for Non-synchronous rotation of Europa. *Nature*, 391:368–370.

Greenberg, R., and Geissler, P. (2002). Europa's dynamic icy crust. *Meteoritics and Planetary Science*, 37:1685–1710.

Greenberg, R., Geissler, P., Hoppa, G., and Tufts, B. R. (2002a). Tidal-tectonic processes and their implications for the character of Europa's icy crust. *Reviews of Geophysics*, 40:1–1.

Greenberg, R., Geissler, P., Hoppa, G., Tufts, B. R., Durda, D. D., Pappalardo, R., Head, J. W., Greeley, R., Sullivan, R., and Carr, M. H. (1998). Tectonic Processes on Europa: Tidal Stresses, Mechanical Response, and Visible Features. *Icarus*, 135:64–78.

Greenberg, R., Hoppa, G. V., Bart, G., and Hurford, T. (2003a). Tidal Stress Patterns on Europa's Crust. *Celestial Mechanics and Dynamical Astronomy*, 87:171–188.

Greenberg, R., Hoppa, G. V., Geissler, P., Sarid, A., and Tufts, B. R. (2002b). The Rotation of Europa. *Celestial Mechanics and Dynamical Astronomy*, 83:35–47.

Greenberg, R., Leake, M. A., Hoppa, G. V., and Tufts, B. R. (2003b). Pits and uplifts on Europa. *Icarus*, 161:102–126.

Greenberg, R., and Weidenschilling, S. J. (1984). How fast do Galilean satellites spin? *Icarus*, 58:186–196.

Hoppa, G., Greenberg, R., Geissler, P., Tufts, B. R., Plassmann, J., and Durda, D. D. (1999a). Rotation of Europa: Constraints from Terminator and Limb Positions. *Icarus*, 137:341–347.

Hoppa, G., Greenberg, R., Tufts, B. R., Geissler, P., Phillips, C., and Milazzo, M. (2000). Distribution of strike-slip faults on Europa. *JGR*, pages 22617–22628.

Hoppa, G., Tufts, B. R., Greenberg, R., and Geissler, P. (1999b). Strike-Slip Faults on Europa: Global Shear Patterns Driven by Tidal Stress. *Icarus*, 141:287–298.

Hoppa, G. V., Randall Tufts, B., Greenberg, R., Hurford, T. A., O'Brien, D. P., and Geissler, P. E. (2001). Europa's Rate of Rotation Derived from the Tectonic Sequence in the Astypalaea Region. *Icarus*, 153:208–213.

Hoppa, G. V., Tufts, B. R., Greenberg, R., and Geissler, P. E. (1999c). Formation of cycloidal features on Europa. *Science*, 285:1899–1902.

Hurford, T. A., and Greenberg, R. (2002). Tides on a Compressible Sphere: Sensitivity of the h_2 Love Number. In *Lunar and Planetary Science Conference Abstracts*, pages 1589–1590.

Hurford, T. A., Greenberg, R., and Frey, S. (2002). Numerical Evaluation of Love's Solution for Tidal Amplitude: Extreme tides possible. *Bulletin of the American Astronomical Society*, 34:893.

Love, A.E. (1944). *A treatise on the mathematical theory of elasticity.* New York, Dover.

Moore, W. B., and Schubert, G. (2000). The Tidal Response of Europa. *Icarus*, 147:317–319.

Pappalardo, R. T., Head, J. W., Greeley, R., Sullivan, R. J., Pilcher, C., Schubert, G., Moore, W. B., Carr, M. H., Moore, J. M., and Belton, M. J. S. (1998). Geological evidence for solid-state convection in Europa's ice shell. *Nature*, 391:365–368.

Pilcher, C. B., Ridgeway, S. T., and McCord, T. B. (1972). Galilean satellites: Identification of water frost. *Science*, 178:1087–1089.

Sarid, A. R., Greenberg, R., Hoppa, G. V., Geissler, P., and Preblich, B. (2004). Crack azimuths on Europa: time sequence in the southern leading face. *Icarus*, 168:144–157.

Sarid, A. R., Greenberg, R., Hoppa, G. V., Hurford, T. A., Tufts, B. R., and Geissler, P. (2002). Polar Wander and Surface Convergence of Europa's Ice Shell: Evidence from a Survey of Strike-slip displacement. *Icarus*, 158:24–41.

Yoder, C. F., and Sjogren, W.L. (1996). Tides on Europa. In *Europa Ocean Conference Abstracts.*

Zahnle, K., Schenk, P., Levison, H., and Dones, L. (2003). Cratering rates in the outer Solar System. *Icarus*, 163:263–289.

NUMERICAL EVALUATION OF LOVE'S SOLUTION FOR TIDAL AMPLITUDE

Amplified tidal effects near singularities

T.A. Hurford
University of Arizona
hurfordt@lpl.arizona.edu

S. Frey
University of Arizona
sfrey@math.arizona.edu

R. Greenberg
University of Arizona
greenberg@lpl.arizona.edu

Abstract The solution for the Love number h of an incompressible, homogeneous sphere is well known. A more general solution for the tidal response of a compressible, homogeneous sphere was also derived by Love, who applied it to two specific cases with parameters relevant for tides on the Earth, showing enhancement of h on the order of a few tens of percent due to compressibility of a sphere. However, the enhancement can be much greater, even arbitrarily large, depending on the rigidity, compressibility and size of the body.

Keywords: Tides, Love numbers

1. Introduction

An extended, 3D body will feel a differential potential exerted on it by a neighboring mass (Figure 1). This tide-raising potential can be written as

$$W_2 = -\frac{GM}{a^3} r^2 P_2(\cos\vartheta), \tag{1}$$

B.A. Steves et al. (eds.), Chaotic Worlds: from Order to Disorder in Gravitational N-Body Dynamical Systems, 307–323.

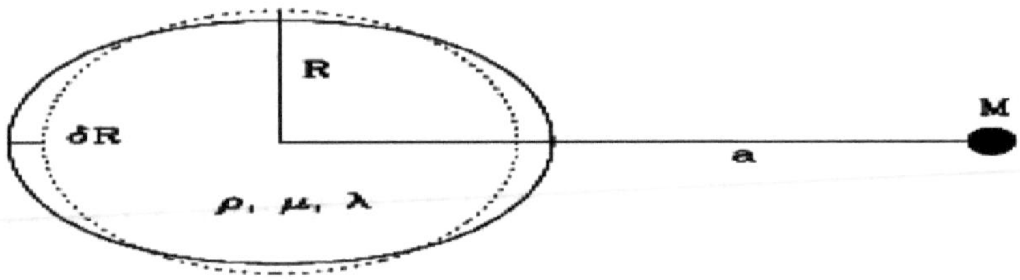

Figure 1. A planet of radius R with density ϱ, rigidity μ and Lamé constant λ is tidally distorted by an amount δR due to a tide-raising mass M at distance a.

at any position (r, ϑ) inside the body defined by the distance from the center of the body, r and an angle, ϑ measured from the axis connecting the centers of the two bodies, where M is the mass of the tide-raising body, a is the distance between the centers of the two bodies, $P_2(\cos\vartheta)$ is the second order Legendre polynomial, and terms of higher order in r/a are assumed negligible. This potential causes the body to deform, raising a tide of height δR on the surface of the body, proportional to the ratio of the original tide-raising potential to the surface gravity of the body, g,

$$\delta R = -\frac{h}{g} W_2(R, \vartheta) \tag{2}$$

where the proportionality constant is known as the Love number h. The Love number h quantifies how much the body will yield to the tide-raising potential .

For the case of a homogeneous incompressible body, Love (1927) showed

$$h_{incompressible} = \frac{5/2}{1 + 19\mu/2\varrho g R}, \tag{3}$$

where μ is the rigidity, ϱ is the density and R is the radius of the body. This result is so well known that it is often cited without derivation or reference.

However, Love (1911) had earlier derived a more general result, valid for arbitrary compressibility. Compressibility is characterized by the Lamé constant λ; λ increases as a material gets more incompressible to a limiting case of $\lambda \to \infty$ for an incompressible body. The equation governing the deformation is re-derived in the appendix, derivation of the Governing Equation.

The problem that Love investigated was the tidal response to the gravitational force of a neighboring point (or spherical) mass, acting on an initially homogeneous sphere. For a compressible sphere, even before a tide raiser is introduced, self-gravity will compress the material until it reaches an equilibrium radial distribution of stress state and density distribution. In the tide-raising problem investigated by Love, he assumed implicitly that the initial condition was a body with uniform density, as well as uniform elastic properties, i.e. it had those properties after spherically symmetrical self-gravitational collapse, but before tide raising. That model is not a realistic representation of any real planet, but following Love we start with that model because it allows an analytic tractable solution. As we shall show, depending on the parameters, that solution can have remarkable properties.

The solution, following Love (see the appendix, Solution to the Governing Equation), is

$$h = \tfrac{5}{2} \quad -\frac{1}{3}A_2(\alpha,\beta)\psi_1(\alpha) - \frac{5}{6}\left(\frac{3\alpha^2+\beta^2}{\alpha^2-\beta^2}\right)A_2(\alpha,\beta)\psi_2(\alpha)$$
$$+\frac{1}{3}B_2(\alpha,\beta)\chi_1(\beta) + \frac{5}{6}\left(\frac{\alpha^2+3\beta^2}{\alpha^2-\beta^2}\right)B_2(\alpha,\beta)\chi_2(\beta), \qquad (4)$$

where α and β can be expressed in terms of the physical properties of the body,

$$\alpha^2 = \frac{\varrho g R}{\mu}\left(\frac{1}{2+\frac{\lambda}{\mu}}\right)\left[\left(16+6\frac{\lambda}{\mu}\right)^{1/2}+2\right], \qquad (5)$$

and

$$\beta^2 = \frac{\varrho g R}{\mu}\left(\frac{1}{2+\frac{\lambda}{\mu}}\right)\left[\left(16+6\frac{\lambda}{\mu}\right)^{1/2}-2\right]. \qquad (6)$$

Note that what we call α and β, Love (1911) called αR and βR, where his α and β had dimensions of inverse length. In (3) the functions $A_2(\alpha,\beta)$, $B_2(\alpha,\beta)$, $\psi_1(\alpha)$, $\psi_2(\alpha)$, $\chi_1(\beta)$ and $\chi_2(\beta)$ are all functions of α and β. These functions are given in Appendix B. Here we have arranged the solution in a form that shows that h is a function of the two ratios, $\varrho g R/\mu$ and μ/λ.

2. Evaluation of *h*

2.1 Love's evaluations

Love's (1911) analysis was motivated by terrestrial geophysics, so he considered parameter values that were plausible as representations of the bulk Earth. For a homogeneous Earth of radius, $R = 6.37\text{x}10^6$ m, and with a density, $\varrho = 5500$ kg/m^3, Love evaluated two cases corresponding to materials with Poisson's ratio of approximately 1/3 and 1/4, respectively, thus bracketing the properties of most geophysical materials (Turcotte and Schubert, 1982).

In the first case Love let $\alpha = 3$ and $\beta = 2$ (1911). According to (5) and (6), in this case $\mu/\lambda = 25/46$ (equivalent to a Poisson's ratio $\approx 1/3$) and $\varrho g R/\mu = 24/5$. Using the values of R, g and ϱ for the Earth, $\mu = 7.16\text{x}10^{10}$ Pa. Love evaluated h (from (3)) and found an 11% enhancement of the incompressible case: Compressibility of the body causes a larger Tidal Amplitude.

Love's (1911) second case has $\alpha \approx 3.3$ and $\beta \approx 2.1$, yielding $\mu/\lambda = 1944/2041$ (Poisson's ratio $\approx 1/4$) and $\varrho g R/\mu = 1107/224$. Again Love found that h is enhanced compared to the Love number of an incompressible case. The enhancement for these given parameter values is 22%.

The value of h can readily be calculated for other values of μ/λ and $\varrho g R/\mu$. Figure 2 shows the solution of the enhancement of h (i.e. $h/h_{incompressible}$) for the two values of $\varrho g R/\mu$ considered by Love, but over a range of values of μ/λ. Recall that $\mu/\lambda = 0$ for an incompressible material. Figure 2 confirms that as $\mu/\lambda \to 0$, the solution (3) approaches the incompressible solution (3), i.e. enhancement of h goes to 1. The two cases evaluated by Love are indicated. Figure 2 shows that the solution seems well behaved as the ratios μ/λ and $\varrho g R/\mu$ vary. Enhancement of h increases over this range as the body becomes more compressible.

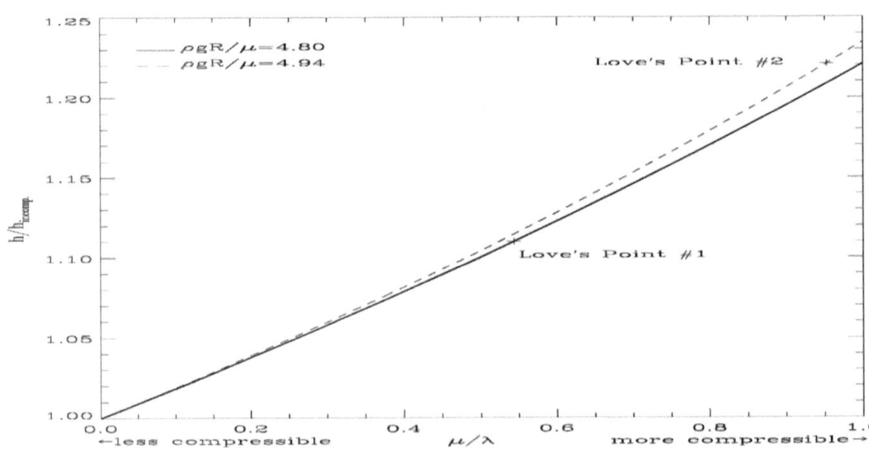

Figure 2. $h/h_{incompressible}$ is shown as a function of μ/λ for the two values of $\varrho g R/\mu$ considered by Love. The + symbols mark the two cases evaluated by Love. For less compressible planets (higher λ), μ/λ is smaller. For the incompressible case, $\lambda \to \infty$,i.e. $\mu/\lambda \to 0$.

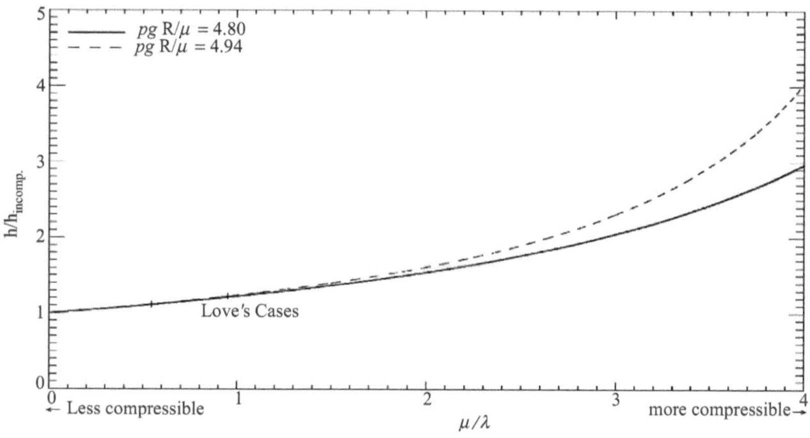

Figure 3. Values of $h/h_{incompressible}$ are shown for the two values of $\varrho gR/\mu$ considered by Love and are extended to higher values of μ/λ than in Figure 2. As in Figure 2, the + symbols mark the two cases evaluated by Love.

2.2 Extending the range of evaluation

Allowing for an even more compressible body, Figure 3 extends results for values of μ/λ up to 4. Although few known materials have such a high μ/λ value (Poisson's ratio of 1/10), the formal trends are interesting. The slopes increase with μ/λ, yielding a rapid increase in h (here up to 4 times that of the incompressible case).

The steepening of the curves depends strongly on $\varrho gR/\mu$ according to Figure 3. The effects of changing $\varrho gR/\mu$ on the solution for h can be explored. If the radius of the planet were increased modestly by $\sim$ 20% above the terrestrial value, while holding ϱ and μ constant, g would increase proportional to R and the ratio $\varrho gR/\mu$ would increase to $\sim$ 7.0. Figure 4 shows evaluation of h for this case over a range of values of μ/λ. Where μ/λ is small, the enhancement $h/h_{incompressible}$ increases sharply with increasing compressibility. As μ/λ approaches a value of 1.442 (not far from reasonable for geophysical material), the enhancement of h becomes very large, approaching infinity. The solution then passes through a singularity, becoming negative for larger μ/λ. According to this result, the shape of the body actually contracts along the axis aligned with the tide-raiser.

This remarkable result, that there exists Love number values that are infinite and/or negative, is not unique to this single set of parameters. Figure 5 shows the enhancement $h/h_{incompressible}$ for various values of $\varrho gR/\mu$. If μ and ϱ are constrained to the reasonable values used by Love, the various values of $\varrho gR/\mu$ correspond to various radii. According to

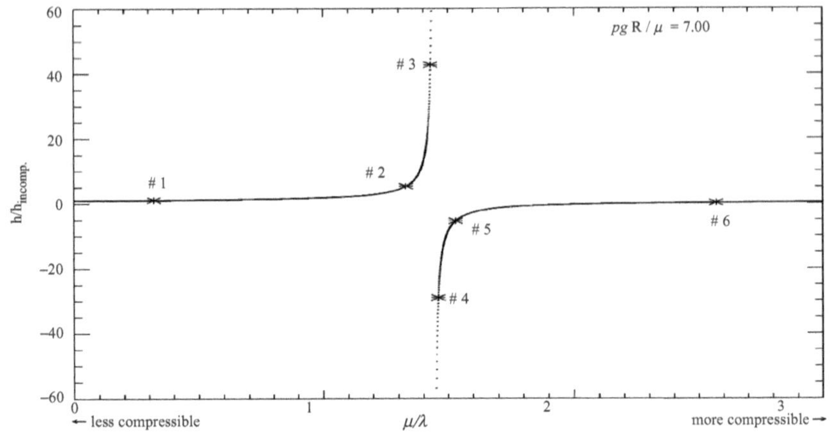

Figure 4. Values of $h/h_{incompressible}$ are shown for $\varrho g R/\mu = 7.0$ as a function of μ/λ. A singularity in the solution is seen at $\mu/\lambda = 1.442$. The * symbols mark cases that are discussed in Section 3.

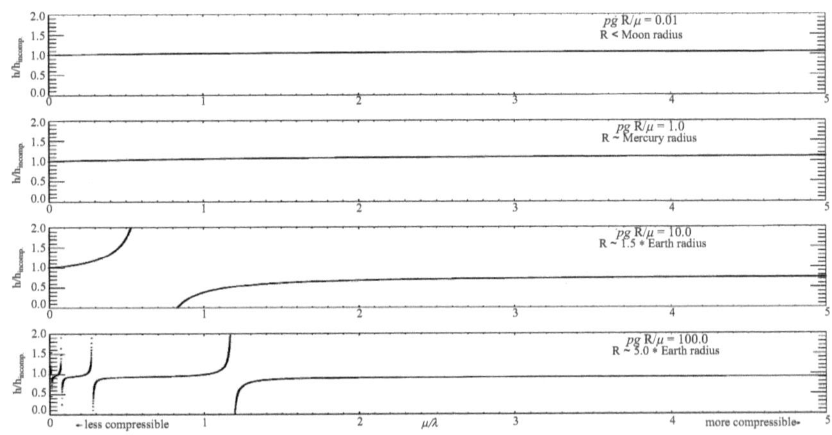

Figure 5. The value of $h/h_{incompressible}$ for various values of $\varrho g R/\mu$ is shown. If μ and ϱ are constrained to values for real materials, the various values of $\varrho g R/\mu$ correspond to various values of R. Solutions for smaller bodies (smaller values of $\varrho g R/\mu$) are well behaved, but elsewhere singularities are seen. The number of singularities increases in the range of μ/λ shown here with $\varrho g R/\mu$.

Figure 5, bodies smaller than the Moon would have their Love numbers enhanced by a few percent over the incompressible values. For bodies only slightly larger than the Earth, singularities in the solution are seen, even for $\mu/\lambda \sim 1$, which is a physically reasonable value. For even larger planets, the number of singularities increases. According to Figure 5,

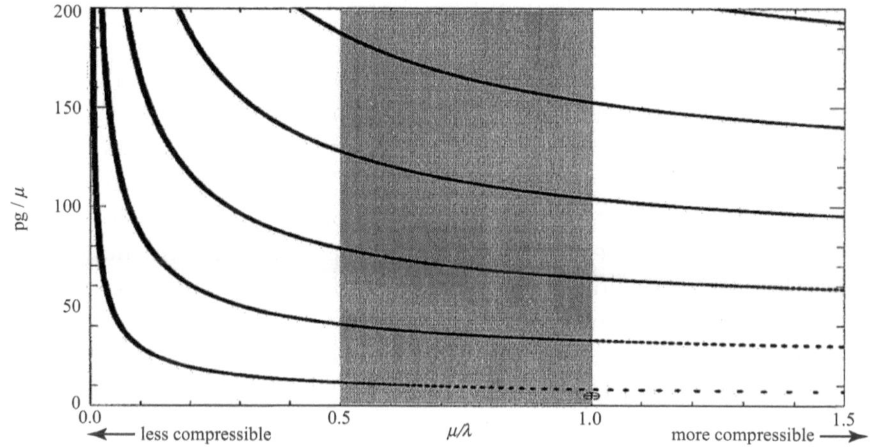

Figure 6. The loci of the singularities in the ($\varrho gR/\mu$,μ/λ) plane are shown. No singularities are observed for small values of $\varrho gR/\mu$ or for $\mu/\lambda = 0$ (rigid and/or incompressible material). The Earth and Venus, represented by the Earth symbol, occupy a region of this space just below the lowest locus line.

a planet with a radius 5 times the Earth's would have singularities at $\mu/\lambda = 0.012$, 0.076, 0.276 and 1.184.

More generally, Figure 6 shows the loci of the singularities of h in the ($\varrho gR/\mu$, μ/λ) plane. In Figure 5, four singularities for the solution when $\varrho gR/\mu = 100.0$ are shown. These same four singularities are shown in Figure 6 as the four locations where the locus lines cross $\varrho gR/\mu = 100.0$. Again in Figure 6, the solution is well behaved for small values of $\varrho gR/\mu$. The shaded region corresponds to values of μ/λ for plausible geophysical materials (Turcotte and Schubert, 1982). Many singularities lie in this range. In Appendix B, we show mathematically the conditions for the singularities.

If we were to adopt values for the parameters that are the rough approximate average values for the Earth or Venus (the two largest rocky planets), the values would plot in Figure 6 near a point (marked by the Earth symbol, $\oplus$), which is just below the lowest singularity locus line. We speculate on possible implications of this result in Section 4.

3. Interpretation

As shown in Section 2, for certain values of $\varrho gR/\mu$ and μ/λ, Love's solution yields tides that could be much larger than tides for the incompressible case, by factors approaching $\pm\infty$. Of course, Love's solution is only valid where the height of the tide is much less than the radius of the body, $\delta R << R$ (see Appendix A). However h may be allowed to become quite large, even while the total height of the tide remains small compared to the radius of the body. A value of $h/h_{incompressible} \rightarrow \infty$ does not necessarily violate the limit $\delta R << R$. It does allow an arbitrarily small tide-raiser to result in a finite, even substantial, Tidal Amplitude. It follows that for any case near a singularity, tidal deformation is predominantly a response to self-gravity of the deformation in the sense that the Earth of the tide-raising body is only a small part of the total potential governing the distortion of any mass element. In that case, the direct tidal potential of the tide-raising body is a minor component of the total potential defining the deformation.

Another surprising result is that the Tidal Amplitude can be negative. In other words, the shortest axis of the deformed body is aligned with the tide-raiser. This result may seem physically counter intuitive, but becomes more understandable if we consider the deformation of the planet throughout its interior. The solution (Appendix B) for the displacement at any location $\vec{u}(r, \vartheta)$ shows that the radial component of $\vec{u}$, $u_r(r, \vartheta)$, at any given r has the same dependence on ϑ as the radial distortion of the surface:

$$u_r(r, \vartheta) = -\frac{h(r)}{g} W_2(R, \vartheta). \tag{7}$$

Remember, the value of h plotted in Figures 2-6 represents only the distortion of the outer boundary of the planet, i.e. $h(R)$.

If we plot the function $h(r)$ for cases near a singularity, and compare these with cases farther from a singularity, we can gain some insight into the character of the distortion in the interior. Consider 6 cases around the singularity shown in Figure 4. At cases #3 and #4, the enhancement $h/h_{incompressible}$ due to compressibility of the surface distortion is a factor of about 40 and -30 respectively. Even though these enhancements are very large the assumption that $\vec{u}$ is small can still be valid if the tide-raiser is small enough or far enough away from the body. The negative enhancements (cases #4 and #5) mean the body is radially shortened along the axis aligned with the tide-raiser.

Consider the radial distribution of the distortion in these cases. Following Love's interest in the Earth's tides, values of the mass M and distance to the tide-raiser a for the Earth-Moon system are used: $M = 5.88\text{x}10^{23}$ kg and $a = 3.84\text{x}10^8$ m. Figure 7 shows $u_r(r)/R$ along the radius parallel to the tide-raiser. According to (7), $h(r)$ is proportional to u_r and $u_r(r)/R$ along the radius perpendicular to the tide-raiser would be $-1/2$ the value. As enhancement of surface displacement increases (i.e. moving from point #1 to #3 in Figure 4), the magnitude of the internal displacement also increases. At #3 the surface displacement is

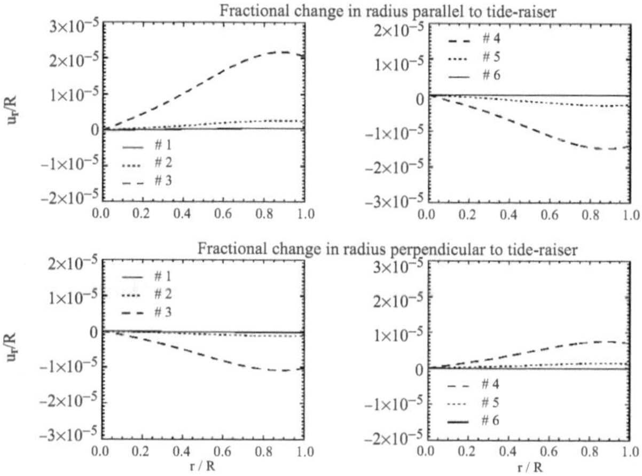

Figure 7. Radial displacements in the interior, normalized to the original radius of the planet, are shown for the six cases marked in Figure 4. For the axes parallel and perpendicular to the tide-raiser, displacements are only in the radial direction.

~ 40 times the incompressible case (Figure 4), however Figure 7 shows the greatest displacements are only around 0.01% of R and these are deep in the interior (near $r = 0.6R$). The same is true of the greatest displacements on the other side of the singularity. Point #4 shows enhancement of ~ -30 times. Even when enhancements become large the assumption that $\vec{u}$ is small can remain valid.

Now consider the density variations in these cases. Using the solution $\vec{u}$, an expression for the change in density, $\Delta\varrho$ can be derived to be

$$\Delta\varrho = -\frac{\varrho r^2}{3gR^3}\left[A_2\alpha^2\psi_2(\frac{\alpha r}{R}) - B_2\beta^2\chi_2(\frac{\beta r}{R})\right]W_2(R,\vartheta). \tag{8}$$

Figure 8 shows the change in density normalized to the original density throughout the body. Since displacement $\vec{u}$ is small throughout, the fractional change in density is $< 1\%$ even in cases where h is enhanced by ~ 40 or ~ -30 times the incompressible value. Since the change in density is related to the change in volume of the body the fact that this change is small also reaffirms that allowing the body to be treated as a linear elastic solid is valid. When the density change becomes large enough it can be imagined that the body would no longer behave elastically.

From this understanding of how the interior deforms and changes density we get insight into how the cases with negative Love number can be physically plausible. For cases as $h \to \infty$, the planet elongates toward the tide raiser (Figure 7a), while the mass density along this axis

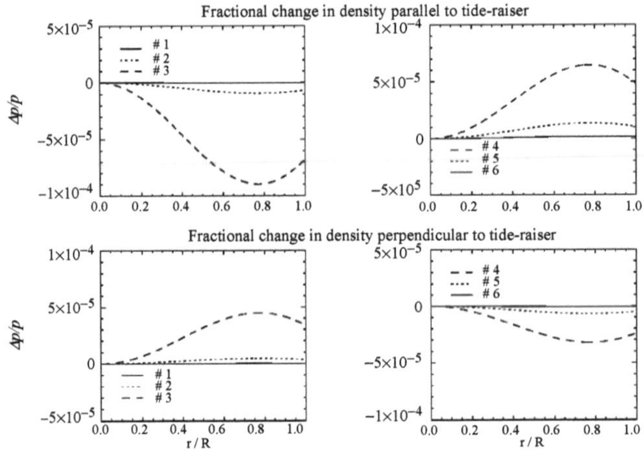

Figure 8. Changes in density in the interior, normalized to the original density of the planet, are shown for the six cases marked in Figure 4.

decreases (Figure 8a). Lowering the density of the material being displaced at the surface corresponds to a relatively small gravitational force from this part of the body. The exact opposite is happening along the axis perpendicular to the tide-raiser, concentrating mass (Figures 7b and 8b). The potential due to the redistribution of mass within the planet, especially the mass concentrated perpendicular to the tide-raiser, will cause the planet to retract along the axis toward the tide-raiser. Thus, for such cases of negative Love numbers, the potential due to the internal redistribution of mass dominates over the original tide-raising potential and over the potential due to the mass displaced at the surface. Along the axis pointing toward the tide-raiser, the material is pulled back by the gravitational force due to the material concentrated perpendicular to the tide-raiser, moving the surface inward relative to the original radius of the planet, leading to a negative Love number.

In order to understand what is happening to the perturbing potential inside the body, we consider the Love number k (defined in Appendix B). Normally k is a dimensionless number that describes the strength of the perturbing potential compared with the tide-raising potential at the surface of the body. However, just as with h, a function for $k(r)$ can be written which describes how the perturbing potentials vary within the body (see Appendix B). When $|k(r)| \sim 1$ the perturbing potentials are as strong as the tide-raising potential . However values of $k(r) > 0$ re-enforce the effects of the tide-raising potential while $k(r) < 0$ oppose the effects of the tide-raising potential . Figure 9 shows profiles of $k(r)$ throughout the body for the four cases from Figure 4. In the

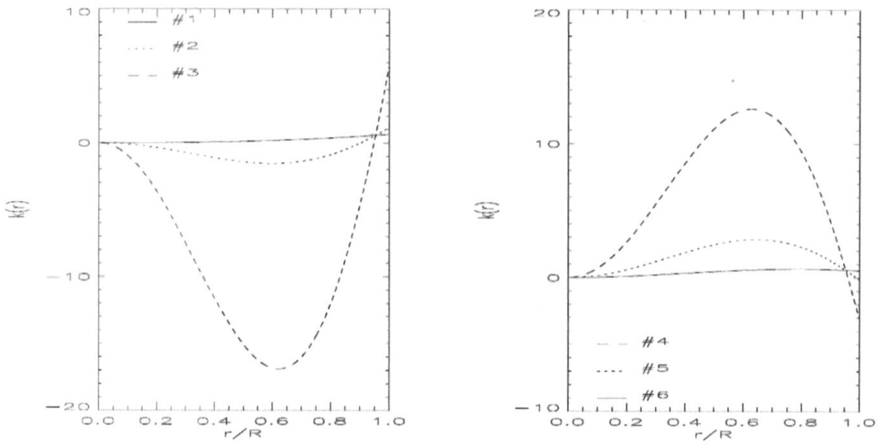

Figure 9. The function $k(r)$ derived in Appendix B, describes how the perturbing potentials vary within the body. When $k(r) = 1$ the perturbing potentials are as strong as the tide-raising potential . Also negative values of $k(r)$ reenforce the effects of the tide-raising potential while positive $k(r)$ counter the effects of the tide-raising potential.

standard case of normal h values (profile #1 in Figure 9) the perturbing potentials are always smaller than the tide-raising potential and work to re-enforce the tide-raising potential . Where enhancement of h becomes significant (profiles #2 and #3), perturbing potentials inside the body become much greater than the tide-raising potential and oppose it. Yet in these cases the pertubing potentials re-enforce the tide-raiser near the surface. This dichotomy pinches mass near the surface, allowing its self-gravity the ability to raise even higher tides at the surface. In fact, features in Figure 9 are manifested in the behavior of u_r shown in Figure 7. Looking at profile #1 in Figure 7a the pullback of u_r/R seen around $r/R \sim 0.9$ corresponds to the change in the perturbing potential, which rapidly switches from opposing the tide-raising potential to re-enforcing it. The peak in the perturbing potential for profile #3 at $r/R \sim 0.6$ also corresponds to a change in concavity of u_r/R at the same location.

For cases of small negative Love number h (#5 in Figure 4), Figure 9 shows that inside the body the perturbing potentials dominate the tide-raising potential and re-enfoce it. However near the surface these potentials over come the tide-raising potential, retracting the surface. Not until the perturbing potentials become smaller in magnitude than the tide-raising potentials does the Love number h become positive once more.

On the surface, in cases closest to the singularity (#3 and #4), Figure 9, shows that k approaches extreme values, just as h does in Figure 4. In fact, singularities in k occur at exactly the same locations in parameter space (e.g. Figure 6) as singularities on h. Mathematically, as shown in Appendix B, both h and k have the same expression in their denominators, which goes to zero at the singularity. Therefore, Figure 6 shows the loci of singularities of $k(r)$, as well as of h.

4. Discussion and Conclusions

Our evaluations show that for certain values of $\varrho g R/\mu$ and μ/λ, the Love number h will be greatly enhanced compared to the incompressible case. This enhancement is most certainly due to the self-gravity of the body sustaining even higher deformation. The governing equation which controls deformation remains valid as long as the total height of the tide and the change in density of the body are small ($\delta R/R$ and $\delta\varrho/\varrho$). It is important to bear in mind that $h \to \infty$ does not necessarily violate that limit. $h \to \infty$ could correspond to a moderate $\delta R/R$ if the tide-raiser is small or far away.

Are any if the singularities near values of parameters that might be relevant to real planets? If we adopt parameters for this artificial model that approximate the bulk properties of the Earth, we would have $\varrho g R/\mu = 5$ and $\mu/\lambda = 1$ (Kaula 1968). In Figure 6, those values plot just below the lowest curve for the locus of a singularity, as indicated by the Earth symbol, $\oplus$. Note that bulk parameters for Venus, would plot close to the same point. If such a body were only a few 10's of percent larger in radius, it would be at a singularity. In principle, an infinitesimal passing body could raise a substantial tide.

The chances that a given body would happen to have parameters near this gravitational instability are vanishingly small. However, suppose a body (similar enough to Love's idealized body that the solution evaluated here applies) were growing and evolving thermally during formation of a planetary system. Its size and bulk physical properties would be changing continuously, possibly crossing over the locus of singularity conditions. At such a time, any passing planetesimal could induce a substantial tide, in essence a self-gravitational instability. It is interesting to speculate whether this effect could limit the size of a growing planet, by only allowing planets to form if they remain below the lowest locus in Figure 6. Certainly it is striking that the two largest terrestrial planets are about the same size as one another and lie just below this boundary.

Also these results show the elastic response to a tide-raising potential. If a given body were to experience a substantial tide, the response would actually take some time. A real body would actually behave visco-elastically and on timescales longer then the Maxwell time for the body these density contrasts would likely drive internal differentiation.

On the other hand, we must bear in mind that no real planet fits the assumptions of Love's classical problem discussed here, with uniform density and uniform elastic properties immediately prior to application of the tide-raising potential. Real planets will certainly not be uniform. Nevertheless, the limits of parameters within which instabilities may occur are not known, and the possibility of instabilities may be worth investigating in greater detail. Such an investigation is currently underway, using numerical and analytical modeling approaches.

Our results could apply to smaller bodies as well, especially as thermal evolution causes them to move through elastic-parameter space. For example, warmer bodies would have lower rigidities and thus the value of $\varrho g R/\mu$ will be greater, possibly allowing larger tides to be raised on their surfaces. While it has long been known that tidal history is important to the thermal evolution of bodies (Peale and Cassen, 1978), it is also likely that thermal history has major effects on tides, especially if it caused bodies to pass through the types of singularities identified here.

Acknowledgements

We would like to thank Jihad Touma, David Stevenson and James Williams for discussion that provided further insight. This work was supported by NASA's Planetary Geology and Geophysics program.

Appendix: Derivation of the Governing Equation

There are several implicit assumptions in Love's (1967) derivation of the partial differential equation that governs the tidal response of a uniform elastic sphere. Here we derive the equation in a way that explicitly illustrates these assumptions.

For a planet in equilibrium, the force of gravity will balance the elastic forces on any mass element inside the planet. With no tide-raiser an elemental cube of volume $(\delta x \delta y \delta z)$ will feel a force of gravity due to the potential of the planet:

$$\vec{F}_g = -\varrho(x,y,z)\delta x \delta y \delta z \vec{\nabla} V_o \tag{A.1}$$

where $\varrho(x,y,z)$ is the density at that location, such that $\varrho(x,y,z)\delta x\delta y\delta z$ is the mass contained in the volume $\delta x\delta y\delta z$ and V_o is the gravitational potential of the planet. It is assumed the planet has been in this condition long enough that shear components to the stress have relaxed away, so the only remaining stress is pressure. The net force on a mass element is due to the gradient of the pressure:

$$\vec{F}_p = -\vec{\nabla} P_o \delta x \delta y \delta z \tag{A.2}$$

where P_o is the pressure perpendicular to each face. In hydrostatic equilibrium, $\vec{F}_g + \vec{F}_p = 0$ or

$$\varrho(x,y,z)\vec{\nabla} V_o = -\vec{\nabla} P_o \tag{A.3}$$

Introduction of a neighboring tide-raising body will disturb this equilibrium. Following an Eulerian approach, each mass element will be displaced an amount $\vec{u}(x,y,z)$, so that the new location will be $x+u_x, y+u_y, z+u_z$. Moreover, each cubic mass element will be distorted during the displacement, each corner of the cube is displaced by a slightly different $\vec{u}$. The new equilibrium written in terms of the conditions at the original location of the mass element ($_o$) is given by

$$(\varrho_o+\Delta\varrho)\vec{\nabla}(V_o+\Delta V) = -\vec{\nabla}P_o - \vec{\nabla}(\varrho_o\vec{u}\cdot\vec{\nabla}V_o) + \mu\nabla^2\vec{u} + (\mu+\lambda)\vec{\nabla}(\vec{\nabla}\cdot\vec{u}), \quad \text{(A.4)}$$

where $\Delta\varrho$ and ΔV are Eulerian perturbations (Dahlen and Trump, 1998). The term $\Delta\varrho\vec{\nabla}\Delta V$ is second order. Furthermore, $\varrho_o\vec{\nabla}V_o = -\vec{\nabla}P_o$ represents the original state of hydrostatic equilibrium of the mass element at its original location. Therefore, the equilibrium equation becomes

$$\varrho_o\vec{\nabla}\Delta V + \Delta\varrho\vec{\nabla}V_o = -\vec{\nabla}(\varrho_o\vec{u}\cdot\vec{\nabla}V_o) + \mu\nabla^2\vec{u} + (\mu+\lambda)\vec{\nabla}(\vec{\nabla}\cdot\vec{u}). \quad \text{(A.5)}$$

ΔV or V' is the sum of the additional potentials, comprising (a) the tide-raising potential, (b) the potential due to the change of mass at the surface from the distortion, and (c) the change in the potential of the body from changes in density due to the redistribution of mass. The change in density $\Delta\varrho$ is $-\varrho(\vec{\nabla}\cdot\vec{u})$. Using this definition and constraining the planet to have uniform density leads to Love's (1911) governing equation:

$$\varrho_o\vec{\nabla}V' - \varrho(\vec{\nabla}\cdot\vec{u})\vec{\nabla}V_o = -\varrho_o\vec{\nabla}(\vec{u}\cdot\vec{\nabla}V_o) + \mu\nabla^2\vec{u} + (\mu+\lambda)\vec{\nabla}(\vec{\nabla}\cdot\vec{u}). \quad \text{(A.6)}$$

Appendix: Solution to the Governing Equation

The equation governing the elastic deformation of a compressible, self-gravitating body is

$$\varrho_o\vec{\nabla}V' - \varrho(\vec{\nabla}\cdot\vec{u})\vec{\nabla}V_o = -\varrho_o\vec{\nabla}(\vec{u}\cdot\vec{\nabla}V_o) + \mu\nabla^2\vec{u} + (\mu+\lambda)\vec{\nabla}(\vec{\nabla}\cdot\vec{u}), \quad \text{(B.1)}$$

where $\vec{u}$ is the displacement at any point inside the body by V', the additional potentials imposed on the body with original potential V_o. Equation (B.1) distorts the body in order to establish an equilibrium between the force of gravity acting on the body and induced stresses inside the body opposing these additional potentials. A 2nd order solution for $\vec{u}$ which satisfies (B.1) is

$$\vec{u} = F_2\vec{\nabla}W_2 + G_2\vec{r}W_2, \quad \text{(B.2)}$$

where F_2 and G_2 in terms of the α and β defined in Section 1 are

$$F_2 = -\frac{R}{g}\left[\frac{5}{4} + \frac{2}{9}\left(\frac{\beta^2}{\alpha^2-\beta^2}\right)A_2\psi_1\left(\frac{\alpha r}{R}\right) + \left(\frac{3\alpha^2+\beta^2}{\alpha^2-\beta^2}\right)A_2\left(\frac{1}{9}\psi_2\left(\frac{\alpha r}{R}\right) - \frac{1}{4}\psi_2(\alpha)\right)\right.$$
$$\left. + \frac{2}{9}\left(\frac{\alpha^2}{\alpha^2-\beta^2}\right)B_2\chi_1\left(\frac{\beta r}{R}\right) - \left(\frac{\alpha^2+3\beta^2}{\alpha^2-\beta^2}\right)B_2\left(\frac{1}{9}\chi_2\left(\frac{\beta r}{R}\right) - \frac{1}{4}\chi_2(\beta)\right)\right] \quad \text{(B.3)}$$

and

$$G_2 = -\frac{R}{gr^2}\Bigg[-\frac{1}{9}\left(\frac{3\alpha^2+\beta^2}{\alpha^2-\beta^2}\right) A_2\psi_1(\frac{\alpha r}{R}) - \frac{5}{9}\left(\frac{3\alpha^2+\beta^2}{\alpha^2-\beta^2}\right) A_2\psi_2(\frac{\alpha r}{R})$$
$$-\frac{1}{9}\left(\frac{\alpha^2+3\beta^2}{\alpha^2-\beta^2}\right) B_2\chi_1(\frac{\beta r}{R}) + \frac{5}{9}\left(\frac{\alpha^2+3\beta^2}{\alpha^2-\beta^2}\right) B_2\chi_2(\frac{\beta r}{R})\Bigg]. \tag{B.4}$$

Both F_2 and G_2 are composed of combinations of A_2 and B_2. The expressions for A_2 and B_2 are

$$A_2 = \tfrac{45}{2} \Big[2(\alpha+\beta)(\alpha-\beta)(\alpha^2+3\beta^2)(-\chi_1(\beta)+5\chi_2(\beta))$$
$$+\alpha^2\beta^2(\alpha^2-3\beta^2)\chi_2(\beta)\Big](\alpha^2-\beta^2)/C_2 \tag{B.5}$$

and

$$B_2 = \tfrac{45}{2} \Big[2(\alpha+\beta)(\alpha-\beta)(3\alpha^2+\beta^2)(\psi_1(\alpha)+5\psi_2(\alpha))$$
$$-\alpha^2\beta^2(3\alpha^2-\beta^2)\psi_2(\alpha)\Big](\alpha^2-\beta^2)/C_2, \tag{B.6}$$

where the denominator C_2 in both A_2 and B_2 is given by

$$\begin{aligned} C_2 \; = \; & -24(\alpha^2+\beta^2)(\alpha+\beta)^2(\alpha-\beta)^2\psi_1(\alpha)\chi_1(\beta) \\ & +\alpha^2\beta^2(\alpha^2+\beta^2)(-5(\alpha^2-\beta^2)+4\alpha^2\beta^2)\psi_2(\alpha)\chi_2(\beta) \\ & -15(\alpha+\beta)(\alpha-\beta)\left((\alpha^2+3\beta^2)^2\psi_1(\alpha)\chi_2(\beta)+(3\alpha^2+\beta^2)^2\psi_2(\alpha)\chi_1(\beta)\right) \\ & -2\alpha^2\beta^2\Big((3\alpha^4+6\alpha^2\beta^2-5\beta^4)\psi_1(\alpha)\chi_2(\beta) \\ & -(5\alpha^4-6\alpha^2\beta^2-3\beta^4)\psi_2(\alpha)\chi_1(\beta)\Big). \end{aligned} \tag{B.7}$$

The functions $\psi_1(x)$,$\psi_2(x)$, $\chi_1(x)$ and $\chi_2(x)$ are related to spherical Bessel functions of real(φ) and imaginary (χ) arguments. They can be written as follows:

$$\begin{aligned} \psi_1(x) &= (x\cos x - \sin x)/x^3 \\ &= \sum_{n=0}^{\infty}(-1)^n x^{2n-2}\left(\frac{2n}{(2n+1)!}\right), \end{aligned} \tag{B.8}$$

$$\begin{aligned} \psi_2(x) &= ((3-x^2)\sin x - 3x\cos x)/x^5 \\ &= \sum_{n=0}^{\infty}(-1)^n x^{2n}\left(\frac{4(n+1)(n+2)}{(2n+5)!}\right), \end{aligned} \tag{B.9}$$

$$
\begin{aligned}
\chi_1(x) &= (x\cosh x - \sinh x)/x^3 \\
&= \sum_{n=0}^{\infty} x^{2n-2}\left(\frac{2n}{(2n+1)!}\right),
\end{aligned}
\tag{B.10}
$$

$$
\begin{aligned}
\chi_2(x) &= ((3-x^2)\sinh x - 3x\cosh x)/x^5 \\
&= \sum_{n=0}^{\infty} x^{2n}\left(\frac{4(n+1)(n+2)}{(2n+5)!}\right).
\end{aligned}
\tag{B.11}
$$

Using this solution the radial component of the displacement is found to be

$$
\begin{aligned}
u_r = -\tfrac{r}{R}\Bigg[\tfrac{5}{2} & -\frac{1}{3}A_2\psi_1(\frac{\alpha r}{R}) - \left(\frac{3\alpha^2+\beta^2}{\alpha^2-\beta^2}\right)A_2\left(\frac{1}{3}\psi_2(\frac{\alpha r}{R}) + \frac{1}{2}\psi_2(\alpha)\right) \\
& +\frac{1}{3}B_2\chi_1(\frac{\beta r}{R}) + \left(\frac{\alpha^2+3\beta^2}{\alpha^2-\beta^2}\right)B_2\left(\frac{1}{3}\chi_2(\frac{\beta r}{R}) + \frac{1}{2}\chi_2(\beta)\right)\Bigg]\frac{W_2(R,\vartheta)}{g}.
\end{aligned}
\tag{B.12}
$$

This expression resembles (equation 2), however it is valid anywhere within the body, not just at the surface. Therefore we define a Love function, $h(r)$ to be

$$
\begin{aligned}
h(r) = \tfrac{r}{R}\Bigg[\tfrac{5}{2} & -\frac{1}{3}A_2\psi_1(\frac{\alpha r}{R}) - \left(\frac{3\alpha^2+\beta^2}{\alpha^2-\beta^2}\right)A_2\left(\frac{1}{3}\psi_2(\frac{\alpha r}{R}) + \frac{1}{2}\psi_2(\alpha)\right) \\
& +\frac{1}{3}B_2\chi_1(\frac{\beta r}{R}) + \left(\frac{\alpha^2+3\beta^2}{\alpha^2-\beta^2}\right)B_2\left(\frac{1}{3}\chi_2(\frac{\beta r}{R}) + \frac{1}{2}\chi_2(\beta)\right)\Bigg].
\end{aligned}
\tag{B.13}
$$

At the surface $u_r(R) = \delta R$ and the Love function $h(R)$ matches the expression for the Love number as given in equation 4, (3). It has been shown that when $C_2 \to 0$, A_2 and $B_2 \to \infty$. Thus $h(r)$ also becomes large under those conditions.

From Love (1911) an expression for the perturbation in potential anywhere inside the body can be written as

$$
\begin{aligned}
V_r' = -\tfrac{r^2}{R^2}\Bigg[\tfrac{3}{2} & +A_2\left(\psi_2(\frac{\alpha r}{R}) - \frac{1}{2}\left(\frac{3\alpha^2+\beta^2}{\alpha^2-\beta^2}\right)\psi_2(\alpha)\right) \\
& +B_2\left(\chi_2(\frac{\beta r}{R}) + \frac{1}{2}\left(\frac{\alpha^2+3\beta^2}{\alpha^2-\beta^2}\right)\chi_2(\beta)\right)\Bigg]W_2(R,\vartheta).
\end{aligned}
\tag{B.14}
$$

As with $h(r)$ a Love function $k(r)$ can be defined anywhere inside the body

$$
\begin{aligned}
k(r) &= \frac{r^2}{R^2}\left[\frac{3}{2} \quad +A_2\left(\psi_2(\frac{\alpha r}{R}) - \frac{1}{2}\left(\frac{3\alpha^2+\beta^2}{\alpha^2-\beta^2}\right)\psi_2(\alpha)\right)\right. \\
&\qquad \left. +B_2\left(\chi_2(\frac{\beta r}{R}) + \frac{1}{2}\left(\frac{\alpha^2+3\beta^2}{\alpha^2-\beta^2}\right)\chi_2(\beta)\right)\right] \qquad \text{(B.15)}
\end{aligned}
$$

which goes to Love's (1911) definition of k at $r = R$.

References

Dahlen, F.A. and Jeroen Tromp. (1998). *Theoretical Global Seismology*, Princeton University Press.

Kaula, W.M. (1968) *An Introduction to Planetary Physics: The Terrestrial Planets*, John Wiley & Sons, Inc.

Love, A.E.H. (1944). *A Treatise on the Mathematical Theory of Elasticity*, New York Dover Publications (republication of work done in 1927).

Love, A.E.H. (1967). *Some Problems of Geodynamics*, New York Dover Publications, (republication of work done in 1911).

Peale, S.J. and P. Cassen. (1978) Contribution of Tidal Dissipation to Lunar Thermal History, *Icarus* 36, 245-269.

Turcotte, D.S. and G. Schubert. (1982) *Geodynamics*, John Wiley & Sons.

ON THE GENERAL SOLUTIONS OF HILBERT-EINSTEIN FIELD EQUATIONS IN VACUUM

C. Marchal
General scientific direction ONERA, BP 72, 92322 Chatillon cedex, France
clbmarchal@wanadoo.fr

Abstract The Hilbert-Einstein field equations describe the structure of space-time in the general theory of relativity. Their simplified form, in a vacuum, was discovered in 1913 by Marcel Grossmann in two joint publications with Einstein, who, for his part, rejected this idea and presented another hypothesis... He reverted to Grossmann's idea only two years later.

The first known solution was that of Schwarzschild that provides the origin of the notion of a black hole. Let us also mention the Kerr solution, plane gravitational waves, the general first order solution, the successive approximations of the two black hole problem: hundreds of rigourous solutions are known today.

The genesis of this discovery called for a series of curious coincidences. The absolute differential calculus was a speciality only of the School of Zürich, and it required an improbable meeting between this School, Albert Einstein and the mathematician Marcel Grossmann, who was also able to think as a pure physicist.

Keywords: General theory of relativity, Field equations, Vacuum

1. The Hilbert-Einstein field equation in vacuum

In the "flat" case of the special theory of relativity the proper time s is given in terms of the space-time displacements by the classical Minkowski expression

$$ds^2 = dt^2 - (dx^2 + dy^2 + dz^2)/c, \tag{1}$$

where c is the velocity of light.

B.A. Steves et al. (eds.), Chaotic Worlds: from Order to Disorder in Gravitational N-Body Dynamical Systems, 325–332.

For the general case we will write as usual : $t = x^0; x/c = x^1; y/c = x^2; z/c = x^3$, (these superscripts are not exponents) and, with the usual Einstein summation convention (summation on the indices that appear both up and down), we obtain:

$$ds^2 = g_{\mu\nu}(x^0, x^1, x^2, x^3)dx^\mu dx^\nu. \tag{2}$$

In the Minkowski case the matrix of the functions $g_{\mu\nu}$ is constant. It has twelve zeros off the diagonal and +1, in the diagonal. In the general case that matrix remains symmetrical: $g_{\mu\nu} = g_{\nu\mu}$. We will also need the inverse matrix $g^{\mu\nu}$ which is given by:

$$[g_{\mu\nu}]^{-1} = g^{\mu\nu}, \qquad g_{\mu\lambda}.g^{\lambda\nu} = \delta^\nu_\mu. \tag{3}$$

The Christoffel brackets are then

$$\Gamma^\lambda_{\mu\nu} = (1/2)g^{\lambda\varrho}[g_{\mu\varrho,\upsilon} + g_{\varrho\upsilon,\mu} - g_{\mu\upsilon,\varrho}, \tag{4}$$

where $g_{\alpha\beta,\gamma} = \partial g_{\alpha\beta}/\partial x^\gamma$, and the components $R_{\mu\upsilon}$ of the Ricci tensor are:

$$R_{\mu\upsilon} = \Gamma^\lambda_{\mu\upsilon,\lambda} - \Gamma^\lambda_{\mu\lambda,\upsilon} + \Gamma^\varrho_{\mu\upsilon}\Gamma^\lambda_{\varrho\lambda} - \Gamma^\varrho_{\mu\lambda}\Gamma^\lambda_{\upsilon\varrho}. \tag{5}$$

Notice that, if Δ is the determinant of the matrix $g_{\mu\upsilon}$, the terms $\Gamma^\lambda_{\mu\lambda}$ and $\Gamma^\lambda_{\varrho\lambda}$ are simply

$$\Gamma^\beta_{\alpha\beta} = \Delta_{,\alpha}/2\Delta. \tag{6}$$

Hence, if we put

$$(1/2)\ln|\Delta| = L, \tag{7}$$

the expression in equation (5) for the function $R_{\mu\upsilon}$ can be simplified to

$$R_{\mu\upsilon} = \Gamma^\lambda_{\mu\upsilon,\lambda} - L_{,\mu\upsilon} + \Gamma^\varrho_{\mu\upsilon}L_{,\varrho} - \Gamma^\varrho_{\mu\lambda}\Gamma^\lambda_{\upsilon\varrho}. \tag{8}$$

This latter expression allows us to verify easily the symmetry of the Ricci tensor, and to notice that if Δ is constant then half of the terms of equation (5) vanish.

In a vacuum the Hilbert-Einstein field equation reduces to

$$R_{\mu\upsilon} + \Lambda g_{\mu\upsilon} = 0, \tag{9}$$

where Λ is the cosmological constant. If $\Lambda = 0$ we obtain the Grossmann equation

$$R_{\mu\upsilon} = 0. \tag{10}$$

2. Some classical solutions

The Schwarzschild solution was the first known solution of the Grossmann equation (beside the Minkowski ds^2 of course).

Scharzschild uses spherical coordinates t, r, ϑ, φ (time, distance, colatitude, longitude), and considers the space-time curvature around a spherical body of mass M, obtaining the following expression for ds^2

$$ds^2 = F(r)dt^2 - (1/c^2).[dr^2/F(r)] + r^2 d\vartheta^2 + r^2 \sin^2 \vartheta d\varphi^2, \tag{11}$$

with

$$F(r) = 1 - (\frac{2m}{r}), \tag{12}$$

where $m = \frac{GM}{c^2}$ is the relativistic radius of the mass M.

If M = 0, we find again the Minkowski ds^2 of equation (1), in spherical coordinates. If $\Lambda = 0$, the function $F(r)$ becomes

$$F(r) = 1 - (\frac{2m}{r}) - (\Lambda \frac{r^2}{3c^2}). \tag{13}$$

For $\Lambda = 0$ the two most famous alternative representations of the Schwarschild ds^2 are the following

- The Robertson ds^2. Robertson uses a radial distance ϱ related to the r of Schwarzschild by

  $$r = \varrho + m + (m^2/4). \tag{14}$$

  He also uses three Cartesian coordinates given by the usual Eulerian relations

  $$x = \varrho \sin\vartheta \cos\varphi, \quad y = \varrho \sin\vartheta \sin\varphi, \quad z = \varrho \cos\vartheta. \tag{15}$$

  Then, for the same ds^2, we obtain the simple Robertson form

  $$ds^2 = f(\varrho)dt^2 - [\frac{g(\varrho)}{c^2}](dx^2 + dy^2 + dz^2), \tag{16}$$

  with

  $$f(\varrho) = [\frac{(2\varrho - m)}{(2\varrho + m)}]^2 \quad ; \quad g(\varrho) = [1 + (\frac{2m}{\varrho})]^4. \tag{17}$$

- The Painlevé ds^2. This involves only one transformation of coordinates: the Painlevé time τ that is given by:

  $$\tau = t + (\frac{2m\varepsilon}{c}) \ln[\frac{(r - 2m)}{2m}], \tag{18}$$

with $\varepsilon = \text{constant} = \pm 1$, which gives

$$ds^2 = d\tau^2 - (\frac{1}{c^2})[dr^2 + r^2 d\vartheta^2 + r^2 \sin^2 \vartheta d\varphi^2] - (\frac{2m}{r})[d\tau + \varepsilon/c)dr]^2 \quad (19)$$

The first two terms give the usual Minkowski ds^2 and the remainder has no singularity at $r = 2m$; that singularity in the Schwarzschild ds^2 is only an artificial singularity.

For $\Lambda \neq 0$, the Painlevé form remains simple

$$ds^2 = d\tau^2 - \frac{1}{c^2}[dr^2 + r^2 d\vartheta^2 + r^2 sin^2\vartheta d\varphi^2] - [\frac{2m}{r} + \frac{\Lambda r^2}{3c^2}][d\tau + \frac{\varepsilon}{c})dr]^2, \quad (20)$$

with

$$\tau = t + h(r); dh/dr = \varepsilon(1-F)/cF; F = 1 - (\frac{2m}{r}) - \frac{\Lambda r^2}{3c^2}). \quad (21)$$

The Kerr solution generalizes the Schwarzschild ds^2 to a rotating black hole, with an angular momentum A and the corresponding length $a = \frac{A}{Mc}$, that always satisfies $|a| \leq m$.

$$\begin{aligned} ds^2 &= dt^2 - 2mr[dt - \frac{a \sin^2 \vartheta d\varphi}{c})]^2 D^{-1} - \frac{(r^2 + a^2) \sin^2 \vartheta d\varphi^2}{c^2} \\ &\quad - \frac{D}{c^2}[d\vartheta^2 + \frac{dr^2}{E}], \end{aligned} \quad (22)$$

with $D = r^2 + a^2 \cos^2 \vartheta$, $E = r^2 - 2mr + a^2$.

A Painlevé form exists for this Kerr ds^2. Let us define the new variables t_n and φ_n by

$$dt_n = dt + (\frac{2mr\varepsilon dr}{Ec}); \quad \varphi_n = d\varphi + [\frac{2mra\varepsilon dr}{E(r^2 + a^2)}], \quad (23)$$

with $\varepsilon =$ constant $= \pm 1$.

The parameters t_n, r, ϑ and φ_n can be called "spheroidal parameters" since for them the Minkowski ds^2 is written

$$ds_M^2 = dt_n^2 - (\frac{1}{c^2})(r^2 + a^2 cos^2\vartheta)[d\vartheta^2 + \frac{dr^2}{(r^2 + a^2)}] + (r^2 + a^2) \sin^2 \vartheta \varphi_n^2. \quad (24)$$

It is easy to verify that, with $\varrho^2 = (r^2 + a^2) \sin^2 \vartheta$ and $\mathrm{z} = r \cos \vartheta$, the four parameters t_n , ϱ, z, φ_n are usual cylindrical parameters, with the corresponding ds_M^2.

With D given by equation (22), the Kerr ds^2 is then

$$ds^2 = ds_M^2 - 2mr[cdt_n - a\sin^2\vartheta d\varphi_n D\varepsilon dr/r^2 + a^2)]^2/Dc^2. \qquad (25)$$

The case $a = 0$ in equations (22)-(25) gives the Schwarzschild case of equations (11)-(19) with $t_n = \tau$ and $\varphi_n = \varphi$.

A third classical solution of the Grossmann equation is the "plane gravitational wave", with u = ct - x:

$$c^2ds^2 = c^2dt^2 - dx^2 - dy^2 - dz^2 - f(u, y, z)(cdt - dx)^2. \qquad (26)$$

These waves move in the x direction at the velocity of light, and the function f has a wide freedom; it is only subjected to the contraint

$$\frac{\partial^2 f}{\partial y^2} + \frac{\partial^2 f}{\partial z^2} = 0. \qquad (27)$$

Notice that the expression equation (26) has the Painlevé form: a Minkowski ds^2 minus the product of a space-time function by the square of a sum of differentials. Furthermore these three sums of differentials given in equations (19), (25) and (26) have the following common point: their ds^2 can be written

$$c^2ds^2 = c^2dt^2 - dx^2 - dy^2 - dz^2 - f(t, x, y, z)(cdt - \alpha dx - \beta dy - \gamma dz)^2, \quad (28)$$

with, for the three functions α, β and $\gamma(t, x, y, z)$

$$\alpha^2 + \beta^2 + \gamma^2 = 1. \qquad (29)$$

This remarkable property has been observed in many of the other rigorous solutions of the Grossmann equation.

It is possible to generalize the plane gravitational waves to the case of a non zero cosmological constant Λ.

We will use the cylindrical space-time coordinates t, x, r, ϑ with $y = r\cos\vartheta$, $z = r\sin\vartheta$ and $u = ct - x$, and we will obtain

$$c^2ds^2 = f(u, r, \vartheta)du^2 - 2g(r)dxdu - dr^2 h(r)d\vartheta^2. \qquad (30)$$

The two functions $g(r)$ and $h(r)$ are imposed; they are only functions of Λ with the length $K = 2c|3Lambda|^{-0.5}$ and with the following

$$\text{For } \Lambda > 0,\ g(r) = \cos(r/K)^{4/3}, \qquad h(r) = K^2 g(r)\tan^2(r/K), \quad (31)$$

$$\text{For } \Lambda < 0,\ g(r) = \cosh(r/K)^{4/3}, \qquad h(r) = K^2 g(r)\tanh^2(r/K).$$

On the contrary the function $f(u,r,\vartheta)$ retains a wide freedom and is only subjected to the following condition that generalizes equation (27)

$$2\Lambda\frac{f}{c^2}+\frac{\partial^2 f}{\partial r^2}+\left[\frac{\partial^2 f}{\partial\vartheta^2}+0.5\frac{(\frac{\partial f}{\partial r})(\frac{\partial h}{\partial r})}{h(r)}\right]=\left(\frac{\partial g}{\partial r}\right)\left[\frac{\partial(f/g)}{\partial r}\right]. \quad (32)$$

Another beautiful result is the general first order solution of the Grossmann equation. We will choose the velocity c as our unit of velocity and obtain

$$g_{\mu v}=f_{\mu v}+f_{v,\mu}+\text{second order}+\begin{vmatrix}1 & A_{,23} & B_{,13} & C_{,12}\\ A_{,23} & -1 & -C_{,03} & -B_{,02}\\ B_{,13} & -C_{,03} & -1 & -A_{,01}\\ C_{,12} & -B_{,02} & -A_{,01} & -1\end{vmatrix}. \quad (33)$$

The four functions f_μ and their partial derivatives correspond to the modifications of referentials. The three "intrinsic" functions A, B, and $C(x^0, x^1, x^2, x^3)$ satisfy the conditions

$$(A+B+C)_{,0123}=0; \Box A=0; \Box B=0; \Box C=0, \quad (34)$$

where the sign $\Box$ is the expression of the "D'Alembertian": $\Box F = F_{,00}-F_{,11}-F_{,22}-F_{,33}$. For instance, for the Robertson form of the Schwarzschild ds^2, we obtain

$$\left.\begin{aligned} &\varrho=[x^2+y^2+z^2]^{1/2}\ ;\ f_0=\tfrac{-mt}{\varrho}\ ;\ f_1=0.5mLn[\tfrac{(\varrho-x)}{(\varrho+x)}]\\ &f_2=0.5mLn[\tfrac{(\varrho-y)}{(\varrho+y)}]\ ;\ f_3=0.5mLn[\tfrac{(\varrho-z)}{(\varrho+z)}]\\ &A=mt\tan^{-1}(\tfrac{x\varrho}{yz})\ ;\ B=mt\tan^{-1}(\tfrac{y\varrho}{xz})\ ;\ C=mt\tan^{-1}(\tfrac{z\varrho}{xy})\end{aligned}\right\} \quad (35)$$

3. Short historical background

A series of improbable coincidences is at the origin of the general theory of relativity.

1820-1830. Birth of non-Euclidian geometries (Gauss, Lobatchevski, Bolyai).

1826-1866. Bernhardt Riemann.

1867. Publication of Riemann's general theory of "Riemannian spaces".

1829-1900. Elwin Bruno Christoffel, **professor at the Zürich polytechnicum** (1862-1869). "On the transformation of homogemneous quadratic forms" (1869).

1853-1925. Gregorio Ricci-Curbastro. Tensorial Analysis and absolute differential calculus.

1878-1936. Marcel Grossmann, **student of Minkowski, friend of Einstein and successor of Christoffel,** after 1907, **at the Zürich polytechnicum. Reports on the works of Christoffel and Ricci-Curbastro.**

1907 and 1911. **Einstein looks for a generalization of relativity.**

1912. **Einstein becomes professor at the Zürich Polytechnicum** (after two years at Prague University and the refusal of a post at Utrecht University). **In Zürich he asks for the help of Grossmann in his researches on relativity.**

1912. **With mathematical and physical considerations, Grossman reaches the conclusion that the real physical space-time is a Riemannian space and that in vacuum it must obey the equation $\mathbf{R}_{\mu\nu} = 0$. This is the simplest covariant solution for a space-time with gravitation. (Publication in June 1913; see references.**

1913. **In the same publications, Einstein gives several reasons to refute the Grossmann solution and presents his own (non-covariant) solution.** There is a falling out between the two friends; they will never work together again. The non covariant solution of Einstein gives the Newtonian curvature of light beams. The astronomers try unsucessfully to detect that curvature on their old pictures of solar eclipses.

1914 March. Einstein goes back to Berlin.

1914 Spring. Erwin Freundlich (Berlin-Babelsberg Observatory) prepares for the observation of the total solar eclipse of August 21, 1914. He receives 2000 Marks (of that time) from the Royal Prussian Academy of Sciences, and then 3000 Marks from the firm Krupp. He buys the best optical equipment for parallactic measures, those coming from the specialized optical firms of Iena (and much better than Eddington's equipment of the eclipse of 1919).

1914, July 25. Freundlich and his team arrive in Theodosia (Crimea). They meet the other teams (Russian, Italian, Spanish, British, French, American) and, proudly, prepare their supermaterial.

1914 August 1. Germany declares war on Russia. Freundlich and his team are placed under house-arrest, they give their material to be kept in Theodosia and are sent to Odessa August 5. They will be exchanged later.

1914 November 30. The report of the Spanish team emphasizes the good meteorological conditions of observation of the eclipse at Theodosia.

1914 November. The German optical material arrives at the University of Odessa.

1915 September. Einstein leaves Berlin for two weeks and goes to Zürich. Notice, of course, the very bad wartime communications between outside and Germany, but Switzerland was neutral...

1915 October. Back to Berlin, Einstein forsakes his one year research on magnetism and hastily returns to the study of general relativity.

1915 November 4, 11, 18. Einstein publishes three successive, and contradictory, notes on general relativity in the publications of the Royal Prussian Academy of Sciences. **These notes are based on the Grossmann equation, that gives a deviation of light beams twice larger than the Newtonian one.** Unfortunately he never gave credit to Grossmann for his equation.

1915 November 16. Hilbert presents his own results on general relativity at Göttingen. He had sent a letter to Einstein on the subject and had invited him. Einstein refuses to come, he answers November 18 **and modifies the text of his third note that was already in press...**

The remainder is simple...

References

Einstein. A, (1913). *Physikalische Grundlagen einer Gravitation theorie.* Vierteljahrschrift der Naturforschung Gesellschaft in Zrich, Bd 58, 8, Seite 284-290

Einstein. A, Grossman. M, (1913). *Entwurf : einer verallgemeinesten Relativitt Theorie und einer Theorie der Gravitation. Zeitschrift fr Mathematik und Physik* . Seite 62 225-244

Grossman. M, (1913). *TMathematische Begriffbildungen zur Gravitationtheorie. Vierteljahrschrift der Naturforschung Gesellschaft in Zrich.* Bd 58, 8, Seite 291-297

List of Participants

Thiery Baertschiger	Departement de Physique Théorique Université de Génève, Geneva, Switzerland thiery.baertschiger@physics.unige.ch
Phillippe Baille	Mathematics Department, Royal Military College of Canada, Kingston, Ontario, Canada baille-p@rmc.ca
Boris Bardin	Department of Theoretical Mechanics, Moscow Aviation Institute, Moscow, Russia boris.bardin@mailcity.com
Rory Barnes	Department of Astronomy, University of Washington Seattle WA, USA rory@astro.washington.edu
Irina Bashkirtseva	Department of Mathematics Ural State University, Ekaterinburg, Russia Irina.bashkirtseva@usu.ru
Peter Berczik	Main Astronomical Observatory Ukranian National Academy of Sciences, Kiev, Ukraine berczik@mao.kiev.ua
Ivano Bertini	Department of Astronomy University of Padova, Padova, Italy bertini@pd.astro.it
Valerio Carruba	Department of Astronomy Cornell University, Ithaca NY, USA vc27@cornell.edu
Alessandra Celletti	Università di Roma Tor Vergata, Italy celletti@mat.uniroma2.it

Caner Cicek	Department of Physics, Science Faculty, Canakkale Onsekiz Mart University, Canakkale, Turkey canerk@bornova.ege.edu.tr
Guido Ciraolo	Dipartimento di Sistemi ed Informatica, facolta di Ingegneria Universita degli studi di Firenze, Firenze, Italy ciraolo@arcetri.astro.it
Josep Cors	Matematica Aplicada III Universitat Politecnica de Catalunya, Manresa, Catalunya, Spain cors@eupm.upc.es
Maria Cosentino	Dipartimento di Matematica Universita di Pisa, Pisa, Italy cosentino@mail.dm.unipi.it
Petra Csomos	Department of Astronomy Eötvos University, Budapest, Hungary cosmos@ludens.elte.hu
Omer Degirmenci	Department of Astronomy and Space Sciences, Faculty of Science, Ege University, Yzmir, Turkey omerd@astronomy.sci.ege.edu.tr
Ahmet Erdem	Department of Physics, Faculty of Arts and Sciences Canakkale Onsekiz Mart University, Canakkale, Turkey aerdem@bornova.ege.edu.tr
Balint Erdi	Department of Astronomy Eötvos University, Budapest, Hungary b.erdi@astro.elte.hu
Sylvio Ferraz-Mello	IAG, Universidade de São Paulo, São Paulo, Brasil sylvio@usp.br
Claude Froeschlé	Observatoire de Nice, Nice, France claude@obs-nice.fr
Barbara Funk	Institut of Astronomy University of Vienna, Vienna, Austria funk@astro.univie.ac.at
Frederic Gabern	Department de Matematica Aplicada i Analisi Universitat de Barcelona, Barcelona, Spain gabern@mat.ub.es
Antonio Giorgilli	Dipartimento di Matematica e Applicazione Università di Milano Bicocca, Milano, Italy antonio@matapp.unimib.it

Justnya Golebiewska	Obserwatorium Astronomiczne Uniwersytet im. Adama Mickiewicza, Poznan, Poland jg@amu.edu.pl
Krysztof Gozdziewski	Torun Centre for Astronomy N. Copernicus University, Torun, Poland K.Gozdziewski@astri.uni.torun.pl
Richard Greenberg	University of Arizona, Tucson, USA greenber@pirlmail.lpl.Arizona.edu
Giovanni Federico Gronchi	Department of Mathematics University of Pisa, Pisa, Italy gronchi@dm.unipi.it
Massimiliano Guzzo	Università degli Studi di Padova, Padova, Italy guzzo@math.unipd.it
Markus Gyergyovitis	Institut of Astronomy University of Vienna, Vienna, Austria A9502592@unet.univie.ac.at
John Hadjedemetriou	University of Thessaloniki, Thessaloniki, Greece hadjidem@physics.auth.gr
Yoshida Haruo	Division of Celestial Mechanics National Astronomical Observatory of Japan, Mitaka, Japan h.yoshida@nao.ac.jp
Martin Hendry	Department of Physics and Astronomy University of Glasgow, Glasgow, UK martin@astro.gla.ac.uk
Marian Jakubik	Department of Interplanetary Matter Astronomical Institute of the Slovak Academy of Sciences, Tatranska Lomnica, Slovak Republic mjakubik@ta3.sk
Suren Khachatryan	Department of Physics Yerevan State University, Yerevan, Armenia skhachat@aua.am
Iwona Kruk	Instytut Astronomiczny Uniwersytet Wrocawski, Wrocaw, Poland kruk@astro.uni.wroc.pl
Ugo Locatelli	Università di Roma Tor Vergata, Italy ugo@arcetri.astro.it
Andrzej J Maciejewski	Institute of Astronomy, University of Zielona Gora, Zielona Gora, Poland maciejka@astro.ia.uz.zgora.pl

Zoltan Mako	Department of Mechanics and Astronomy Babes-Bolyai University, Cluj-Napoca, Romania zmako@math.ubbcluj.ro
Christian Marchal	Direction Scientifique Générale ONERA Chatillon, France marchal@onera.fr
Mario Melita	Astronomy Unit, School of Mathematical Sciences Queen Mary and Westfield College, University of London London, UK M.D.Melita@qmul.ac.uk
Philip James Message	Department of Mathematical Sciences, University of Liverpool, UK sx20@liverpool.ac.uk
Andrea Milani	Dipartimento di Matematica, Università di Pisa, Pisa, Italy milani@dm.unipi.it
Merce Olle	Department Matematica Applicada I Universitat Politecnica de Catalunya, Barcelona, Spain Merce.olle@upc.es
Nuno Pereira	Department of Mathematics Escola Superior de Tecnologia e Gestao de Beja, Beja, Portugal nsap@estig.jp.beja.pt
Giuseppe Pucacco	Department of Physics University of Rome "Tor Vergata", Rome, Italy pucacco@roma2.infn.it
Sean Raymond	Department of Astronomy University of Washington, Seattle, WA, USA Raymond@astro.washington.edu
Philippe Robutel	Astronomie et Systèmes Dynamiques IMC, Paris, France robutel@imcce.fr
Lev Ryashko	Mathematical Department Ural State University, Ekaterinburg, Russia lev.ryashko@usu.ru
Slawomir Rybicki	Faculty of Mathematics and Computer Science Nicolaus Copernicus University, Torun, Poland Slawomir.Rybicki@mat.uni.torun.pl
Zsolt Sandor	Department of Astronomy Eötvos University, Budapest, Hungary Zs.sandor@astro.elte.hu

Christian Schwarz	Insutitut of Astronomy University of Vienna, Vienna, Austria schwarz@astro.univie.ac.at
M Shoaib	School of Computing and Mathematical Sciences Glasgow Caledonian University, Glasgow, UK safridi@gmail.com
Ioannis Sideris	Department of Physics Northern Illinois University, DeKalb, IL, USA sideris@nicadd.niu.edu
Milos Sidlichovsky	Astronomical Institute Academy of Sciences of the Czech Republic, Prague, Czech Republic sidli@ig.cas.cz
Carles Simó	Department de Matematica Aplicada i Analisi, Universitat de Barcelona, Barcelona, Spain carles@maia.ub.es
Charalampos Skokos	Department of Mathematics, CRANS University of Patras, Patras, Greece hskokos@cc.uoa.gr
Daniel Snow	Analysis Division, Headquarters Air Force Space Command Colorado Springs, CO, USA dan.snow@peterson.af.mil
Denise Snow	Analysis Division, Headquarters Air Force Space Command, Colorado Springs, CO, USA denise.snow@peterson.af.mil
Tomasz Stachowiak	Department of Astronomy, Astronomical Observatory Jagiellonian University, Krakow, Poland toms@oa.uj.edu.pl
Bonnie Steves	School of Computing and Mathematical Sciences, Glasgow Caledonian University, Glasgow, UK bst@gcal.ac.uk
Aron Suli	Department of Astronomy Eötvos University, Budapest, Hungary a.suli@astro.elte.hu
Andras Széll	School of Computing and Mathematical Sciences, Glasgow Caledonian University, Glasgow, UK a.szell@gcal.ac.uk
Marek Szydlowski	Astronomical Observatory Jagiellonian University, Krakow Poland uoszydlo@cyf-kr.edu.pl

Giacomo Tommei	Department of Mathematics "L. Tonelli" University of Pisa, Pisa, Italy tommei@mail.dm.unipi.it
Kleomenis Tsiganis	Observatoire de la Cote d'Azur CNR, Nice, France tsiganis@obs-nice.fr
Giovanni Valsecchi	Istituto di Astrofisica Spaziale e Fisica Cosmica, CNR, Roma, Italy giovanni@ias.rm.cnr.it
Benjamin Villac	Department of Aerospace Engineering The University of Michigan, Ann Arbor, Michigan, USA bvillac@umich.edu
George Voyatzis	Department of Physics University of Thessaloniki, Thessaloniki, Greece voyatzis@auth.gr
Jörg Waldvogel	Swiss Federal Institute of Technology ETH, Zurich, Switzerland waldvoge@sam.math.ethz.ch
Piotr Waz	Centre for Astronomy Nicolaus Copernicus University, Torun, Poland Piotr.waz@astri.uni.torun.pl

Index

MIX
Papier aus verantwortungsvollen Quellen
Paper from responsible sources
FSC® C105338

If you have any concerns about our products,
you can contact us on
ProductSafety@springernature.com

In case Publisher is established outside the EU,
the EU authorized representative is:
Springer Nature Customer Service Center GmbH
Europaplatz 3, 69115 Heidelberg, Germany

Printed by Libri Plureos GmbH
in Hamburg, Germany